RISK MANAGEMENT APPLICATIONS IN PHARMACEUTICAL AND BIOPHARMACEUTICAL MANUFACTURING

Wiley Series in
BIOTECHNOLOGY AND BIOENGINEERING

Significant advancements in the fields of biology, chemistry, and related disciplines have led to a barrage of major accomplishments in the field of biotechnology. The Wiley Series in Biotechnology and Bioengineering will focus on showcasing these advances in the form of timely, cutting-edge textbooks and reference books that provide a thorough treatment of each respective topic.

Topics of interest to this series include, but are not limited to, protein expression and processing; nanotechnology; molecular engineering and computational biology; environmental sciences; food biotechnology, genomics, proteomics, and metabolomics; large-scale manufacturing and commercialization of human therapeutics; biomaterials and biosensors; and regenerative medicine. We expect these publications to be of significant interest to practitioners both in academia and industry. Authors and editors were carefully selected for their recognized expertise and their contributions to the various and far-reaching fields of biotechnology.

The upcoming volumes will attest to the importance and quality of books in this series. I would like to acknowledge the fellow co-editors and authors of these books for their agreement to participate in this endeavor. Lastly. I would like to thank Ms. Anita Lekhwani, Senior Acquisitions Editor at John Wiley & Sons, Inc. for approaching me to develop such a series. Together, we are confident that these books will be useful additions to the literature that will not only serve the biotechnology community with sound scientific knowledge, but also with inspiration as they further chart the course in this exciting field.

ANURAG S. RATHORE
Amgen Inc.
Thousand Oaks, CA, USA

Titles in series

Quality by Design for Biopharmaceuticals: Principles and Case Studies / Edited by Anurag S. Rathore and Rohin Mhatre

Emerging Cancer Therapy: Microbial Approaches and Biotechnological Tools / Edited by Arsenio Fialho and Ananda Chakrabarty

Risk Management Applications in Pharmaceutical and Biopharmaceutical Manufacturing / Edited by A. Hamid Mollah, Mike Long, and Harold S. Baseman

RISK MANAGEMENT APPLICATIONS IN PHARMACEUTICAL AND BIOPHARMACEUTICAL MANUFACTURING

Edited by

A. HAMID MOLLAH
MIKE LONG
HAROLD S. BASEMAN

A JOHN WILEY & SONS, INC., PUBLICATION

Copyright © 2013 by John Wiley & Sons, Inc. All rights reserved

Published by John Wiley & Sons, Inc., Hoboken, New Jersey
Published simultaneously in Canada

No part of this publication may be reproduced, stored in a retrieval system, or transmitted in any form or by any means, electronic, mechanical, photocopying, recording, scanning, or otherwise, except as permitted under Section 107 or 108 of the 1976 United States Copyright Act, without either the prior written permission of the Publisher, or authorization through payment of the appropriate per-copy fee to the Copyright Clearance Center, Inc., 222 Rosewood Drive, Danvers, MA 01923, (978) 750-8400, fax (978) 750-4470, or on the web at www.copyright.com. Requests to the Publisher for permission should be addressed to the Permissions Department, John Wiley & Sons, Inc., 111 River Street, Hoboken, NJ 07030, (201) 748-6011, fax (201) 748-6008, or online at http://www.wiley.com/go/permission.

Limit of Liability/Disclaimer of Warranty: While the publisher and author have used their best efforts in preparing this book, they make no representations or warranties with respect to the accuracy or completeness of the contents of this book and specifically disclaim any implied warranties of merchantability or fitness for a particular purpose. No warranty may be created or extended by sales representatives or written sales materials. The advice and strategies contained herein may not be suitable for your situation. You should consult with a professional where appropriate. Neither the publisher nor author shall be liable for any loss of profit or any other commercial damages, including but not limited to special, incidental, consequential, or other damages.

For general information on our other products and services or for technical support, please contact our Customer Care Department within the United States at (800) 762-2974, outside the United States at (317) 572-3993 or fax (317) 572-4002.

Wiley also publishes its books in a variety of electronic formats. Some content that appears in print may not be available in electronic formats. For more information about Wiley products, visit our web site at www.wiley.com.

Library of Congress Cataloging-in-Publication Data is available.

Printed in the United States of America

ISBN: 9780470552346

10 9 8 7 6 5 4 3 2 1

CONTENTS

Preface		vii
Contributors		xi
About the Authors		xiii
1	**Background and Introduction** *Harold S. Baseman and A. Hamid Mollah*	1
2	**Risk Management Tools** *Mark Walker and Thomas Busmann*	17
3	**Risk Management: Regulatory Expectation, Risk Perception, and Organizational Integration** *Mike Long*	49
4	**Statistical Topics and Analysis in Risk Assessment** *Mike Long*	75
5	**Quality by Design** *Bruce S. Davis*	89
6	**Process Development and Clinical Product Manufacturing** *Karen S. Ginsbury*	101

7	**Points to Consider for Commissioning and Qualification of Manufacturing Facilities and Equipment**	129
	Harold S. Baseman and Michael Bogan	
8	**Process Lifecycle Validation**	179
	A. Hamid Mollah and Scott Bozzone	
9	**Aseptic Processing: One**	227
	James P. Agalloco and James E. Akers	
10	**Aseptic Processing: Two**	243
	Edward C. Tidswell	
11	**Pharmaceutical Product Manufacturing**	275
	Marlene Raschiatore	
12	**Biopharmaceutical Manufacturing**	325
	Ruhi Ahmed and Thomas Genova	
13	**Risk-Based Change Control**	367
	William Harclerode, Bob Moser, Jorge A. Ferreira, and Christophe Noualhac	
Index		**387**

PREFACE

With the introduction of FDA's 21st century GMP and ICH initiatives (such as Q8 Pharmaceutical Development, Q9 Quality Risk Management, and Q10 Pharmaceutical Quality System), drug manufacturing entered a new era of risk management. Although regulatory agencies are encouraging the use of risk management in pharmaceutical and biopharmaceutical product manufacturing, regulatory guidance and comprehensive literature on how to use and implement risk management is limited and in need of further development. This book will fill this large void and assist both industry as well as agency to implement a compliant and effective risk management approach. The book has been prepared to provide the readers with some points to consider for managing risks to product quality incurred during the manufacture of biopharmaceutical and pharmaceutical products, including

- Industry trend towards to use of QRM for manufacturing control
- Regulatory expectations
- Use of limited resources and cost control
- Maintaining and assuring product quality
- Process and quality improvement

Over the years, the authors and editors of this book have presented to and met with numerous people at many companies, plants, and organizations in the pharmaceutical industry, asking about and providing instruction in quality risk management. While audiences have generally become more aware of the topic, including the expectation and perhaps even the benefit of its use, it has not been clear whether the aspects of its implementation have been fully appreciated and taken advantage of. We saw many instances where resources and time were spent to complete hundreds of risk assessments, but never utilized into manufacturing operations. Identification of risk without action has minimal benefit and can be viewed as shifting a company from ignorance to avoidance. This has led the editors of this book to believe that there was a strong need for further explanation and education on approaches for the practical implementation of quality risk management for pharmaceutical and biopharmaceutical processes. Over the past half-decade or more, companies and individuals have explored, developed, implemented, benefited, and refined methods for risk management, assessment, review, and decision making that utilize approaches, models, systems, techniques, schemes and plans—some more effective, some less. This book creates this link by exploring risk management of manufacturing processes through a collection of chapters written by some of the leading experts in pharmaceutical and biopharmaceutical product manufacturing. The editors chose topics that represented the most significant challenges in the industry at the time of writing.

The ultimate goal of the risk management process is to bring focus and effort to the issues in an organization that imparts the highest risk to product quality and/or patient safety. The degree of formality and rigor applied should be commensurate with the complexity and/or criticality of the element or issue. There are many tools that can be applied, which are new to many in the industry. For risk assessors, there can be confusion with the level and detail of an assessment as well as the potential subjectivity, which can permeate an assessment. These issues are compounded with the uncertainty of the regulatory framework and application of risk. Hence, it is quite important to use proven and compliant risk management approaches to assure acceptable results.

This book was written by authors who have used risk management to improve processes, investigate failures, design operations, validate processes, and increase overall quality and productivity of their respective operations. This book is not a collection of history or fundamentals, nor is it a book of theoretical desired states. Although the book does explore and present some background and introduction, it is primarily focused on the practical presentation of points to consider and methods to help the reader make better decisions based on risk and to help manage that risk. This book is written by people who lived through the introduction of this topic to our industry and participated in its implementation. Our authors want to communicate best practices to help the reader better implement and benefit from its use without experiencing the unnecessary burden and redundancies that often accompany activities in this industry. This book provides examples of risk management in areas from the process development to sterile fill operations.

Risk management should be a method for making our jobs more doable, adding to our understanding of processes, and helping us make better decisions. It should not be just a checklist item, another corporate directive, of little value. To the contrary, the objective of this book is to help the reader understand, appreciate, use, and benefit from risk management. We hope you will find it useful and enjoyable.

We express our gratitude to all the authors for contributing to this book. Thanks to Anurag Rathore for his encouragement and persuasion to complete this book. We also thank Lynn Torbeck at Torbeck and Association, Kevin O'Donnel at the IMB (Irish Medicines Board), Emma Ramnarine and David Reifsnyder at Genentech, Penny Butterell and Simon Smith at Pfizer, and Brian Turley and Walter Henkels at Valsource for providing feedback.

Thank you,
THE EDITORS

CONTRIBUTORS

James P. Agalloco, Agalloco & Associates, Belle Mead, NJ

Ruhi Ahmed, Regulatory Affairs, Ultragenyx, Novato, CA

James E. Akers, Akers Kennedy & Associates, Kansas City, MO

Harold S. Baseman, ValSource LLC, Jupiter, FL

Michael Bogan, Integrated Commissioning & Qualification, LLC, Hope Valley, RI

Scott Bozzone, Pfizer Inc., Peapack, NJ

Thomas Busmann, Focus Compliance & Validation Services, Knoxville, TN

Bruce S. Davis, Bruce Davis (Global Consulting), Haslemere, Surrey, UK

Jorge A. Ferreira, Jacobs Engineering Group, Conshohocken, PA

Thomas Genova, Johnson & Johnson, Raritan, NJ

Karen S. Ginsbury, PCI Pharmaceutical Consulting Israel, Petach Tikvah, Israel

A. Hamid Mollah, Quality Engineering and Validation, XOMA Corporation, Berkeley, CA

William Harclerode, Forest Labs, Harborside Financial Center, Jersey City, NJ

Mike Long, ConcordiaValSource, Wayland, MA

Bob Moser, Jacobs Engineering, Conshohocken, PA

Christophe Noualhac, Product Development, Halo Pharmaceutical, Whippany, NJ

Marlene Raschiatore, Johnson & Johnson, Glen Mills, PA

Edward C. Tidswell, Baxter Healthcare Corp, Sterility Assurance Department, Round Lake, IL

Mark Walker, Focus Compliance & Validation Services, Knoxville, TN

ABOUT THE AUTHORS

James Agalloco, President, Agalloco & Associates, is a pharmaceutical manufacturing expert with more than 40 years of experience. He worked in organic synthesis, pharmaceutical formulation, pharmaceutical production, project/process engineering, and validation during his career at Merck, Pfizer, and Bristol-Myers Squibb. Since the formation of A&A in 1991, Jim has assisted more than 100 firms with validation, sterilization, aseptic processing, and compliance. He has edited/coedited four texts, authored/coauthored more than 40 chapters, published more than 100 papers, and lectured extensively on numerous subjects. He is a past president of the Parenteral Drug Association (PDA), a current member of USP's Microbiology Expert Committee, the Editorial Advisory Board for Pharmaceutical Technology and Pharmaceutical Manufacturing.

Ruhi Ahmed is currently the Senior Director in the Regulatory Affairs department at Ultragenyx Pharmaceutical, Inc. She has previously worked in the Regulatory Affairs at Watson Pharmaceuticals and BioMarin Pharmaceutical Inc. Dr. Ahmed graduated in 2003 from the University of Southern California (USC) with a PhD in Molecular Pharmacology and Toxicology. She also has an MS degree in Regulatory Science from the USC and an MA degree in Biochemistry from the University of Texas at Austin. Dr. Ahmed has earned her RAC certification for both the European Union and the United States. As a member of PDA's Task Force on Quality Risk Management, Dr. Ahmed has participated in coauthoring PDA Technical Report #44 on "Quality Risk Management for Aseptic Processes." She is currently leading a PDA Task Force on assessing general applications of Quality Risk Management for Biopharmaceutical APIs.

James E. Akers, PhD, is the President of Akers Kennedy & Associates, Inc., located in Kansas City, Missouri. Dr. Akers has over 25 years experience in the pharmaceutical industry and has worked at various director-level positions within the industry and for the past decade as a consultant. Dr. Akers' areas of expertise include aseptic processing, contamination control, sterilization, biological products, and advanced aseptic technologies. Dr. Akers served as President of the PDA from 1991 to 1993 and as a member of the PDA Board of Directors from 1986 to 1999. Currently, he is the Chairman of the USP Committee of Experts Microbiology and has served on a number of PDA Task Forces.

Hal Baseman has over 30 years of experience in pharmaceutical operations, validation, and regulatory compliance. Hal has held positions in executive management and technical operations at several manufacturing and consulting firms. He is a long-time member and a frequent presenter for PDA and ISPE (International Society for Pharmaceutical Engineering). Hal has held positions as a member of the PDA Board of Directors, the Cochair of the PDA Science Advisory Board, the Coleader of the PDA Validation Interest Group, the Cochair of the PDA Aseptic Process Simulation Task Force, and the Cochair of the PDA Quality Risk Management Task Force. Hal Baseman holds an MBA from the LaSalle University and a BS degree from the Ursinus College.

Michael Bogan is the President of Integrated Commissioning & Qualification Consultants Corp. During his 20-year career in the life sciences industry, Michael also held various leadership roles in project management, process development, and manufacturing. Michael has also held a position with Amgen as the Senior Manager of Validation, who is responsible for validation activities at various plants. He has presented on the subject of commissioning and qualification strategies to industry trade associations, including the ISPE. He holds his degree from the Franklin Pierce College, with additional study in Biology at the Northeastern University.

Scott Bozzone is the Senior Manager in the Quality Systems and Technical Services, Validation for Pfizer, Inc. based in Peapack, New Jersey. He has been at Pfizer for 26 years and prior to that worked several years at Revlon Health Care Group (Armour/USV). In his current position, he is responsible for validation site support and leads Pfizer's global Validation Community of Practice. He also serves on Pfizer's Quality Risk Management training team. He has a doctoral degree in Industrial Pharmacy from the St. John's University in New York.

Thomas Busmann is a Vice President of Focus Compliance & Validation Services. He is a registered professional chemical engineer with over 30 years of experience serving numerous clients in the pharmaceutical, biopharma, medical device, and chemical industries to assist them in achieving regulatory compliance with their processes and projects. Areas of expertise include risk management, verification of process control systems, computer system validation, process validation, utility equipment validation, training, and hazard and safety analysis.

Bruce Davis runs his own Consultancy in Quality by Design (QbD) and engineering. He carries out in-house and external training in this area and also helps run a UK university QbD course. He is a professional engineer with a wide international knowledge of the pharmaceutical industry, having previously worked at AstraZeneca, where his responsibilities included managing international engineering, facilitating QbD, and leading a global change process for enhanced qualification.

He is a past chair of the ISPE Board of Directors and, for them, leads a cross-company case study for practical implementation of QbD. He is the Secretary to ASTM E55.03 on General Pharmaceutical Standards.

Jorge A. Ferreira is the Technology Manager with Jacobs Engineering Group. Before this, he held project management, engineering, and operations positions with Bristol-Myers Squibb. Mr. Ferreira has been involved in the pharmaceutical industry for more than 25 years with expertise in engineering design, construction, and plant operations, including supervision and management responsibilities for aseptic manufacturing operations. His background includes clean utilities, formulation, component preparation, sterilization and depyrogenation, aseptic filling, lyophilization, inspection, packaging, and warehousing logistics. He holds a BS in Mechanical Engineering from the New Jersey Institute of Technology, New Jersey.

Thomas Genova, PhD, is a Johnson & Johnson Quality Fellow supporting the development of combination products and risk management activities. He began his career qualifying ethylene oxide and gamma irradiation processes. His career evolved to include steam and dry heat processes, filtration operations, and aseptic filling processes. At the Temple University, he taught graduate classes in *Sterilization Processes* and *High Purity Water Systems*. He was a member of the PhARMA Water Quality Committee, the Technical Advisory Group of ISO/TC 209 Cleanrooms and Associated Environments, and the PDA Risk Management Task Force. Tom is an ASQ Certified Quality Engineer (CQE) and a Six Sigma Master Black Belt. He holds a PhD in Immunotoxicology from the Rutgers University.

Karen S. Ginsbury, BPharm, MSc, MRPS, is a London-trained pharmacist who has worked for the past 25 years in the field of pharmaceutical quality assurance and compliance. Karen also has a Master of Science degree in Microbiology from the Birkbeck College, University of London. Currently, Ms. Ginsbury is the President and the CEO of PCI Pharmaceutical Consulting Israel Ltd., a consultancy company working with multinational as well as small start-up companies, and has acquired considerable expertise in the field of Investigational Products. Also, she is currently chairing the PDA's Task Force in preparing a Technical Report on GMP Points to Consider for Investigational Products, and she has particular knowledge in the application of risk management to the field of product and process development throughout the product lifecycle.

William Harclerode has been with Forest Labs since 2006 in the Corporate Quality Risk Management Department. He is currently responsible for providing leadership and support for implementation of quality risk management activities across the product supply chain. Before that, he worked as Validation Manager and Production Manager at Abbott Labs and Knoll Pharmaceutical Company. He holds an MBA from the Fairleigh Dickinson University, a BS in Chemical Engineering from the Drexel University, and a BS in Biology from the St. Lawrence University.

Dr. Mike Long has two decades of experience leading product, process development and validation efforts on a wide range of pharmaceutical, medical device, and combination products. He is a frequent speaker/writer on topics such as Risk Management, Quality Systems, QbD, Process Validation, and Process Robustness. He is an active member on industry committees including the PDA's Science Advisory Board. Mike has instructed graduate courses in Data Analysis and in Risk Management and Quality Systems. Mike is a Master Black Belt with a BS from the Worcester Polytechnic Institute, an MS from the Tufts University, and a doctorate from the Northeastern University.

A. Hamid Mollah is currently the head of Quality Engineering and Validation at XOMA. He held positions at Genentech Inc. and Baxter BioScience and has 14 years of pharmaceutical and biopharmaceutical industry experience in the areas of process development, manufacturing, validation, and regulatory affairs. Before joining the pharmaceutical industry, he worked on fermentation process development, design, scale-up, optimization, and project management for 7 years. He received his PhD in Biochemical Engineering from the Imperial College of Science, Technology, and Medicine in London, England. Hamid has given talks at various conferences (PDA, IBC, IQPC) and is published in a number of peer-reviewed journals, including articles on risk assessment and risk-based validation. He is an RAPS-certified regulatory affair professional (RAC) and ASQ CQE.

Bob Moser has been working at Jacobs Engineering for the past 11 years, most recently, as a Senior Engineer and Supervisor. Mr. Moser is a chemical engineer with 25 years experience in quality, environmental, safety, and risk management within the chemical, pharmaceutical, and general manufacturing industries. He has broad domestic and international (Europe and Asia) experience. His background includes leadership of FMEA (Failure Modes and Effects Analysis) processes for both quality and reliability. He holds a BS in Chemical Engineering from the Pennsylvania State University. His designations include

Professional Engineer (PE), Certified Safety Professional (CSP), and Associate Risk Manager (ARM).

Christophe Noualhac is currently the Project Manager at Halo Pharmaceutical. He is responsible for directing and coordinating activities associated with pharmaceutical development projects. Before joining Halo Pharmaceutical in 2009, he worked as Validation Manager at PAR Pharmaceuticals and Technical Support at Abbott Laboratories. He holds a Master Degree in Pharmaceutical Chemistry from the University of Toulouse (France) and a Master Degree in Applied Physics and Chemistry from the University of Montpellier (France).

Marlene Raschiatore is the Director of Global Regulatory Compliance for Johnson and Johnson, Consumer Sector. During her career, she has held quality, compliance, and regulatory affairs positions with pharmaceutical product manufacturing companies, including Elkins-Sinn Inc. and Wyeth Pharmaceutical. She was an author and contributor on the PDA's Technical Report No. 44, Risk Management for Aseptic Processing. Ms. Raschiatore holds a Juris Doctor from the Beasley School of Law and a Bachelor of Science in Medical Technology from the Temple University in Philadelphia, Pennsylvania.

Edward C. Tidswell, BSc, PhD, is the Senior Director of Sterility Assurance for Baxter Healthcare located north of Chicago, Illinois. His organization provides support for more than 40 facilities globally, encompassing the entire breadth of microbiological control, sterilization, and sterility assurance across a diversified healthcare company. Dr. Tidswell is a leading authority on risk, aseptic, and sterile manufacture. In 2004, he received the Parenteral Society's George Sykes Memorial Award for his contribution to pharmaceutical risk assessment. In June 2010, Dr. Tidswell joined the USP expert committee on Microbiology & Sterility Assurance.

Mark Walker is a Vice President of Focus Compliance & Validation Services (Focus CVS) located in Knoxville, Tennessee. Focus CVS is dedicated to providing regulatory compliance, risk assessment, validation, training, and process engineering services to pharmaceutical, biotechnology, and medical device industries. Mr. Walker has more than 25 years experience in providing consulting, project management, and business development services. His expertise includes risk assessment, regulatory compliance, computer systems validation, GMP and validation training, process validation, document management, local area network system design, database system design and implementation, and business development services to the pharmaceutical, medical device, and biotech industries. Mr. Walker currently serves on the Association for the Advancement of Medical Instrumentation's (AAMI) medical device software work group.

1

BACKGROUND AND INTRODUCTION

Harold S. Baseman and A. Hamid Mollah

Companies wishing to manufacture and distribute regulated health care products to the population of the United States must comply with the U.S. Food and Drug Administration (FDA) regulations, better known as *Current Good Manufacturing Practices* (CGMPs). 21 CFR 211.100 of U.S. CGMPs states "There shall be written procedures for product and process control designed to assure that drug products have identity, strength, quality, and purity they purport or are represented to possess ..."[1]. Regulations are a legal requirement and this CFR, among others, mandates that companies must take active steps to assure product quality.

Companies and individuals working for health care industries have an obligation to provide products that are safe and effective to their customers, users, and patients. The regulations codify this obligation, thus making it a legal requirement; but the obligation to provide safe and effective products is also a moral and ethical obligation that goes beyond the legal regulatory requirements. People working for pharmaceutical companies also have a duty of loyalty to operate for the welfare of the company. In other words, they have an additional obligation to operate efficiently and earn optimal profits within the framework of regulatory requirements and ethics. Companies and individuals must be able to align these legal requirements and business obligations.

The failure to provide safe and effective products will likely result in loss of business as well as other legal consequences. However, in recent years, it seems that the industry has faced pressure and challenges to balance these requirements and obligations. It has become more difficult to remain in compliance, serve

---

Risk Management Applications in Pharmaceutical and Biopharmaceutical Manufacturing, First Edition. Edited by A. Hamid Mollah, Mike Long, and Harold S. Baseman.
© 2013 John Wiley & Sons, Inc. Published 2013 by John Wiley & Sons, Inc.

customers, and be competitive. Companies have struggled with balancing regulatory requirements, scientific elements of product development and manufacture, and maintaining a productive business situation.

The pharmaceutical and biopharmaceutical industries are facing financial pressure because of the high cost of drug development and manufacturing as well as generic competition. There are business drivers and regulatory expectations for innovative approaches to speed up pharmaceutical product development and licensure, optimally use resources, and to assure continued product quality and patient safety. The industry must apply comprehensive risk management and innovative approaches to product life cycle not only to enhance patient safety but also to improve business outcomes. Hence, it is critical to understand appropriate risk management tools and approaches that would be acceptable to regulatory agencies. Other industries, including closely related ones such as the medical devices and food industries, have adopted a more structured approach to this subject than we have traditionally used. The application of risk management to medical devices is expected by medical device regulatory bodies [2–4]. Hazard analysis and critical control points (HACCP) is used in the food industry to identify potential food safety hazards, so that key actions, known as *critical control points* (CCPs), can be taken to reduce or eliminate the risk of the hazards being realized [5].

In the summer of 2002, the FDA announced an initiative to "enhance and modernize" pharmaceutical regulation. In the fall of 2004, it published the final report on *Pharmaceutical cGMPs for the 21st Century—A Risk-Based Approach*. This paper represented an attempt to "enhance and modernize" pharmaceutical regulation. It not only speaks of product quality and patient safety but also of the need for innovation and the cost of drug development and manufacture [6].

The paper offered initiatives and recommendations with the following objectives in mind:

1. Encourage the early adoption of new technological advances by the pharmaceutical industry.
2. Facilitate industry application of modern quality management techniques, including implementation of quality systems approaches, to all aspects of pharmaceutical production and quality assurance.
3. Encourage implementation of risk-based approaches that focus the attention of both industry and agency on critical areas.
4. Ensure that regulatory review, compliance, and inspection policies are based on the state-of-the-art pharmaceutical science.
5. Enhance the consistency and coordination of FDA's drug quality regulatory programs, in part, by further integrating enhanced quality systems approaches into the agency's business processes and regulatory policies concerning review and inspection activities.

The reference to risk-based approaches mentioned in (3) is of particular interest to the subject of this book. Facing limited resources, the agency recognized that to best serve public interest, decisions on resource allocation, focus, and prioritization should be based on risk to patient safety and public safety. Those in the industry are impacted by the approach. For instance, a firm manufacturing over-the-counter (OTC) oral dosage products and having a relatively clean compliance record would likely be inspected less often or receive less attention than a firm aseptically manufacturing sterile injectables and having a more problematic compliance record.

The prioritization of resources based on risk to public safety make sense and it led to better productivity and effectiveness. It was logical that the agency would expect the industry to employ similar approach to make resource- and focus-related decisions. Firms are encouraged to use risk to product quality and patient safety as a criterion for decision making.

Risk management and assessment are not new. People use risk assessment as a way to help make decisions every day. When you walk across the street, drive through a yellow light, or order a meal—you employ a level of risk assessment, weighing the impact of a hazard and the likelihood of the hazard happening against anticipated benefit. Companies do the same in many aspects of corporate functioning from financial decisions, to investments, to plant locations, and product development. If their objective is to serve their customer, then it makes sense that they would employ this type of decision making to manufacturing and response to patient needs and safety.

In 2005, the ICH (International Conference for Harmonization) issued Q9 Guidance on quality risk management. ICH Q9 was later issued in 2006 as Guidance for Industry by the FDA and adopted by the EU as Annex 20 of the European GMPs in 2011. The guidance remains optional for pharmaceutical product manufacturers in the United States and Europe [7]. However, references to risk assessments and criticism for not employing such measures have appeared in FDA warning letters dating back to 2006 [8]. Regulatory citations indicated that companies face questions on how decisions related to product quality were made, if assessments of the risk of process steps and changes to product quality were not employed. If a company's obligation is product quality and patient safety, it should take such risks into account when making manufacturing decisions. How else could it make these decisions?

In the spring of 2005, at the PDA (Parenteral Drug Association) annual meeting in Chicago, the leaders of the Process Validation Interest Group asked its members for their topic of most interest or concern. The overwhelming answer was risk management. The leaders then asked how many of those individuals were currently utilizing or were aware of efforts within their respective organizations to utilize risk management. Only a few raised their hands. This was not unexpected. ICH Q9 was being issued and reviewed. Papers presented at the PDA annual meeting spoke about the need for risk management.

One person in the meeting noted that their risk assessment efforts were unsuccessful, as they were subject to criticism from local regulators, because of the misuse of the risk management. The misuse apparently involved using risk assessment to identify process-related risk, but then failing to take steps to mitigate that risk. The objective of risk management, as discussed later, is not just to identify risk, but to mitigate and reduce risk, thus improving the manufacturing process.

The outcome of the 2005 meeting was an initiative by the PDA Science Advisory Board to create a task force of industry professionals to investigate and develop a model for the use of risk management for aseptic processes. This would later become the basis of PDA *Technical Report No. 44 Quality Risk Management for Aseptic Processes*, as well as later efforts on companion documents and reports. The task force was made up of 15 individuals from sterile drug manufacturing within 15 different organizations and companies. Only a few had direct experience with formal risk management and that experience had largely come from the medical device industry. The use of formal risk assessment and management techniques for pharmaceutical and biopharmaceutical manufacturing appeared to be a work in progress at best.

In 2008, the PDA published Technical Report No. 44. The technical report presented concepts and a program for evaluating the risk of process failure in making decisions for the manufacture of sterile drug products using aseptic processing. One point presented in TR 44 was that aseptic processing was not necessarily risky. The hazards associated with aseptic processing were significant. However, if well controlled, the risk should not necessarily be high. In other words, determining the risk was the objective of risk management—rather, process improvement through control and mitigation were the key objectives [9].

Since 2004, more and more FDA guidance has included recommendations for risk management and assessments. In the 2008 draft version of the FDA Guidance for Industry on the General Principles of Process Validation, the FDA included a modest level of references to risk assessments in the text. Some industry comments questioned the apparent "lack" of focus on risk in the document. When asked, FDA representatives responded that they felt risk management principles and methodology were so prevalent in the fabric of industry operation that it was not necessary to emphasize it in the guidance. The number of references to risk management and assessment nearly doubled in the 2011 final version [10–12].

Throughout the next several years, industry standards, guidance, and technical reports were prepared to address risk-based decision-making. In 2001 through 2011, the ISPE (International Society of Pharmaceutical Engineering) published a series of industry guides, employing risk-based methods for design and qualification of pharmaceutical manufacturing facilities and processes, including Volume 5 of its *Facilities Baseline Guides: Commissioning and Qualification* (with revisions in progress) and the *ISPE Guide: Science and Risk-Based Approach for the Delivery of Facilities, Systems, and Equipment*. These guides presented methods for qualifying pharmaceutical manufacturing facilities incorporating risk to product-quality-based decision criteria. The baseline guide introduced the concept of evaluating systems based on their relative impact to product quality [13].

In 2007, the ASTM (American Society for Testing and Methodology) issued E2500-07, the *Standard Guide for Specification, Design, and Verification of Pharmaceutical and Biopharmaceutical Manufacturing Systems and Equipment*. E2500-07 discussed a risk- and science-based approach to the qualification or verification of equipment used to manufacture and test pharmaceutical products. It was an effort to use risk to product quality and patient safety as important factors when deciding what to qualify, how to qualify, when to qualify, and who should be involved in the qualification and approval effort [14].

When the ASTM committee E55 was assembled to create and review what would become E2500-07, they discovered that while many, companies recognized that quality risk management was an important tool for making product manufacturing decisions, few, had real experience or input into practical means to accomplish this in an effective manner. As such, the committee was faced with creating desired state approaches rather than reflecting tried and true best practices. Throughout related meetings and discussions, it appeared that most companies had some appreciation for the need to manage risk to product quality and as a part of that to take steps to assess and document the assessment of that risk. However, it also appeared that companies did not always utilize risk management techniques optimally or effectively in making decisions. One is reminded of a company visited not long after the 2005 PDA meeting. The company had a vigorous risk management program, complete with corporate directives, policies, procedures, and a risk management department. They had volumes of carefully filled out risk assessments, which were placed in binders and displayed. When asked what these risk assessments were used for, the response was to assess risk. The assessment forms were meticulously filled in, reviewed, and approved. After that, they were placed in binders and placed on a shelf. Whether the information was used to help make any decisions was not apparent. This illustrates the misconception that the objective of risk management is to merely assess or categorize risk, rather than using it to provide information to help make informed decisions and improve the process.

The objective of risk management should be to improve the process by reducing or mitigating risks. There needs to be a clear link between risk management principles as described in guidances such as ICH Q9 and other guidances and practical manufacturing activities. The book offers the reader multiple perspectives and approaches to risk management and assessments. The chapters in this book emphasize that quality risk management of pharmaceutical manufacturing processes is an important topic because of the following:

1. It is a regulatory expectation, in that it helps to assure product quality and ensure patient safety. The FDA, European Medicines Agency (EMA), and many other regulatory agencies strongly recommend the use of risk assessment and the consideration of risk to patient safety and product quality in making decisions related to product development, manufacture, and distribution.
2. It is a good business practice. Properly used, risk management should help assure product quality as well as promote efficient utilization of resources and prioritization of efforts. It should help companies reduce redundant and non-value added efforts, while allowing them to focus on efforts that have optimal impact on product quality.

3. It is a logical and effective means of obtaining useful information needed to make sound quality and business decisions. Risk management represents an organized method for obtaining, analyzing, and communicating useful information.

The regulatory environment emphasizes the use of enhanced knowledge of product performance over a range of material attributes, manufacturing process options, and process parameters to identify risks to patient safety and product quality. Risk analysis and risk management are acceptable and effective ways to minimize patient risk and determine appropriate levels of validation and controls. The use of quality risk management does not obviate company obligation to comply with regulatory requirements. However, effective quality risk management can facilitate better and informed decisions, can provide regulators with greater assurance of a company's ability to deal with potential risks, and might affect the extent and level of regulatory inspections. The effective and consistent analysis of risks associated with manufacturing processes and quality systems typically leads to more robust decisions and yields greater confidence in outcomes. The ultimate goal of the risk management process of an organization is to bring focus and effort to the issues that impart the highest risk to product quality and/or patient safety. Risk management outputs will potentially serve as reference documents to support product development and control strategy discussions in regulatory filings.

1.1 RISK MANAGEMENT OF PHARMACEUTICAL AND BIOPHARMACEUTICAL MANUFACTURE

The efficient manufacturing of quality pharmaceutical products presents a challenge in the present day business environment. If not properly controlled, these challenges can represent risk to product quality and in turn patient safety. There are several reasons why the business environment presents unique challenges, including the following:

1. *The need to understand and comply with evolving regulatory requirements and expectations.* In the past, the FDA represented the benchmark or standard source for drug product and manufacturing regulations. Most foreign regulators utilized the principles presented by the FDA in their own regulations. However, as European and other non-U.S. markets developed, companies faced unique interpretations and presentations of regulations and requirements from other regulatory bodies. At times, as one regulatory body would attempt to modernize its regulations or guidance, other countries would lag behind, creating the potential for misinterpretation and misalignment of regulatory expectations or focus.
2. *The use and integration of innovative technologies with existing manufacturing methodology.* New technologies such as automated processes, PAT

(process analytical technology), rapid microbiological analysis, and single-use manufacturing systems offer the potential for process improvement and risk reduction by eliminating process variation, human intervention, and more reliable product and process testing/monitoring. However, these technologies may bring with them new or additional risks associated with understanding the limitations of the technology.

3. *Adapting existing manufacturing methodology to new products and dosage forms.* Sometimes, tried and true approaches that have worked for older technologies will not work as well for newer technologies. An example may be the use of aseptic process simulations or media fill tests designed to demonstrate aseptic technique proficiency, used to assure aseptic processing control in relatively intervention-free automated or isolator systems.

4. *The retooling of facilities and the transfer of products and technology as a result of consolidation of plants and assets.* There are likely to be physical, procedural, and cultural differences between facilities that need to be considered in order to effectively manufacture products.

5. *The loss of knowledgeable staff through attrition and reorganization.* Even the best controlled processes, with the most well-written procedures, are subject to a certain level of "tribal knowledge." Efficient and effective manufacturing depends in part on the dissemination of information, much of which is learned from experience. However, if that experience or the people who have it are no longer with the company, then what will replace that experience-based knowledge?

6. *The need to better understand the interdependencies and variability of materials, technology, and product on more complex processes.* More complex products and dosage forms, as well as combination products, present new process development and manufacturing challenges.

7. *The need to maximize productivity and minimize cost.* Quality may be the number one factor in pharmaceutical manufacturing, but controlling cost and resources has taken on a major level of importance in most modern manufacturing operations. This often leads to LEAN manufacturing methods, doing more with less, automation, and streamlined operations and workforces. All are for the good, but established methods of quality assurance may also need to adapt to this new business environment.

8. *The need to control processes to achieve consistent quality and maintain product supply from and across multiple plants and locations.* As seen in the 2009 heparin issue, where products and supplies originated in emerging growth countries, there may be a need to examine the effectiveness of existing methods and procedures for quality control and quality assurance including audit, training, monitoring, and testing regiments [15]. In addition, more complex products and inspection techniques may result in or uncover quality issues with critical supplies, such as glassware and product contact containers. Improved methods of identifying and addressing these issues may be necessary. Where inspection is not enough, quality by design

(QbD) and other ways to identify and mitigate risk may be one answer to supplier and supply-related quality issues.

1.2 A PRACTICAL GUIDE TO RISK MANAGEMENT

It is important to have a clear concept on various terms used in risk management. The concept of risk has two components: (i) probability of occurrence of harm and (ii) the consequence of that harm (i.e., severity). A hazardous situation is a circumstance in which people, property, or environment is exposed to one or more harm(s)[16] [ISO 14971]. Risk analysis involves the estimation of risk(s) for each hazardous situation or failure mode. Harm, in the context of this book, is damage to health, including the damage that can occur from loss of product quality or availability. Severity is a measure of the possible consequences of a hazard. Hazard is a potential source of harm (i.e., an immediate output from the product/process/system that directly causes harm). Risk is the possibility of suffering harm or loss. More specifically, risk is the relationship between impact of a hazard and the probability of that hazard occurring to such an extent as to result in harm. Risk to product quality is the combination of the severity or the impact of an unwanted event and the likelihood that event will occur to a degree which will adversely affect product quality. Some examples of hazards, harm, and risk are listed here:

1. Hazards
   - Product not sterile or impure
   - Product subpotent or superpotent
   - Product contaminated
   - Product mislabeled
   - Product unsealed or improperly sealed
   - Product missing or unusable product
   - Ineffective product
   - Lack of product supply
   - Noncompliance
   - Product rejection
   - Inefficient process
   - Misuse of product
   - Poor process yield
   - Failure to receive product approval
2. Harm
   - Injury to patient
   - Disruption of product supply

ICH Q9 explains what quality risk management is, how it can be applied to pharmaceuticals, and how it can provide a common language with an agreed process for the pharmaceutical industry and regulators. In ICH Q9 risk management models, "risk" is defined as "the combination of the probability of occurrence of harm and the severity of that harm." While the combination of probability and severity is helpful in reflecting the level of risk importance, detectability often influences the decision-making process in manufacturing risk management.

ICH Q9 places focus on risk to patient safety due to a product defect or loss of quality along the supply chain. ICH Q9 defines risk management as *a systematic application of quality management policies, procedures, and practices to the tasks of assessing, controlling, communicating and reviewing risk*. Risk management is then a process by which sources of risk are recognized and steps taken to mitigate, reduce, or eliminate the chance of harm. If the objective of risk management is to avoid harm, then one way to meet this objective is to provide a means to make decisions based on relative risk [17].

Figure 1.1 of ICH Q9 presents a logical flow for managing risk [17]. The flow can be broken down into distinct steps and substeps.

1. *Risk assessment*—understand the process.
   - *Risk identification—identify process/quality hazards and potential harm it might cause.*
   - *Risk analysis—determine what event or condition could cause the hazard.*
   - *Risk evaluation—rank or score the relative risk of the hazard, in an effort to recognize when improvement has occurred.*
2. *Risk control*—react to the outcome.
   - *Risk reduction—improve the process through mitigation of the risk.*
   - *Risk acceptance—decide if the process risk has been reduced to an acceptable level or if further mitigation or evaluation is necessary.*
3. *Risk communication*—interact with interested parties to relay risk-related information in order to implement mitigation-related changes or communicate residual or remaining risk.
4. *Risk review*—follow up and periodically reassess to determine if changes have been implemented and if changes are effective.

Risk management is a method to

- recognize and address potential weakness in the process, in an effort to assure objectives are met
- identify potential hazards
- assess likelihood of occurrence
- decide if process risk is acceptable
- communicate risk
- reduce risk

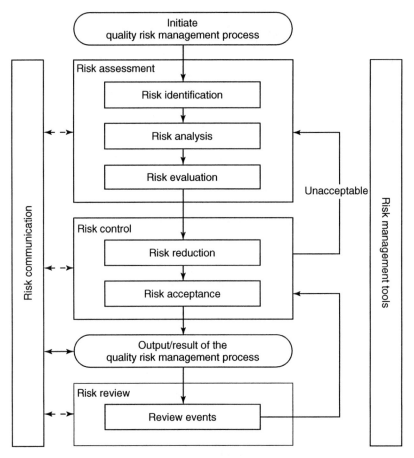

Figure 1.1 Overview of a typical quality risk management process.

- improve the process
- provide information needed to help make decisions

Risk management approaches should

- focus on risk to patient safety
- result in improved process understanding
- result in improved process
- be planned, logical, and documented
- add value
- avoid "checklist" approach
- should support and be consistent with the validation program
- should be documented

1.2.1 Additional Points to Consider

People who are designing manufacturing processes and respective control strategies need to realize the relative risk inherent in these process steps. In this case, it is the relative risk of process failure that could adversely affect product quality and patient safety. The unwanted effect on product quality is the loss of quality attributes. Quality attributes are those elements or functions of the product that define it.

For pharmaceutical products, these attributes include the following:

Strength—potency, efficacy, effectiveness
Safety—does not cause harm, contamination, loss of sterility
Purity—free of foreign substances, contamination
Identity—what it purports to be, lot number, expiry date

In the pharmaceutical industry, we are primarily concerned with risk to patient safety or public welfare. Loss of product quality leads to loss of patient safety. If the failure or unwanted event is found and removed before it can affect the patient, then there is no harm. If there is no harm, then there is no risk. Therefore, detection becomes an important element in determining relative risk.

Risk assessments and management techniques should be used to provide the information needed to make sound decisions, but they should not be used to make the decision itself. In other words, setting a predetermined decision based on a numeric scoring from a risk assessment model and then blindly following that outcome without further analysis and thought can lead to problems and biased assessments.

For risk management to be useful, risk assessments must be as accurate as possible. Therefore, objectiveness and unbiased assessment and analysis are key. It is essential that companies take care of the following:

1. Avoid preconceived decisions or results.
2. Do not use risk assessment to *validate* a position or to justify a questionable process.
3. Use diverse assessment team(s) to assure objectivity.
   - Use experienced facilitator if necessary.
   - Allow for adequate time, but do not overdo it.
4. Pay attention to the outcome.
   - Aseptic process is an example.
5. Document for future reference.
6. Pay attention to results.
7. Plan for follow through and feedback.
8. Plan for communication.

Mitigation of risk involves taking action and making changes. These changes may not mitigate all risk and may add other risks—or residual risks. Residual risks are risks that remain after mitigation changes are made. It is important to recognize and address these residual risks. This does not mean avoiding mitigation changes because of residual risk.

Choose the risk assessment method that best fits the complexity and impact of the decision to be made. One method may not fit all applications. Avoid unnecessarily complex or burdensome methods when simple ones will accomplish the same result. However, choose a method that will provide enough information and is as objective as possible.

1.3 OVERVIEW OF THE BOOK

Companies have struggled with the best way to effectively use and demonstrate the use of risk-based information gathering, evaluation of information, and decision-making processes, often doing too little or too much. In an effort to assist the reader with the development and use of effective risk-based approaches for pharmaceutical and biopharmaceutical product manufacturing, this book will provide guidance, including some divergent views on practical and pragmatic ways to incorporate risk into their operations. While the contents of this book are not meant to include all of the methods and areas where QRM can and should be incorporated, it provides several approaches and examples that can be used as a framework for developing and implementing risk-based decision-making methods for other processes and process steps.

The first chapters are designed to familiarize the reader with the subject of risk management in the context of pharmaceutical and biopharmaceutical manufacturing and present basic concepts and ideas for developing and utilizing an effective risk management program. To that end, Chapter 1 provides an introduction and background to risk management for pharmaceutical and biopharmaceutical products and processes. Chapter 2 introduces the reader to general information on the development and use of a risk management program. It shows widely used risk assessment tools and methods, many of which are used in subsequent chapters. Chapter 3 provides additional insight into the nuances of the risk management program, including regulatory expectations, a high level overview of the cognitive and social aspects of risk, as well as thoughts on developing an objective risk management program and an effective organizational culture. Chapter 4 presents views and a commentary on the use of statistical analysis to assist with the planning of risk-based approaches and with useful analysis of the resulting information.

The next chapters present specific, yet not exclusive, areas of pharmaceutical and biopharmaceutical product development and validation where risk management techniques can be used, along with programs and approaches for doing so. As such, Chapter 5 discusses the use of QRM in QbD aspects of product and process development. The QbD approach is where quality is designed from the outset, using a science and risk basis, as opposed to a traditional approach, where normally end-product testing used to check quality requirements have been met. Chapter 5 shows key steps for QbD and how to apply them.

Chapter 6 presents approaches to use and disseminate risk-based information in making decisions related to early product development and clinical product manufacture. It shows the integration of science and risk management to allow for successful product development, such as basis for design and product control strategy. It uses a block diagram for tablet manufacture, and an Ishikawa (fishbone) diagram, breaking the process into the 6Ms to identify risk factors that need to be considered and possibly controlled in designing the manufacturing and control process.

Chapter 7 discusses methods for using QRM to commission and qualify equipment and facilities utilized in the manufacture of products, including the evaluation and leveraging of information. It presents important considerations for the use of QRM in making decisions needed to plan, develop, and conduct more effective commissioning and qualification efforts. It shows some of the areas where risk assessment can effectively be used to help develop and implement a sound, efficient qualification program using simple but effective tools.

Chapter 8 then picks up the use of risk in developing and implementing a sound and effective process validation life cycle. It shows how risk-based approaches can be applied during process characterization and validation, for the identification of CPPs for new and existing processes, risk prioritization, technology transfer, process changes, and in defining review schedule. Cleaning validation and cross-contamination risks are also discussed in this chapter.

The next chapters present several specific areas of product manufacture where risk assessment can be used. Chapters 9 and 10 provide two views on the use of risk assessment for the evaluation and improvement of one of the most challenging and complex manufacturing processes—aseptic processes. Risk assessment in the context of aseptic pharmaceutical manufacture is the principal focus in Chapter 9, with particular description and explanation of a quantitative, statistical tool of risk analysis permitting a more exacting evaluation of risk. It describes aseptic processing hazards, such as intrinsic hazards, extrinsic hazards, risk of endotoxin, and models for microbial ingress. Chapter 10 reviews contemporary thinking relative to aseptic processing risk assessment and mitigation. Formalized risk assessment described in this chapter and its essential counterpart risk mitigation will play an increasing role in the design, operation, and maintenance of aseptic operations.

Chapters 11 and 12 present views on the use of risk for better understanding and operation of drug and biopharmaceutical manufacture. Chapter 11 focuses on the areas of risk that a drug company may encounter in pharmaceutical manufacturing, specifically addressing oral solid and liquid formulations. Common risks associated with the manufacturing process for a solid tablet outlined in a fishbone diagram are identified. A case study illustrates how to apply risk management principles to identify and mitigate risks that could affect product quality and patient safety using the HACCP process. Chapter 12 discusses applications of risk management in critical areas of biopharmaceutical production, such as raw material supply, cell banking, fermentation, cell culture, purification, scale-up of production process, and distribution issues associated with cold chain.

Finally, Chapter 13 provides a risk-based approach to controlling processes and process-related changes. Integrating QRM into the change control system is essential to maintain risk management as a "living" process, but it can be especially challenging because change control covers many areas in manufacturing and most of the product life cycle. Chapter 13 provides some practical methods and tools that can be used to integrate QRM into an existing change control system.

1.4 FINAL THOUGHTS

Risk management can be a useful tool in controlling processes, assuring product quality, prioritizing resources, and understanding processes. Risk assessment can be helpful for obtaining the information needed to make manufacturing decisions in a challenging business environment. The consideration of risk in making product-quality-related decisions is not only a logical business practice but also a regulatory expectation.

Effective quality risk management can facilitate better and more informed decisions, can provide regulators with greater assurance of a company's ability to deal with potential risks, and might affect the extent and level of direct regulatory oversight. Proper documentation is important to achieve this goal. However, the level of effort, formality, and documentation of the quality risk management process should be commensurate with the level of risk. The application of risk management practices enables manufacturers to design processes that proactively identify, mitigate, and/or control risks. Risk management practices may be implemented via a well-designed risk assessment plan and activities.

The use of risk management, when properly planned and implemented, can provide valuable information. That information, when properly considered, can then lead to better product quality-related decisions. Those decisions, if properly implemented, can assure product quality and improved processes. Process improvement can better ensure patient safety, which is the objective of quality risk management.

REFERENCES

1. Department of Health and Human Services, U.S. Food and Drug Administration, Code of Federal Regulation, Title 21, Chapter 211, Section 100 (CFR 211.100 of cGMPs), 2008.
2. Medical Device Directives, Annex 1—Essential Requirements, European Commission.

REFERENCES

3. FDA, Medical Devices: Current Good Manufacturing Practices (CGMP) Final Rule; Quality Systems Regulation, Preamble, Federal Register, 61(195), 52,602–52,662, October 7, 1996.
4. 21 CFR 820.30 (g).
5. 21CFR120, Hazard Analysis and Critical Control Point (HACCP) System.
6. Department of Health and Human Services, U.S. Food and Drug Administration, Pharmaceutical cGMPs for the 21st Century—A Risk-Based Approach Final Report—Fall 2004, September 2004.
7. Department of Health and Human Services, U.S. Food and Drug Administration, Guidance for Industry, Q9 Quality Risk Management, 2006.
8. Department of Health and Human Services, U.S. Food and Drug Administration, Warning Letter to Bell-More Laboratories, Inc., January 5, 2007.
9. Technical Report 44, Quality Risk Management of Aseptic Practices, PDA Journal of Pharmaceutical Science and Technology, 2008 Supplement Volume 62, No. S-1.
10. Department of Health and Human Services, U.S. Food and Drug Administration, Guidance for Industry on the General Principles of Process Validation, January 2011.
11. Department of Health and Human Services, U.S. Food and Drug Administration, Guidance for Industry: Sterile Drug Products Produced by Aseptic Processing, September 2004.
12. PDA Submits over 400 Comments on FDA Draft Revised Validation Guidance, PDA Letter, Volume XLV, Issue #3, March 2009.
13. International Society for Pharmaceutical Engineering ISPE Baseline®, Pharmaceutical Engineering Guide Series, Volume 5—Commissioning and Qualification, First Edition, March 2001.
14. ASTM (American Society for Testing and Methodology) E2500-07, Standard Guide for Standard Guide for Specification, Design, and Verification of Pharmaceutical and Biopharmaceutical Manufacturing Systems and Equipment.
15. Heparin Issue Reference China may be Source of Tainted Heparin, Associated Press, March 5, 2008, www.msnbc.msn.com.
16. ISO 14971:2000—Application of Risk Management to Medical Devices.
17. Department of Health and Human Services, U.S. Food and Drug Administration, Guidance for Industry, Q9 Quality Risk Management, 2006.

2

RISK MANAGEMENT TOOLS

Mark Walker and Thomas Busmann

This chapter provides an overview of the quality risk management methodology to be considered for the application of quality risk management tools and procedures to identify, mitigate, and communicate risks within the pharmaceutical and biological manufacturing industries.

In general terms, risk management is a process. In its simplest form, risk management follows a typical quality approach.

Risk management plan	Write down what you are going to do.
Risk assessment	Do what you said you would.
Risk management report	Document what you did, what you discovered, and what you concluded.
Risk monitoring and control plan	Determine methods for monitoring and procedures for control.
Residual risk	Identify the risk that could not be removed and determine whether it is acceptable.

2.1 APPLICABILITY

The tools discussed in this chapter do not represent an exhaustive or complete list of all tools available within the risk management toolbox, but are a cross-section of some of the more widely used tools. Companies may choose to

---

Risk Management Applications in Pharmaceutical and Biopharmaceutical Manufacturing,
First Edition. Edited by A. Hamid Mollah, Mike Long, and Harold S. Baseman.
© 2013 John Wiley & Sons, Inc. Published 2013 by John Wiley & Sons, Inc.

standardize on using certain tools exclusively or may leave it to the project team to select an appropriate tool. In certain cases, a company may choose an informal risk analysis process to meet their risk management requirements, whereas in other circumstances, a very rigorous, quantitative, and involved risk analysis will be required. In utilizing risk management tools for managing the risks of the quality of a product (drug, biotechnological products, combination products, etc.), the risks to the patient should be the most important consideration for which tool or tools to use. The product manufacturer has the best information and the most at stake, so they are the best positioned to identify the scope and selection of the tools for their risk management approach. Also, keep in mind there is no "one size fits all" approach to risk management. Each situation must be evaluated on the basis of facts present at the time in the specific project that is being evaluated.

2.2 RISK MANAGEMENT

Risk management is an ongoing process of minimizing risks throughout a project, process, or product's life cycle to maximize its benefit and reduce its risk to individuals or the environment. Risk information will emerge throughout the product life cycle. This information combined with risk management tools makes it easy for the application of quality risk management principles. When the risk cannot be eliminated, it must be determined at each decision point if the currently identified risk level can be reasonably reduced further or not and if the current level of risk is acceptable to the company.

2.3 RISK MANAGEMENT PROCESS

The quality risk management process includes the assessment, control, communication, and review of risks associated with a product, project, or process (i.e., drug) throughout its life cycle. The use of risk management tools can provide documented and reproducible methods for accomplishing the risk management process. The flow chart shown in Figure 2.1 is an example of a typical quality risk management process [1].

2.3.1 Risk Assessment

As you see in Figure 2.1, the first process is assessing the risk. This has three components: risk identification, risk analysis, and risk evaluation. Each of these is discussed later. This is a systematic process to understand risk: (i) what can go wrong, (ii) how likely is it, and (iii) what are the impacts?

A risk assessment is conducted on the basis of historical experience, analytical methods, knowledge, and sometimes intuition.

2.3.1.1 Risk Identification In order to address risk, the hazards leading to harm must be identified. This process requires subject matter experts knowledgeable and experienced with the product or process to "brain-storm," use checklists,

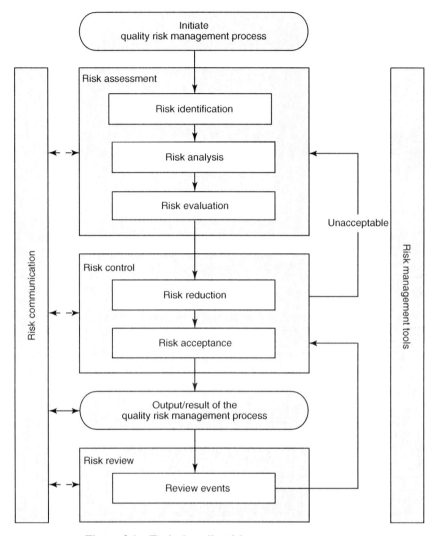

Figure 2.1 Typical quality risk management process.

review historical records, gather shared corporate knowledge, and systematically evaluate the risk posed by the system under evaluation. Each of the identified risks must be evaluated to identify those risks that are realistic and therefore must be addressed. As the process summarized in Figure 2.1 is followed, the next steps are discussed below.

2.3.1.2 Risk Analysis Using the information that is available at the time of the investigation, each identified hazard is systematically analyzed qualitatively, quantitatively, or semiquantitatively to examine estimated risk under various conditions:

- intended use
- unintended use
- normal and fault conditions.

This process may be inductive or deductive and will be as exhaustive as resources allow. The process examines the likelihood of the occurrence and severity of the harm as the result of the hazard. In some cases, the process looks at the detectability of the harm as the result of a hazard in estimating the risk. At the end of this, the identified risk will be scored for comparison with risk acceptance criteria as described next.

2.3.1.3 Risk Evaluation In a quantitative, qualitative, or semiquantitative process, each estimated risk is compared with defined risk criteria to determine the significance of the risk. This allows the risk to be categorized as acceptable or unacceptable. For those risks that are acceptable, the company is willing to accept the risk. In some cases, it may not be possible to completely eliminate risk. For those that are determined to be unacceptable, risk control must be employed to achieve risk acceptance.

2.3.2 Risk Control

Risk control is the process through which decisions are reached to implement protective measures to reduce risk or maintain risk within specified levels. It is a balance between benefits, risks, and resources. Through this process, it must be evaluated whether there are new risks introduced as a result of the risk control. The objective is to reduce the risk to make it acceptable. If the risk cannot be reduced to an acceptable level, the product or process should be abandoned or more data must be obtained which allows the risk evaluation to be reexamined.

2.3.2.1 Risk Reduction Risk reduction is the process of identifying mitigating approaches to address the identified risk. Risk can be reduced by mitigating the severity of harm, reducing the likelihood of harm, or improving the detectability of hazards. When applying risk reduction, the priority of approach should be to make changes inherent in the design to remove the hazard without introducing new hazards. This may include protective measures or controls (i.e., barriers, alarms), instituting administrative controls, and, only if no other approach is identified, including information regarding risk (labeling). It is important to include validation and verification procedures for the mitigations in the qualification of the product, process, or project.

2.3.2.2 Risk Acceptance As applied to the risk management process, risk acceptance is the establishment of the level of risk that the entity is willing to allow in the product, process, or project. This decision to accept risk is dependent upon many parameters, some of which include an entity's overall risk tolerance, the intended purpose for product or project, the ability to mitigate the risks, and the severity of the harm. Risk acceptance is decided and documented on a case-by-case basis. It can be a formal decision or it can be passive. It may be automatic acceptance based on meeting predefined criteria.

2.3.3 Risk Review/Communication

Decisions reached regarding risk identification, evaluation, mitigation, and acceptance need to be communicated throughout the entity. This most often is performed through the report that documents the process, findings, and conclusions of the risk management process. This communication step is important for management to understand the risk associated with the product, process, or project. Depending on the product, process, or project, the communication step may also be important for health care professionals, patients, caretakers, regulators, stakeholders, or other workers involved in understanding the associated risks identified and the mitigations employed. It is also important to document and communicate the process so that during evaluation of residual risk later in the project life cycle, information is available to assess the effectiveness of the mitigating actions and to understand what was known, what was considered, and what decisions were made during the risk assessment process to accept the risk.

It is important to note that writing the risk assessment report is not the end of the risk management process. Throughout the product life cycle, the risk assessment must be monitored and reviewed to ensure that mitigating actions remain effective. Also, as additional information is collected or as events unfold, it will be necessary to reevaluate the risk to determine if new hazards are identified and to verify that the new information or events do not change the risk conclusions previously established.

2.4 RISK ANALYSIS/ASSESSMENT TOOLS

Decisions are necessary as part of the quality risk management process. These decisions are based on probability, severity, and, sometimes, the detectability of risk. There are various risk management tools that can be used for the risk management process. Risk management tools can be arranged into three categories: (1) risk analysis/assessment tools, (2) basic facilitation tools, and (3) decision-making, and statistical tools.

In using a risk analysis/assessment tool, the objective is to identify hazards and analyze and evaluate the risk. Once the risks are identified, risk control is used to reduce risk or to manage risk to acceptable levels. To determine the appropriate risk analysis/assessment tool, it is necessary to have a well-defined description of the risk. It is helpful in defining risk with the use of three questions: what might go wrong, what is the probability or likelihood that the event will occur, and how severe are the consequences of the event?

Risk assessment can be categorized as inductive risk assessment (looking forward in time), deductive risk assessment (looking backward in time), or both.

Inductive risk assessment is forward looking and considers planned activities. This approach tries to predict risk in a proactive manner. Examples of events where this is appropriate include transfer from design to production, scale-up of the production, changing materials, or processes, or a move of the facility or production line. The purpose of deductive risk assessment is to identify the cause of quality issues to reduce associated risks. Examples of events where this is appropriate include sudden change in yield, contamination event, or regulatory mandate. The most common scenario where performing an inductive and deductive risk assessment would be an event requiring assessment of what went wrong (look back), but a full assessment is necessary to consider what else might go wrong (look forward).

There are many risk analysis/assessment tools that can be used for your risk assessment process. Table 2.1 lists many of the common risk management tools. The following sections discuss many of the major tools that are included in the table. However, this is not to be considered a complete list of tools that are available. The objective is to discuss many of the most common tools that are available.

2.4.1 Preliminary Hazard Analysis (PHA)

2.4.1.1 Description A preliminary hazards analysis (PHA), also called a *screening risk assessment*, is a tool that requires experience or knowledge of a hazard to identify future hazards, hazardous situations, and events that might cause harm. The tool identifies potential hazards or hazardous events, and ranks those events, considering the possibility of a hazard or hazardous event occurring, and their severity (qualitatively) and identifies safeguards. This tool might be helpful when there is little information available or in the early development of a process, product, or facility design. This tool is often a precursor to other more detailed risk analysis. However, the PHA has input to every aspect of the quality risk management process as shown in Figure 2.2. The tool can be used early in product life cycle to examine areas of potential risk to assist in prioritizing resources. PHA is typically used as a high-level tool early in the life of a project, product, process, or device [2,3]. Additional references defining in more detail how to conduct a PHA are listed at the end of the chapter.

2.4.1.2 Benefits A PHA can be performed by small teams. It is applicable to any activity, system, or risk assessment application. During the early product development stages, a PHA is easy to implement as it allows modifications to be made with less effort, thus reducing surprises and decreasing development and design time.

2.4.1.3 Limitations It is necessary to have the ability to foresee hazards and identify them for a PHA. Without having the knowledge and experience from persons involved with the process, hazards are not easily identified. In addition, it is difficult to identify the interaction of hazards to each other using this tool.

RISK ANALYSIS/ASSESSMENT TOOLS

TABLE 2.1 Advanced Risk Management Tools

Risk Management Tool	Description/Attributes	Potential Applications
Fault tree analysis (FTA)[a]	• Method used to identify all root causes of an assumed failure or problem • Used to evaluate system or subsystem failures one at a time, but can combine multiple causes of failure by identifying causal chains • Relies heavily on full process understanding to identify causal factors	• Investigate product complaints • Evaluate deviations
Hazard operability analysis (HAZOP)[a]	• Tool assumes that risk events are caused by deviations from the design and operating intentions • Uses a systematic technique to help identify potential deviations from normal use or design intentions	• Access manufacturing processes, facilities, and equipment • Commonly used to evaluate process safety hazards
Hazards analysis and critical control points (HACCP)[a]	• Used to identify and implement process controls that consistently and effectively prevent hazard conditions from occurring • Bottom-up approach that considers how to prevent hazards from occurring and/or propagating • Emphasizes strength of preventative controls rather than ability to detect • Assumes comprehensive understanding of the process and that critical process parameters (CPPs) have been defined before initiating the assessment.	• Better for preventative applications rather than reactive • Great precursor or complement to process validation • Assessment of the efficacy of CPPs and the ability to consistently execute them for any process

(continued)

TABLE 2.1 (*Continued*)

Risk Management Tool	Description/Attributes	Potential Applications
Failure modes effects analysis (FMEA)/failure modes effects criticality analysis (FMECA)[a]	• Assesses potential failure modes for processes, and the probable effect on outcomes and/or product performance • Once failure modes are known, risk reduction actions can be applied to eliminate, reduce, or control potential failures • Highly dependent upon strong understanding of product, process, and/or facility under evaluation • Output is a relative "risk score" for each failure mode • FMECA is an extension to FMEA to include the means of ranking the criticality of the failure mode	• Evaluate equipment and facilities; analyze a manufacturing process to identify high risk steps and/or critical parameters
Preliminary hazard analysis (PHA)	• Also referred to as a screening risk assessment • Used early in product life cycle to examine areas of potential risk to assist in prioritizing resources • It is applicable to any activity, system, or risk assessment application • Usually requires additional follow-up or in-depth analysis	• Used early in the development of a process, product, or facility design • Precursor to other more detailed risk analysis • Used when there is little information available

[a] Source: Final draft, quality risk management principles, and industry case studies, Product Quality Research Institute, December 28, 2008 [9].

The quality of results of a PHA is highly dependent upon the team and leader. It is often the case that with a PHA, additional follow-up or in-depth analysis is required.

2.4.1.4 Example Table 2.2 presents an example of a PHA using a table to record information. This is representative of a PHA for a batch chemical processing step. Figure 2.3 is an example of a risk matrix that would be used in defining level of risk. Refer to Chapter 6 for additional examples.

RISK ANALYSIS/ASSESSMENT TOOLS

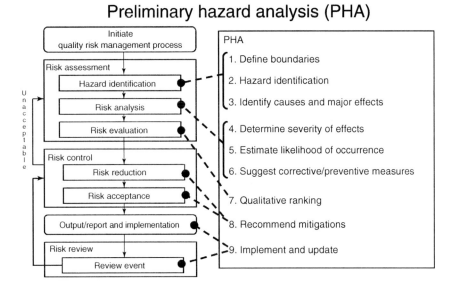

Figure 2.2 How preliminary hazard analysis fits in the risk management process.

2.4.2 Failure Mode Effects Analysis (FMEA) and Failure Mode Effects and Criticality Analysis (FMECA)

2.4.2.1 Description A failure mode effects analysis (FMEA) is a systematic inductive risk analysis of a system to identify the potential failure modes, their causes and effects. An FMEA may also be the procedure by which each potential failure mode in a system is analyzed to determine the results or effects thereof on the system and to classify each potential failure mode according to its severity. An FMEA requires product and process knowledge. It can be applied to a manufacturing operation to analyze the effect on product or process [4].

In addition to FMEA, there is also failure modes, effects, and criticality analysis (FMECA). FMECA is an extension to the FMEA to include a means of ranking the "criticality" of the failure modes to allow prioritization of controls or mitigations. It is not uncommon that the terms FMEA and FMECA are used interchangeably. However, by definition, FMEA stops with scoring severity and probability of occurrence. If it involves matrices (for ranking and prioritization), risk priority number (RPN), or ranking, the failure analysis is technically FMECA.

The primary elements of an FMEA or FMECA include the following:

1. Define the question/system boundaries.
2. Identify potential failure modes.
3. Identify failure effects (and causes).
4. Determine severity of effects.
5. Estimate likelihood of failure mode occurrence.

TABLE 2.2 Preliminary Hazards Analysis Worksheet

Preliminary Hazards Analysis—(Example)
Batch Chemical Process

Hazard No.	Hazard	Cause	Major Effects	Severity of Harm/Probability	Risk Category	Corrective/Preventive Measures Suggested
1.0	Explosion	Static electricity igniting volatile vapors in tank	Significant damage to property and/or injury to possible fatality of personnel	Critical/Remote	High	Inert tank with nitrogen gas
2.0	Tank over pressurization	Failure of relief valve	Damage to tank	Negligible/Occasional	Low	No action required
3.0	Operator inhalation of toxic fumes	Operator did not follow procedure and opened tank while chemicals were present inside	Injury to possible fatality of personnel	Critical/Occasional	High	Implement mechanical controls that lock tank when chemicals are present
4.0	Shock	Faulty wiring	Nonlethal to lethal exposure of electricity to personnel	Critical/remote	High	Install GFI to electrical system

RISK ANALYSIS/ASSESSMENT TOOLS

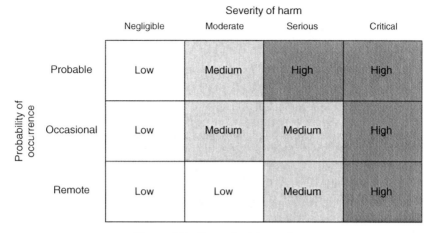

Figure 2.3 Example risk matrix.

6. Identify current controls.
7. Rank and prioritize (using a criticality analysis (matrix or RPN method)).
8. Recommend mitigations.
9. Implement recommendations and update the analysis.

Figure 2.4 shows how the process fits into the risk management program.

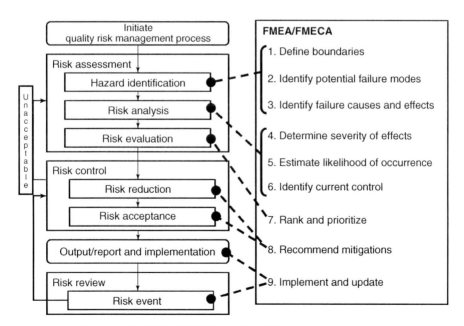

Figure 2.4 FMEA/FMECA quality risk management process.

2.4.2.2 Benefits When it is important to document a company's knowledge of a system or process, it is beneficial to use an FMEA. When risks in a well-defined system need to be prioritized, use FMECA. It is also beneficial to use FMEA/FMECA when there is little past experience with failure modes, for example, a new process or when it is desirable to identify potential risks early in the development of a product or process. FMEA or FMECA is also a good choice when it is important to identify potential process problems, process controls, or to prospectively identify what could go wrong before the scale-up of a process. This will allow the opportunity to mitigate problems before they occur.

2.4.2.3 Limitations An FMEA or FMECA is not a good tool for analyzing interactions between failure modes/causes. FMEA is not a good tool for reactive or retrospectively initiated risk management processes for pharmaceutical manufacturing.

2.4.2.4 Examples Table 2.3 provides an example of an FMEA/FMECA analysis of a chemical reaction vessel. Additional references to include examples on how to conduct a FMEA/FMECA include *The Quality Toolbox Second Edition* [5] and *The Basics of FMEA 2nd Edition* [6]. You can also see examples in Chapters 6, 7, 8, and 13.

2.4.3 Hazard and Operability Analysis (HAZOP)

2.4.3.1 Definition Hazard operability analysis (HAZOP) is based on a theory that assumes that risk events are caused by deviations from the design or operating intentions. It is a systematic brainstorming technique for identifying hazards using the so-called guide-words. Guide-words (e.g., no, more, other than, part of, etc.) are applied to relevant parameters (e.g., flow, temperature, pressure) to help identify potential deviations from normal use or design intentions. It always uses a team of people with expertise covering the design of the process or product and its application [7].

A HAZOP is a very thorough and creative process requiring experienced persons knowledgeable in the system to be involved. The process requires looking at deviations from the design or operating conditions. The system to be analyzed must be defined. The objective of the HAZOP is to identify potential hazards in the system. Hazards involved may include both those essentially relevant only to the immediate area of the system and those with a much wider sphere of influence. Identifying the causes of the operational disturbances and production deviations help identify safeguards and recommendations [7].

2.4.3.2 Benefits The primary benefit of the HAZOP is the identification of critical operations within the system. Identifying these critical operations allows for the opportunity to redesign to remove the hazard, incorporate safety devices to reduce the harm from the hazard, or include warning devices that allow for notification of potential hazards.

TABLE 2.3 Example FMEA/FMECA Worksheet

Item or Process Step	Potential Failure Mode	Potential Failure Effect(s)	Severity	Potential Cause(s)	Occurrence	Current Controls	Detection	RPN	Recommended Action	Responsible	Target Date	Severity	Occurrence	Detection	RPN
												"After" Action Taken			
1	Overpressure of reaction vessel	Possible overpressure of Vessel YY with popping of rupture disk to floor	4	Level indicator fails on tank causing valve to be left open too far	4	None	8	128	Change supplier, 100% inspection of level indicators	AB1	8/1	4	2	2	16
2	Loss of lot due to filling line failure	Inability to transfer material to tank	10	Process control valve fails closed or block valve closed or set too far closed	4	None	4	160	Add process flow meter	JD1	8/1	10	2	2	40
3	Overfill of reaction vessel	Possible overfill of Tank XX if vessel is overfilled	2	HMI location and controls for PW are located on media prep	2	None	8	32	None						0
4	Loss of lot due to temperature excursion	Inaccurate reading from temperature gauge	8	Gauge not correctly calibrated	2	Calibration every 2 yrs	8	128	Calibrate quarterly	AW2	7/1	8	2	2	32

The example above is a traditional FMECA worksheet. Although FMEA and FMECA are referred to interchangeably. FMECA adds Criticality to the FMECA process. Criticality is the combination of frequency of occurrence and severity.

2.4.3.3 Limitations

Conducting a HAZOP requires a significant amount of information and can be very time consuming. The process only considers hazards from a single component. The process does not consider interactions between multiple hazards. As the process only looks at single components, there is a possibility that some likely hazards may not be identified. The process is highly dependent upon the ability and experience of the leader and team members.

2.4.3.4 Examples

Figure 2.5 shows how the HAZOP fits into the risk management program.

Table 2.4 provides an example template for conducting HAZOP analysis. Figure 2.6 presents a process flowchart for conducting the HAZOP.

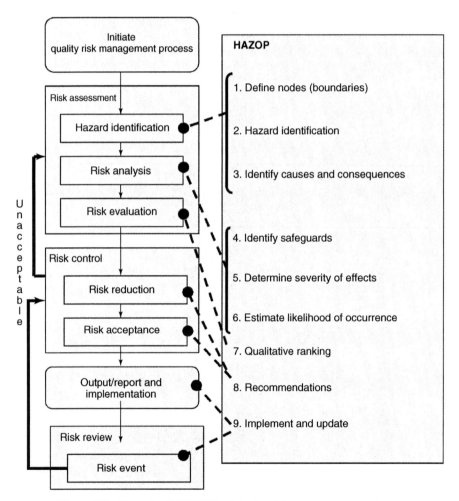

Figure 2.5 How the HAZOP fits into the risk management program.

TABLE 2.4 Example Template for Conducting HAZOP Analysis

Node:
10

Process Stream:
Purified Water System

Drawings:
PFD 1
P&ID 2

Design Conditions/Parameters:
Supply water by batch to process vessel at one of several steps—initialization, quench, final purge, cleaning, makeup

Comments:
Evaluation conducted as part of Design Review.

| Deviation | Causes | Consequences | Safeguards | Risk Matrix | | | Recommendations | Responsibility | Remarks |
				S	L	RR			
No flow	Flow Valve fails closed when open is required	Inability to supply water when needed	1. Valve position indication 2. Flow indication FI-YY on line 3. Operator surveillance	2	4	8	No recommendations—Safeguards deemed adequate	N/A	None
No flow	Loss of Pressure on supply line	Inability to supply water when needed	1. Pressure measurement PI-XX 2. Operator surveillance	2	3	6	No Recommendations—Safeguards deemed adequate	N/A	None
No flow	Loss of supply Pump	Inability to supply water when needed	1. Indication of pump operation 2. Operator surveillance	6	4	24	Install spare supply pump	Engineering	Include as part of Capital Request

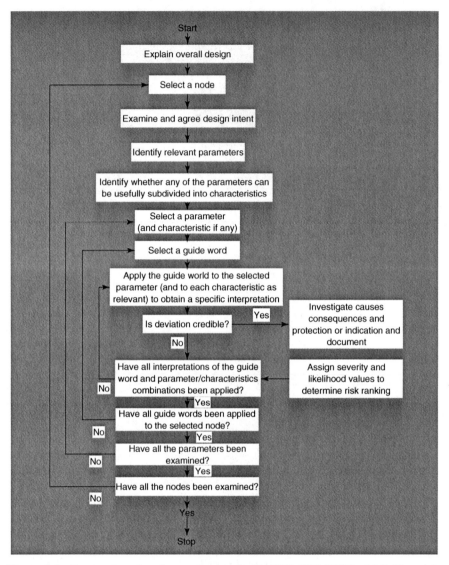

Figure 2.6 Process flowchart for conducting the HAZOP. IEC 61882 ed.1.0. Copyright © 2001 IEC Geneva, Switzerland. www.iec.ch [7].

2.4.4 Fault Tree Analysis (FTA)

2.4.4.1 Description Fault tree analysis (FTA) is a deductive technique that assumes failure of the functionality of a product or process. From the failure of the functionality, the process looks at possible causes and links them down (in the form of a tree of fault modes) to the desired level of detail or system

RISK ANALYSIS/ASSESSMENT TOOLS

function. This technique can be used for analyzing hazards already identified via other techniques. An FTA can be quantitative if data on component failure rates are available. In addition, an FTA can reveal combinations of events leading to failure. It is useful both for risk assessment and in developing monitoring programs [8].

2.4.4.2 Benefits FTA is an effective tool for evaluating how multiple factors can impact a system. It provides a visual representation of the failures, which is useful in the analysis. An FTA can be used prospectively or retrospectively and includes qualitative or quantitative data. It can be used in many phases of a project, to include product development, design engineering, operations, process expansion or modification, or as part of an incident investigation. This technique also helps in identifying common cause events.

2.4.4.3 Limitations An FTA can be very tedious and time consuming. It requires a fair amount of training, skill, and experience of the persons involved in the process. An FTA is very narrow in focus.

2.4.4.4 Examples Figure 2.7 provides a simple example of an FTA for a mixing motor.

2.4.5 Hazard Analysis and Critical Control Point (HACCP)

Hazard analysis and critical control point (HACCP) is a systematic, proactive risk management tool that focuses on manufacturing processes. The process relies on the identification of critical control points, and maintains control by operating within critical limits to prevent hazards from occurring. There are seven steps of HACCP: (i) conduct the hazard analysis, (ii) determine critical control points, (iii) establish critical control limits, (iv) establish monitoring procedures and monitor the critical control points, (v) establish corrective actions, (vi) establish verification procedures, and (vii) establish documentation procedures and keep records. HACCP allows flexibility on how to conduct the previously mentioned steps. For example, HACCP states that it is necessary to establish monitoring procedures, but it does not say what the procedures are or how often to use them. This process is commonly used for addressing specific chemical, physical, and biological hazards. HACCP ensures quality without relying on end-product testing and can be used for planning ahead for correction of problems when prevention fails [9].

2.4.5.1 Benefits It is beneficial to use HACCP when in-product testing is difficult, costly, and/or time consuming. The other advantage of HACCP is the ability to identify potential risks early in the development or during a scale-up of a process or product so that they can be effectively managed.

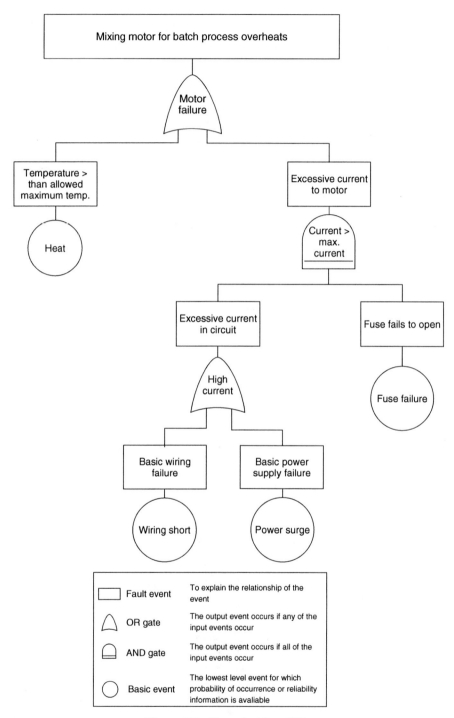

Figure 2.7 Example of an FTA.

RISK ANALYSIS/ASSESSMENT TOOLS

2.4.5.2 Limitations It is not a good idea to use HACCP when there is little experience with the process. HACCP is not a good tool for analyzing interactions between failure modes or causes. This process is also limited when there is a need to address overall process or product reliability or other risks other than for safety and quality.

2.4.5.3 Examples Figure 2.8 presents how HACCP fits into the risk management process. Figure 2.9 shows a simple example of decision making as part of conducting a HACCP analysis.

2.4.6 Risk Ranking and Filtering (RRF)

2.4.6.1 Description Risk ranking and filtering (RRF), also called "risk ranking," is a tool for comparing and ranking risks. This tool is helpful for managing a large set of diverse risks that are difficult to compare. The ranking is performed after each identified risk factor has been assigned a composite risk score. This is used to sort the risks relative to each other. The filtering of risks is performed by using weighting factors, cutoff values for scores, or other criteria. This process can be used to fit the risk ranking into management or policy objectives. The overall process of conducting RRF includes (i) identifying risk factors (hazards), (ii) grouping risk factors into categories, (iii) assigning a score to each risk factor, and (iv) ranking and filtering the risk factors. Figure 2.10 shows how RRF fits into the risk management program.

2.4.6.2 Benefits It is beneficial to use a RRF risk assessment tool when there are many risks to be managed and they are diverse and difficult to compare. This process can be used where both qualitative and quantitative risks are to be managed.

2.4.6.3 Limitations RRF does not add value when the list of identified hazards or risk factors is small, or the scope is very focused. RRF is not a good tool for reactive/retrospectively initiated risk management processes for specific incidents.

2.4.6.4 Examples Tables 2.5 and 2.6 are examples of risk ranking and RRF, respectively. The RRF has a cutoff where any process with a rank of >32 will be addressed.

2.4.7 Other Risk Analysis Tools

Although not as commonly used as the previously mentioned tools, the following risk analysis tools are also available for the risk management process.

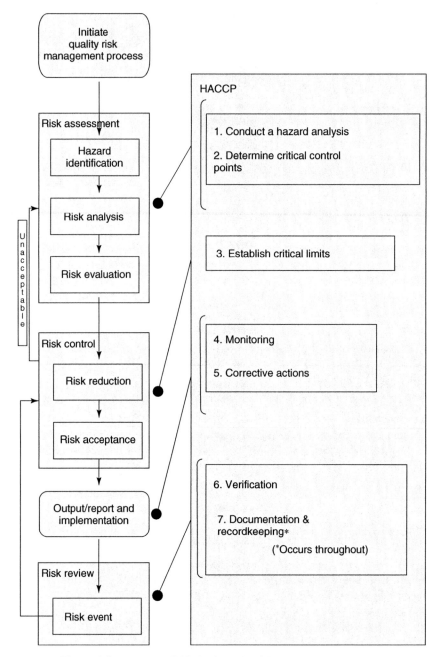

Figure 2.8 How HACCP fits into the risk management process.

RISK ANALYSIS/ASSESSMENT TOOLS

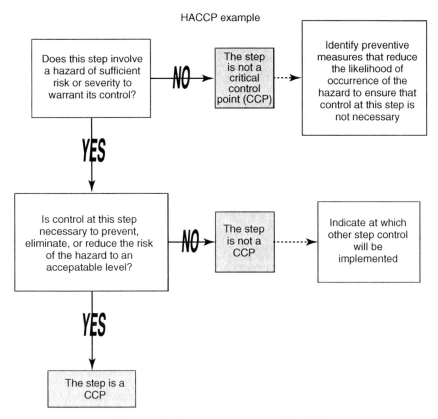

Figure 2.9 Simple example of a HACCP.

2.4.7.1 Layer of Protection Analysis (LOPA) Layer of protection analysis (LOPA) is a more simplified risk assessment tool that provides more information than the traditional qualitative process hazard analysis. It provides a semiquantitative risk assessment that looks at the effectiveness of individual protection layers and then combines the protection layers to evaluate against risk criteria. LOPA is not a hazard identification technique. It is necessary to identify the hazards from another tool such as HAZOP. LOPA provides a semiquantitative risk analysis, which is beneficial when there is a concern that a qualitative risk analysis may be insufficient. A LOPA may be more cost-effective and less time consuming than other quantitative risk analysis tools. LOPA does not identify hazards. It is used with other risk assessment tools.

2.4.7.2 Event Tree Analysis (ETA) Event tree analysis (ETA) is similar to FTA, but starts with the initiator and expands to the consequences (fault trees start with the effect and drill down to causes). ETA is an inductive analysis that determines all possible outcomes from an event. In a complex system, an ETA is

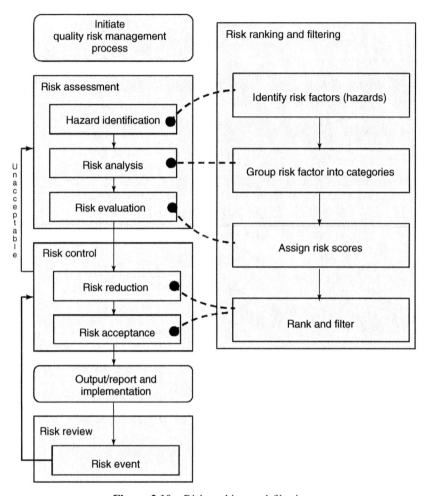

Figure 2.10 Risk ranking and filtering.

very useful for identifying all potential hazards that may not have been identified from a PHA or HAZOP.

2.4.7.3 Checklist Analysis A checklist analysis is a systematic evaluation against preestablished criteria in the form of a checklist. The process generates qualitative lists of conformance and nonconformance determinations, with recommendations for correcting nonconformance. The best use of checklists is when there is an existing process or standard. For example, there is a design review for a product, and there is an ASTM (American Society for Testing and Materials) standard for certain characteristics that the product needs to meet. A checklist would be created so that at the end of the risk assessment the checklist would be used for conformance to the standard.

TABLE 2.5 Risk Ranking of the Previous FMEA Example

Item or Process Step	Potential Failure Mode	Potential Failure Effect(s)	Severity	Potential Cause(s)	Occurrence	Current Controls	Detection	RPN	Recommended Action	Responsible	Target Date	"After" Action Taken			
												Severity	Occurrence	Detection	RPN
2	Loss of lot due to filling line failure	Inability to transfer material to tank	10	Process control valve fails closed or block valve closed or set too far closed	4	None	4	160	Add process flow meter	JD1	8/1	10	2	2	40
1	Overpressure of reaction vessel	Possible overpressure of Vessel YY with popping of rupture disk to floor	4	Level indicator fails on tank causing valve to be left open too far	4	None	8	128	Change supplier, 100% inspection of level indicators	AB1	8/1	4	2	2	16
4	Loss of lot due to temperature excursion	Inaccurate reading from temperature gauge	8	Gauge not correctly calibrated	2	calibration every 2 yrs	8	128	Calibrate quarterly	AW2	7/1	8	2	2	32
3	Overfill of reaction vessel	Possible overfill of Tank XX if vessel is overfilled	2	HMI location and controls for PW are located on media prep	2	None	8	32	None						0

TABLE 2.6 Risk Filtering of the Previous Risk Ranking Example

Item or Process Step	Potential Failure Mode	Potential Failure Effect(s)	Severity	Potential Cause(s)	Occurrence	Current Controls	Detection	RPN	Recommended Action	Responsible	Target Date	"After" Action Taken Severity	Occurrence	Detection	RPN
2	Loss of lot due to filling line failure	Inability to transfer material to tank	10	Process control valve fails closed or block valve closed or set too far closed	4	None	4	160	Add process flow meter	JD1	8/1	10	2	2	40
1	Overpressure of reaction vessel	Possible overpressure of Vessel YY with popping of rupture disk to floor	4	Level indicator fails on tank causing valve to be left open too far	4	None	8	128	Change supplier, 100% inspection of level indicators	AB1	8/1	4	2	2	16
4	Loss of lot due to temperature excursion	Inaccurate reading from temperature gauge	8	Gauge not correctly calibrated	2	Calibration every 2 yrs	8	128	Calibrate quarterly	AW2	7/1	8	2	2	32

40

BASIC FACILITATION TOOLS

2.4.7.4 What-If Analysis What-if analysis is a loosely structured systematic technique that uses the question "what-if?" to identify hazard scenarios, consequences, and safeguards.

2.4.7.5 Health Hazard Evaluation (HHE) Health hazard evaluation (HHE) is a hazard evaluation specifically directed at human health and safety. The Food and Drug Administration (FDA) and the National Institute for Occupational Safety and Health (NIOSH) have their own forms and procedures for HHEs. FDA conducts an HHE on recall products, and uses them to determine whether the recall is a Class I, II, or III. NIOSH uses HHEs to learn whether workers are exposed to hazardous materials or harmful conditions.

2.5 BASIC FACILITATION TOOLS

Basic facilitation tools are used to help think about the problem as a risk assessment is being performed. These tools help the team obtain a common understanding of the process being analyzed, and identify hazards and their causes. Table 2.7 is a list of tools that are discussed in this section. There are many facilitation tools for which this section covers only a few of them.

2.5.1 Flowcharts

Flowcharts or block diagrams are graphical representations of a process or sequence of activities. It can be used to facilitate hazard identification by breaking the process down into individual steps that can be analyzed. Having a flowchart provides a visual representation that would assist the team in developing a common understanding of the scope of the system.

2.5.2 Process Mapping

Process mapping provides a clear and simple visual representation of a process. This visual representation facilitates a common understanding for explaining and analyzing the process and its hazards. Process mapping is a prerequisite for some tools for example HACCP.

2.5.3 Check Sheets

Check sheets in the risk assessment assist the team in thinking about hazards. A check sheet can be standardized for certain situations where applications are similar or repetitive. For example, while investigating an out-of-control process, having a list of events based on a historical knowledge of similar scenarios, processes, or products may assist with the analysis of the process. The points to consider in a check sheet are generally qualitative in nature.

TABLE 2.7 Basic Risk Management Tools

Risk Management Tool	Description/Attributes	Potential Applications
Diagram analysis[a] • Flowcharts • Check sheets • Process mapping • Cause/effect diagrams	• Simple techniques that are commonly used to gather and organize data, structure risk management processes, and facilitate decision-making	• Compilation of observations, trends, or other empirical information to support a variety of less complex deviations, complaints, default, or other circumstances
Risk ranking and filtering[a]	• Method to compare and rank risk • Typically involves evaluation of multiple diverse quantitative and qualitative factors for each risk, and weighting factors and risk scores	• Prioritize operating areas or sites for audit/assessment • Useful for situations when the risk and underlying consequences are diverse and difficult to compare using a single tool
5 Why analysis	• The technique of repeatedly asking "why" something occurred	• Facilitates the identification of deep underlying causes of problems and failures
Histograms	• Summarizes the frequency distribution of the data set	• Used in decision-making by identifying outliers of data from a risk assessment
Pareto analysis	• Technique for prioritizing information under the principle that 80% of the problems are produced by 20% of the causes	• Useful in identifying hazards that have the most impact to reducing risk
Control charts	• A tool used to determine whether or not a process is in a state of statistical control	• Used for risk mitigation or monitoring

[a] Source: Final Draft, Quality Risk Management Principles and Industry Case Studies, Product Quality Research Institute, December 28, 2008 [9].

2.5.4 Cause and Effect Diagrams (Fishbone or Ishikawa Diagram)

Cause and effect diagrams (also known as fishbone or Ishikawa diagrams) assist in identifying whether the cause of failures were hazards. The main branches of the diagram represent categories of contributing causes. This facilitation tool is beneficial for associating multiple possible causes with a single effect.

2.5.5 5 Why Analysis

5 Why analysis is the process of repeatedly asking "why" something occurred. The rule of thumb is to ask "why" five times. But it could be more or less than five times. The process is to keep asking "why" until the result is unchangeable or is out of control. The benefit of the 5 Why Analysis is that it facilitates the identification of deep underlying causes of problems and failures.

2.5.6 Histograms

Histograms or bar charts graphically summarize the frequency distribution of the data set. Histograms can be used in decision making by identifying outliers of data from a risk assessment where risk priority numbers are assigned.

2.5.7 Pareto Analysis

Pareto analysis is a decision-making technique that prioritizes information under the principle that 80% of the problems are produced by 20% of the causes. This technique is useful as a benchmark and should not be used as being an accurate measurement of risks, but more for identifying the risks that have the most impact on a system.

2.5.8 Control Charts

Control charts are often used in process quality control to determine whether or not a process is in a state of statistical control. This tool can be used to distinguish real process variations from "noise." If the chart indicates that the process is not in control, it can be used to determine the source of the problem so the process can be controlled back to steady-state. Control charts can also be used for risk mitigation (by improving detection) or monitoring.

2.6 COMPARISON OF RISK ASSESSMENT TOOLS

This section will look at various risk assessment tools and compare them to each other on the basis of:

1. Scope—the scope of the risk analysis to be performed, how broad.
2. Granularity—the level of detail in the analysis and shown in the documented output.
3. Inductive or deductive—Inductive is forward looking and planned, deductive is backward looking.
4. Interactions versus individual risks.
5. Complexity—a rating of complexity from low to high.

TABLE 2.8 Risk Comparison of Risk Assessment Tools

Risk Assessment Tool	Scope	Granularity	Inductive/ Deductive	Interactions vs. Individual Risks	Complexity
PHA	Any	Low	Either	Individual	Low
HAZOP	Focused	High	Inductive	Individual	Medium
HACCP	Any	Variable	Either	Individual	Low
FMEA	Any	High	Inductive	Individual	Medium
FTA	Focused	High	Deductive	Interactions	High
RRF	Broad	Low	Inductive	Individual	Low

Table 2.8 compares PHA, HAZOP, HACCP, FMEA, FTA, and RRF. Granularity is used in this example to indicate the level of detail that the analysis will reach and what will show up in the documented output. As an example, the granularity of HACCP depends on what tool is selected for hazard analysis. Similarly, HACCP does not specify the use of a particular hazard analysis technique, so it uses either an inductive or deductive technique. There will always be exceptions to these generalizations, but these are typical when the tools are applied in the best way. For example, you could do a deductive/retrospective HAZOP, but that is probably not the most effective use of the tool.

2.7 RESIDUAL RISK EVALUATION

Residual risk is the risk that remains after applying risk mitigation measures. A residual risk evaluation looks at the risk remaining after you have performed your risk analysis, applied risk mitigation, and implemented controls. It is important to recognize that risk cannot be eliminated, merely reduced. An effective risk management program must address acceptable levels of residual risk and ongoing processes to continually reduce risk through improvements in processes or procedures.

2.8 SOURCES OF RISK INFORMATION

A number of sources for supporting risk information are available. The references listed here are sources for understanding risk management, risk analysis, risk analysis techniques, and risk control. When executing risk analysis, the following are typical sources for select data and information to assist in the execution of the risk analysis.

- Published standards.
- Scientific and technical data.
- Field data.

- Test data.
- Clinical evidence.
- Use and user data.
- Expert opinion.

In addition, a number of government agencies such as the Department of Energy and National Aeronautics and Space Administration have routinely conducted risk assessments on a number of projects. Some of these studies are available publicly.

2.9 CONCLUSION

This chapter has attempted to introduce a number of risk analysis tools that are appropriate for use in risk management within the pharmaceutical and biological manufacturing industries. It is important to note that there is not one tool that fits all situations. It should also be noted that the selection of an appropriate tool, a particular tool, or the "correct tool" is not as important as following a methodical approach to management of risk and consistently employing risk management throughout the project or product life. As indicated throughout, numerous tools may be used for a particular situation with similar results achieved in identifying hazards and evaluating their respective risk. The authors' advice is to select a tool that works for the application, use it well, do not presume to know the results before they are achieved, be faithful to the process, be robust in documentation, and ensure the results are communicated to the stakeholders. Finally, remember that this is not a one-time process, but rather a life cycle activity.

DEFINITIONS

Risk	Combination of the frequency or probability of occurrence and the consequence of a specified hazardous event. IEC 300-3-9
Harm	Physical injury or damage to health, property, or the environment. ISO/IEC Guide 73
Hazard	Potential source of harm. ISO/IEC Guide 51; ICH Q9
Risk analysis	Systematic use of available information to identify hazards and to estimate risk
Risk assessment	Overall process comprising a risk analysis and a risk evaluation
Risk control	Process through which decisions are reached and protective measures are implemented for reducing risks to, or maintaining risks within, specified levels

Risk evaluation   Compares the estimated risk against defined risk criteria using a quantitative, qualitative, or semiquantitative scale to determine the significance of the risk

Risk management   Systematic application of management policies, procedures, and practices to the tasks of analyzing, evaluating, controlling, communicating, and monitoring risk throughout the product life cycle

ACKNOWLEDGMENT

The author thanks the International Electrotechnical Commission (IEC) for permission to reproduce information from its International Standard IEC 61882 ed.1.0 (2001).

All such extracts are copyright of IEC, Geneva, Switzerland. All rights reserved. Further information on the IEC is available from www.iec.ch. IEC has no responsibility for the placement and context in which the extracts and contents are reproduced by the author, nor is IEC in any way responsible for the other content or accuracy therein.

REFERENCES

1. International Conference on Harmonization, Q9 Quality Risk Management, Guidance for Industry, Rockville, MD: U.S. Department of Health and Human Services Food and Drug Administration, 2006.
2. Ericson II, Clifton A. Hazard Analysis Techniques for System Safety. Hoboken: John Wiley, 2005.
3. Hyatt N. Guidelines for Process Hazards Analysis, Hazards Identification and Risk Analysis. Boca Raton: CRC, 2005.
4. International Electrotechnical Commission,. Analysis Techniques for System Reliability—Procedure for Failure Mode and Effects Analysis (FMEA), Technical Standard. Geneva, Switzerland: International Electrotechnical Commission, 2006.
5. Tague NR. The Quality Toolbox. 2nd ed. Milwaukee: ASQ Quality Press, 2005.
6. McDermott, Robin E., Raymond J. Mikulak, and Michael R. Beauregard. The Basics of FMEA 2nd ed New York: Productivity Press, 2009.
7. International Electrotechnical Commission. Hazard and operability studies (HAZOP studies)—Application guide IEC-61882. Geneva, Switzerland: International Electrotechnical Commission, 2001.
8. International Electrotechnical Commission. IEC 61025—Fault Tree Analysis (FTA), International Standard. Geneva, Switzerland: International Electrotechnical Commission, 2006.
9. T. Frank et al. Quality Risk Management Principles and Industry Case Studies, (Dec 2008), sponsored by the Pharmaceutical Quality Research Institute Manufacturing Technology Committee (PQRI-MTC).

FURTHER READING

1. Risk management—principles and guidelines, ISO 31000:2009
2. Risk management—risk assessment techniques, ISO/IEC 31010: 2009
3. Risk management—vocabulary, ISO 73:2009
4. Medical devices—application of risk management to medical devices, ISO 14971: 2007
5. Risk analysis of technological systems—applications guide, IEC 60300-3-9:1995
6. Project risk management—application guidelines, IEC 62198:2001
7. Pharmaceutical Development Q8, International Conference on Harmonization, published in the Federal Register, 9 June 2009, ICH Q8
8. Quality Risk Management Q9, International Conference on Harmonization, published in the Federal Register, 2 June 2006, ICH Q9
9. Pharmaceutical Quality Q10, International Conference on Harmonization, published in the Federal Register, 8 April 2009, ICH Q10
10. Guidelines for Hazard Evaluation Procedures: with worked examples, 2nd edition, 1992, American Institute of Chemical Engineers (AIChE), Center for Chemical Process Safety (CCPS).
11. Guidelines for Auditing Process Safety Management Systems, 1993, AIChE CCPS.
12. Plant Guidelines for Technical Management of Chemical Process Safety, 1992, AIChE CCPS.
13. The Quality Toolbox, 2nd Edition by Nancy R. Tague.
14. The Basics of FMEA, 2nd Edition by Robin McDermott.
15. Failure Mode and Effect Analysis: FMEA from Theory to Execution, D H Stamatis (1995).
16. The Basics of FMEA, McDermott, Mikulak & Beauregard (1996).
17. MIL-STD-1629A. Procedures for Performing a Failure Mode, Effects, and Criticality Analysis (1980).

3

RISK MANAGEMENT: REGULATORY EXPECTATION, RISK PERCEPTION, AND ORGANIZATIONAL INTEGRATION

MIKE LONG

3.1 INTRODUCTION

Quality risk management (QRM) is an enabling process that supports the product life cycle and the pharmaceutical quality system of an organization. Organizations that have fully functioning mature QRM systems in place have a solid foundation in the regulatory guidance and expectations set down by the global health authorities and administrations. These organizations also have a solid foundation in understanding how those charged with creating, maintaining, and assessing the risks of their medical products approach risk from a cognitive and social standpoint. These foundational elements are embedded throughout the organization through policies and procedures and reinforced via effective training. This chapter provides an overview of these foundational risk management elements and is broken down into three basic sections. The first section outlines the basic regulatory requirements for embedding QRM within a company's quality system and life cycle. The second section provides a high level overview of the cognitive and social aspects of risk that impact the way we react to and control risk. These are important to understand when creating new QRM policies and procedures within an organization. The third section of the chapter discusses

---

Risk Management Applications in Pharmaceutical and Biopharmaceutical Manufacturing,
First Edition. Edited by A. Hamid Mollah, Mike Long, and Harold S. Baseman.
© 2013 John Wiley & Sons, Inc. Published 2013 by John Wiley & Sons, Inc.

logistical requirements and considerations when rolling out a QRM program and the organizational training required for an effective QRM system.

3.2 QRM REGULATORY EXPECTATIONS

Working in this industry, we have a responsibility, within the clinical risk/benefit construct for individual products, to produce safe and effective products. An effective quality management system assists in meeting the responsibility. Risk management is the enabling process in the design and implementation of that quality management system [1]. Global regulations lay out the responsibilities:

The (social) responsibility of a pharmaceutical manufacturer is quite clear:

The holder of a manufacturing authorization must manufacture medicinal products so as to ensure that they are fit for their intended use, comply with the requirements of the Marketing Authorization and do not place patients at risk because of inadequate safety, quality, or efficacy [2].

The responsibility of management within the company is also quite clear:

The attainment of this quality objective is the responsibility of senior management and requires the participation and commitment by staff in many different departments and at all levels within the company, by the company's suppliers, and by the distributors [3].

As is the method by which these quality objectives will be met:

To achieve the quality objective reliably there must be a comprehensively designed and correctly implemented system of Quality Assurance Incorporating Good Manufacturing Practice, and thus Quality Control and Quality Risk Management [3].

So, as detailed earlier, a modern pharmaceutical quality system cannot truly function properly in the absence of an effective, integrated QRM system.

The expectations that regulators have for QRM have been laid out directly through regulations or guidance by agencies, consortiums, and health organizations across the globe.* These are clear indicators of an expectation for having a fully embedded QRM process within the quality system. The days of having a standalone risk assessment procedure, such as for failure modes and effects analysis, are in the past.

Reviewing recent inspection observations is a good way to understand how the expectations of international regulatory bodies on a given subject such as risk management need to be put into practice. Recent observations show the importance of having both a fully embedded system of QRM and appropriately executed risk assessments (see Table 3.1 for examples of observation deficiencies) [5,6].

*Chapter 3 of the EudraLexVol 4 GMPs "Premises and Equipment". From a practical standpoint, you cannot minimize risk without assessing it. You cannot assess risk unless you have standard tools. You cannot have standard risk assessment tools without a system of risk procedures.

TABLE 3.1 Agency-Observed General QRM Deficiencies [7]

System Level (Policy/Procedure) Deficiency	Risk Assessment Deficiency
No consideration given to QRM	Inadequate or no assessment of impact on product quality
Inappropriate application of QRM	
Improper implementation	Lack of evidence supporting decisions
Variable tolerance of risk	Lack of process understanding and/or regulatory requirement
Systematic approach not applied to the review of assessments.	There is a desired outcome and risk management is used to justify it (invalid assumptions—suit the desired outcome)

3.2.1 General System Expectations

There is an expectation that organizations have a high level risk management policy document that defines [8,9]:

- areas of the business where QRM will be applied;
- risk management methods/tools to be used;
- responsibilities of management and individuals engaged in risk assessments;
- who owns the risk decisions;
- how risk is documented and controlled;
- methods and timing for reviewing and communicating risk;
- standard guides on ranking and accepting risk; and
- risk management training and resourcing.

There are additional requirements that lower level tactical procedures exist, providing evidence that risk management is imbedded into the quality management system. Specific guidance on individual risk management tools is expected, as are updated procedures in the following areas [10]:

- deviation management;
- investigations;
- complaints;
- change control;
- validation;
- computer systems;
- premise and equipment design and operation;
- supplier evaluation;
- annual reviews; and
- sampling.

Additional potential areas of application for risk management implementation have been suggested in areas such as [11]:

- multipurpose facilities;
- equipment design and installation;
- process controls;
- HVAC containment strategies;
- cleaning regimes;
- campaign manufacture;
- environmental monitoring plans;
- services and equipment monitoring; and
- maintenance and calibration intervals.

The expectations for individual risk assessments are more detailed and tie back to how well the policies and procedures were written and implemented. Table 3.2 provides a detailed list of risk assessment level expectations and practical considerations for meeting the expectations.

3.2.2 Establishing Risk Communication and Reporting Mechanisms

Risk communication is the *"sharing of information about risk and risk management between the decision makers and others"* and is one of the most important parts of a fully functioning, life cycle approach to QRM [15]. It is a *"continual and iterative processes that an organization conducts to provide, share, or obtain information and to engage in dialog with stakeholders regarding the management of risk"* [16]. Without robust risk communication and reporting, integrated risk management is not possible.

The process requires multiple stakeholders to engage in multilateral discussion regarding the *"existence, nature, form, likelihood, significance, evaluation, acceptability, and treatment of the management of risk"* [17]. This process is an input into the decision to accept or reject a risk, although not the acceptance process itself.

Setting up clear flows for the communication of risk up and down an organization is important for many more reasons than just meeting the expectations of international guidance documents. Effective communication of risk:

- encourages accountability and ownership of risk;
- ensures information from risk management exercises is available at right levels of the organization and at the right time;
- promotes a mature culture of risk management; and
- creates a level of risk awareness that may assist in reducing the overall residual risk of an organization.[†]

[†]This idea that making people aware of a risk can help reduce that risk is a concept that directly relates to risk communication activities during Quality Risk Management work. This is one reason why Risk Communication is so important, and any action that serves to help reduce risk should be considered. Dr. Kevin O'Donnell, Irish Medicines Board, 2010.

TABLE 3.2 Regulatory Expectations and Practical Considerations When Executing Risk Assessments [12–14]

Regulatory Expectations when Executing Risk Assessments	Practical Considerations
Clearly identify the process being assessed and what it is attempting to achieve.	Properly plan and scope risk assessments. Questions to ask: • Is the assessment ad hoc/for cause (such as a deviation/failure/change), or a life cycle/living assessment (manufacturing operations) • At what part of the process will the assessment start and where will it end?
Be based on systematic identification of possible risk factors	Procedures must exist which provide guidance on the risk assessment tool being used (e.g., FMEA) and have standard ranking indications (e.g., a value of 1 for occurrence has a descriptive indicator such as "less than 1 in 1,000, 000")
Take full account of current scientific knowledge	Instead of an "I think" approach, use of an "I know" approach using existing literature, clinical data, using experiments.
Be conducted by people with experience in the risk assessment process and the process being risk assessed	Risk assessments must be conducted by a team that has been fully trained on the organizations risk procedures. Facilitators should have advanced training on the tools and facilitation techniques.
Use factual evidence supported by expert assessment to reach conclusions	Severity and occurrence rates should be determined by clinical, development, and manufacturing data guided by subject matter expertise.
Do not include any unjustified assumptions	Do not guess. It is okay to say "We don't know" and collect the information, so accurate evaluations can occur
Identify all reasonably expected risks—simply and clearly along with a factual assessment and mitigation where required	Do not shortchange the risk assessments or omit items. Additionally, do not list hazards that clearly cannot occur. These are common errors that are easily rectified by a good facilitator and with proper review during the assessment execution phase. Remember, risk assessments are tools to assist in the continuous improvement cycle. The better the data that enters the assessment, the more robust the improvement cycle.
Be documented to an appropriate level and controlled/approved	Organizations must have policies and procedures which detail how risk assessments are approved, controlled, and reviewed. The level and formality of risk assessments can vary, although they all need to be documented.

(*continued*)

TABLE 3.2 (*Continued*)

Regulatory Expectations when Executing Risk Assessments	Practical Considerations
Ultimately be linked to the protection of the patient	The requirement is taken directly from ICH Q9 "the protection of the patient by managing the risk to quality should be considered of prime importance." Risk assessment activities should not be executed with the initial intention to reduce activities (validation, sampling, etc.). Properly executed assessments will naturally provide opportunities to reduce activities for low risk items. Items that require additional work, when completed, provide for reduction in variation and lower cost of quality, increasing profitability.
Contains objective risk mitigation/reduction plans.	Organizations must have procedures that provide standard ways to determine if hazards require risk reduction or can be accepted in their current state. The plans must be documented, approved (they can exist as a part of the original assessment), and reviewed as required.

Companies should ask the following questions when setting up or evaluating their risk communication process.

- What mechanisms exist for the company to quickly respond to risk?
- Are these mechanisms two way?
- Are there robust lines for the communication of risk robust both internally as well as externally?
- Are there policies that facilitate external risk communications with the public, regulators, and notified bodies?
- Does the company have a risk register, or similar method, to collect associated product and process risk as a method to discuss the site's overall residual risk?

3.2.3 Risk Registers

There is an expectation that pharmaceutical manufacturing sites have specific high level risk overview documents known as *risk registers, or risk master plans*. The initial expectation was clearly and formally laid out in the MHRA's Good Manufacturing Practice (GMP)—Quality Risk Management: frequently asked questions:

> ...*a risk register (or equivalent title document) should list and track all key risks as perceived by the organisation and summarise how these have been mitigated. There*

QRM REGULATORY EXPECTATIONS 55

should be clear reference to risk assessments and indeed a list of risk assessments conducted should be included or linked to the register. A management process should be in place to review risk management–this may be incorporated into the quality management review process [18].

(See Table 3.3 for the entire list of questions and answers.)

International agencies are also ensuring that this expectation is being met through audit observation such as:

"There was no risk register to facilitate the management, monitoring, and review of formal risk assessments . . ." [19]

The expectation for the use of a high level risk summary document such as a risk register arises from the formal inclusion of the review of risk assessments during inspections. It is now a well-understood requirement that all organizations have a system for risk management. The evolution of the use of risk management in the industry has brought it to a point where a collection of risk assessments performed on the site need to be managed and reviewed as a whole and not just as individual assessments. Without a high level risk summary document for a site, managing (review and communication) of the highest risk items from dozens of risk assessments becomes unwieldy.

A formal risk register or risk master plan can be seen as mutually beneficial for both the industry and the regulators. It provides a summary document for the regulators to review during inspections, and it provides the management of a manufacturing site a living document that summarizes the high risk items for their site (see Fig. 3.1).

This high level document will summarize the significant risks of a manufacturing site and should provide a brief explanation of the mitigation of those risks or the current plan to which these risks are being reduced. It is also expected that the risk register or risk master plan has links to, or has listed within, the individual formal risk assessments that have been performed at the site. There is also an expectation that the risk register will include an explanation of the manufacturing site's risk review process, or how often the risk register is reviewed and residual risk updated.

The risk register will look very similar in form to an FMEA, although some of the information from the detailed assessments does not need to be transferred to the register.

3.2.4 Audit of the Risk Management Systems

International health authorities and agencies will set aside time to audit a company's QRM program. These audits will generally target three areas (see Table 3.4 for a detailed list of typical audit questions).

TABLE 3.3 MHRA's List of Frequently Asked QRM Questions [42]

MHRA QRM Frequently Asked Question	MHRA Response
1. Do all inspections cover the QRM process?	Yes, QRM (QRM) is a requirement of Chapter 1 of the EU GMP Guide Part I. All manufacturing authorization holders, third-country manufacturing sites, blood establishments, active pharmaceutical ingredient manufacturers must have a system for QRM. Inspectors will review the QRM system as part of the Quality Systems section of the inspection (along with complaints, recalls, deviations, and product quality reviews, etc.). Additionally, inspectors may review specific risk assessments when encountered during inspection. Inspectors will allocate time commensurate with their perceived significance of the risk and if necessary request the company to produce a formal summary of the risk assessment, key decisions, and conclusions, or take copies of risk assessments for further consideration outside the inspection.
2. How will deficiencies be categorized?	As with other areas of inspection, deficiencies will be categorized dependent on the significance of the findings. Typically, complete lack of a system should be classed as a major deficiency, while lesser deviations within a system would be classed as other. Critical deficiencies may reference QRM where risk assessments have inappropriately supported release of products that pose a threat to patient safety. QRM deficiencies may be grouped with other quality systems deficiencies under a quality systems heading. As always, factual statements of what are seen as deficiencies will be clearly recorded.
3. Should a company have a procedure to describe how it approaches QRM related to manufacture and GMP?	Yes, the procedure should be integrated with the quality system and applied to planned and unplanned risk assessments. It is an expectation of Chapter 1 that companies embody QRM. The standard operating procedure (SOP) should define how the management system operates and its general approach to both planned and unplanned risk management. It should include scope, responsibilities, controls, approvals, management systems, applicability, and exclusions.
4. Is it acceptable to link QRM with cost-saving measures?	The expectation of QRM is to assess risks to the medicinal product and patient and manage these to an acceptable level. It is appropriate for companies to assess their control systems to implement the optimum controls to ensure product quality and patient safety. If this can be achieved in a more cost-effective manner while maintaining or reducing risk to the product and patient, then this is acceptable. However, inappropriate risk assessment and mitigation in order to achieve cost savings is not appropriate.

5. Should sites have a formal risk register and management process?	Yes, a risk register (or equivalent title document) should list and track all key risks as perceived by the organization and summarize how these have been mitigated. There should be clear reference to risk assessments and indeed a list of risk assessments conducted should be included or linked to the register. A management process should be in place to review risk management—this may be incorporated into the quality management review process.
6. What tools are acceptable to use in QRM?	There is no definitive list, although a number of examples are given in Annex 20 (and ICH Q9). In some cases, combinations of tools or other approaches may be used. The important criterion is that the tool used should support the key attributes of a good risk assessment (see the following text).
7. Do formal tools and a full report have to be issued for every risk assessment?	As stated in Chapter 1 of the EU GMP guide, "…the level of effort, formality, and documentation of the QRM process is commensurate with the level of risk." As such expectations of inspectors will be pragmatic regarding the degree of formality that is required, however, appropriate evidence should be available of what has been done and as such a written output must be retained. Inspector's pragmatism will be directly related to the nature of the risk with increasingly more formality and detail required for more significant risk (risk being the probability of occurrence of harm and the severity of that harm, often supplemented by the ability to detect the potential harm occurring).
8. What are the key attributes of a good risk assessment?	The following key attributes should be observed (mindful of the risk significance addressed in the previous question): • Clearly identify the process being assessed and what it is attempting to achieve, that is, what the harm/risk is and what the impact could be on the patient • Be based on systematic identification of possible risk factors • Take full account of current scientific knowledge • Be conducted by people with experience in the risk assessment process and the process being risk assessed • Use factual evidence supported by expert assessment to reach conclusions • Do not include any unjustified assumptions • Identify all reasonably expected risks—simply and clearly along with a factual assessment and mitigation where required • Be documented to an appropriate level and controlled/approved • Ultimately be linked to the protection of the patient • Should contain objective risk mitigation plans.

(continued)

TABLE 3.3 (*Continued*)

MHRA QRM Frequently Asked Question	MHRA Response
9. What is the difference between a planned and unplanned risk assessment?	A planned risk assessment is one that is conducted in advance of conducting an activity, either before any activity is conducted or before further activity is conducted. This would often allow quality to be built in to activities and risk reduced (quality by design), for example, design of facilities for manufacture of cytotoxic products or organization/design of a label-printing room. An unplanned risk assessment is one that is conducted to assess the impact of a situation that has already occurred, egg impact of a deviation from normal ways of working.
10. Should we expect there to be no risk to patient safety as a conclusion to a risk assessment?	In reality, there is always a degree of risk in all situations but mitigation controls should minimize the likelihood to an acceptable level of assurance. The degree of risk tolerated very much depends on the circumstances, the proximity to the patient, and other controls that may follow the process being assessed before the product is used by the patient. It should be expected that risk mitigation plans are identified and implemented where any risk to patient safety is posed. Companies should take a holistic view and be mindful that critical issues often occur where multiple failures in systems occur together, so mitigation plans should be sufficiently robust to tackle such potential. Inspectors will be assessing if risk assessments underrate either the likelihood, consequences, or detection of occurrences in order to make it appear that there is minimal risk to the patient. The factual evidence behind statements may be challenged.
11. Are any areas out of bounds for risk assessment?	It would be unacceptable for risk assessment to conclude that statutory, regulatory, or GMP requirements should not be followed or are not appropriate, for example, risk assessment could not conclude that it was appropriate for licensed products to be released by someone who was not a qualified person (QP). Otherwise, risk assessments can be used within GMP systems as a tool to identify, quantify, and minimize risk to patient safety.
12. How should risk assessments be controlled?	Risk assessments should be controlled within a defined document management system. If risk assessments are conducted to justify controls for an ongoing process, then the assessments should be subject to change control and periodic review, egg line clearance risk assessment. Frequency of review should be appropriate for the nature of the process. Such risk assessments should be seen as

13. Do risk assessments have to be supported by factual evidence or can they just use professional judgment?	living documents that are visible and subject to change as required. Risk assessments that were conducted as one off activities to assess a situation that will not recur need not be controlled in a "live" manner but must be documented, approved, and retained for assessment of a temperature excursion on storage of a batch of starting material. Such "one off" activities should be controlled as live documents if any conclusions are to be used in any future excursions. Ultimately, these may then need to be reviewed in light of experience or developments.
There should be factual evidence recorded to support any conclusions drawn from egg plant design details in controlling cross contamination—an unsupported assumption that the plant must be suitably designed as we have used it for 10yr or we have had an standard operating procedure (SOP) for 5yr so it must be suitable is a weak approach that may be unfounded and must be challenged by those conducting risk assessments. Professional judgment should be used in the interpretation of factual evidence but must be subject to justification.
14. Scoring in risk assessments is subjective, is there danger that risk assessments may be manipulated to draw desired conclusions?	The scoring system and trigger points for mitigating action are subjective. However, as important as the scores in risk assessments is the rationale for the score. If supported by factual evidence, it should be more obvious what mitigating action is required—the mitigating action is as important as the score assigned. Companies should not score risks in a blinkered manner without considering the factual causes, likelihood of detection, and consequences. Inspectors will be alert to improper use of risk assessments to condone poor practice or exclude patient risk.
15. Is it acceptable to allow external consultants to participate in site risk assessments?	It may be appropriate for consultants to provide support for risk assessments where they can provide specific expertise or knowledge. Their role in the risk assessment should be clear. The reason for delegation and resultant accountability must be understood. Inspectors will expect sites to demonstrate that delegation was effective and that appropriate skill, knowledge, local knowledge, and local accountability was appropriate for the life cycle of the risk assessment. A technical agreement may be appropriate with the consultant where GMP responsibility is assumed.
16. Is it acceptable to allow contract staff to participate in site risk assessments?	It would be usual for contract staff, for example, contract QPs, to lead or participate in risk assessments. The extent of involvement as responsibility/accountability must be documented in the technical agreement between the individual and the organization.

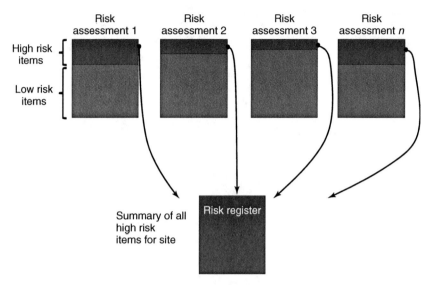

Figure 3.1 Risk register. (*See insert for color representation of the figure.*)

- The first will be a review of the QRM policy that defines what the QRM program entails for the specific organization.
- The second area will be a review of specific procedures, which will dive deeper into the "how" of risk management.
- The third target area for QRM audit will be the evaluation of the actual risk assessments themselves.

The time that is actually spent on QRM during an audit will be dependent on the risk of the product and the processes being used, the maturity and clarity of the QRM program and procedures, effectiveness of the risk review and communication processes, and the thoroughness of the individual risk assessments reviewed.

3.3 PROBLEMS OF SUBJECTIVITY AND UNCERTAINTY DURING QRM EXERCISES

It is said that all politics are local. In a sense, risks are as well; they are personal and relative.‡ We bring our individual life story to each and every risk assessment engagement, all the while being bombarded by external stimuli.

---

‡You know the story that all risks are relative—the risk depends on which one of your relatives is taking the medicine (your wonderful loving grandmother or your crazy uncle Al)!

TABLE 3.4 Auditing of the Risk Management Process [20,21]

Risk Management Audit Level	Area of Scrutiny/Typical Audit Questions
Policy/System	• The areas of application of QRM should be appropriately defined in the company quality management system • There should be an appropriate number of personnel with relevant qualifications, experiences, and training. Their responsibilities should be clearly defined • Senior management is involved in the identification and implementation of QRM principles within the company
Procedures	• Procedures exist for the Assessment Control (incl. risk reduction and risk acceptance) and Communication of risk in various areas within the quality management system • Are there Regular Reviews of Risks? • RM procedures are oriented toward patient safety • The formality of approach should be commensurate with the level of patient risk • Show a logical approach to selection of methods and tools
Assessments	• Whether the QRM performed was integrated into the Quality System? • Was the process and documentation transparent and traceable? • Were the formalized documented procedures followed? • Was a risk problem or question well defined? • Did the completed process address the risk problem/question? • How was the decision made and documented? • Was there appropriate communication throughout the process? • Was a systematic approach applied? • Were the selected methods/tools suitable? • The key risks should have been adequately identified and analyzed, with all relevant data having been generated and/or considered. • All data reviewed must be from a reliable database. • The risk acceptance criteria must be adequate for the specific situation in question; • The risk-based decision(s) must be considered to be well informed, science based and comprehensible. They must be concordant with the preset acceptance criteria • The assessments link to product quality

There are two stories I use in my classes about assessing risk and the impact of outside influences. I call them "The tale of two risk assessments" and "Everyone is a speed demon."

A Tale of Two Risk Assessments

Consider the following:
- You are on a material review board/disposition team. It is the end of the quarter, and a batch of your highest volume product has been quarantined for potential quality issues. The team has also been notified that their business has met all of its financial goals for the quarter.

Question: *Does this financial information influence the team's disposition (which is a risk assessment process)?*

Now consider:
- The following quarter the material review board/disposition team must decide the disposition of a batch of the exact same product for the same quality issue. Again it is the end of the quarter, but this time the team knows but the business unit *is not* meeting its financial goals.

Question: *Does this change in the external context in which the risk of the batch is reviewed change our internal risk calculus? Why?*

There is only one change, one piece of information that is different between the two scenarios of this real-life example. Whether we like to admit it or not, our decision-making process on the risk of releasing the product or not will be different in each of these occasions. Even if we make the same deposition decision on the lots (accept both, reject both), the context in which those decisions are made will be different.

Everyone is a Speed Demon

The legal speed limit on most major highways in the United States is 65 mph. When you drive on these highways, it seems that most folks are driving above the speed limit to a different degree. In the graduate classes I teach or in industry talks I give, as a part of our risk-management-related discussion, I ask the students/audience: *"at what speed above the posted 65 MPH speed limit do you think you will get a ticket?"*; the vast majority answer *"above 75 MPH."*

Question: *What is the "true" speed limit?*

There are internal and extra influences that are always present when we assess risk. We apply our internal risk "sensor" based on the external context we see applied. These are always influencing our risk decisions. They cannot be completely eliminated. We can create training programs and procedures that assist in minimizing their influences in our risk management systems if we have better understanding of how our risk-making decisions are influenced.

TABLE 3.5 Dual Process Cognitive System Characteristics

Intuitive System (System 1) "Gut"	Reflective/Reasoning System (System 2) "Head"
Automatic	Controlled
Effortless	Effortful
Rapid	Slow
Associative	Deductive

3.3.1 Cognitive Operations When Assessing Risk: "This is Your Brain on Risk"

An individual's decision making is governed by two main systems, traditionally called *intuition* and *reason*. They are formally categorized as *system 1* and *system 2* in what is called *dual-process theory* (see Table 3.5). System 1 is our ancient intuitive process that is quick and effortless. System 2 is our rational or reflective process that is controlled, slow, calculating, and effortful. It is sometimes easier to think of these in the common vernacular such as when folks say, "I used my *gut* instinct" or "I followed my *head*, not my heart on this decision" [22,23].

The systems generally work interactively, although both systems are not "applied to every problem a person confronts nor (does) each system have an" exclusivity on the problem. The systems overlap, but the amount of overlap "differ(s) depending on the individual reasoner's knowledge, skill, and experience" [24].

These systems assist, monitor, and control our judgments under uncertainty (a fancy way to say: our decisions about what is risky and what is not). How we make the decisions are controlled by what are called *heuristics and biases*. Heuristics are rules of thumb and internal automatic settings (see Table 3.6 for a select list of heuristics). They are unconscious controls that can be both inherent as well as learned through experience [25]. Biases are tendencies we all use to weigh and adjust our initial intuitive judgments. They are based on many factors including our culture. Biases are cues available to us in different forms that cause us to give too much or not enough weight to a given piece of information when making risk-based decisions. The systems generally work quite well, although they can also be a significant source of error in our risk decision-making ability and can cloud our perceptions of risk because of our heuristics and biases.

The dual processes of decision making and the underlying heuristics and personal and group biases that assist in their governance have a profound impact on the implementation of the risk management system. It also means that defining risk can be quite personal and there is generally a significant difference between what the general population perceives as a high risk compared to what thesubject matter experts perceive (called lay–expert bias).

There are a number of common-sense factors that impact and direct our perceptions or risks. One of the key factors in the perception of a risk is whether

TABLE 3.6 Select List of Heuristics [26–28]

Heuristic (Rule of Thumb)	Description
Anchoring and adjustment	Estimates are made based on an initial input (conscious or unconsciously accepted) that our minds "anchor" or grab onto. We adjust our decision/risk rating up or down from that anchor.
Affect	An instinctual feeling about the "goodness or badness" (or risk) of something. Can also be called the *"Good–Bad Rule"*
Availability	We estimate the likelihood of an event based on how easy it is for us to think of an example of it occurring. If we can easily think of an example, our estimate of the likelihood of that example occurring increases. Can also be called the *"Example Rule"*

TABLE 3.7 Risk Perception Influence Factors [30]

Factors Causing *Higher* Risk Perception	Factors Causing *Lower* Risk Perception
Uncontrollable	Controllable
High visceral dread	Low dread factor
Global/catastrophic	Localized /non-catastrophic/low impact
Impact on future generations	Low impact on future generations
Steadily increasing risk	Decreasing risk
Risk not easily mitigated or reduced	Risk is easily reduced
Fatal risk effects	Repairable harm
Risk is unseen, unknown to those exposed with delayed effects	Observable, exposed are aware, effects of risk are immediate
New risk	Old risk
Effects of risk not know to science	Effects of risk are known

Adapted from Slovic (1987).

a person engages or voluntarily exposes himself or herself to the risk (called voluntary–involuntary bias). This is not the only factor that impacts the perceived risk/benefit balance and a person's ability to accept a risk. Factors such as catastrophic potential, the amount of visceral dread it creates within a person, and the impact on future generations also weigh heavily in our measurements of risk (see Table 3.7 for a detailed list of risk perception influence factors). In fact, the discussion and debate about the riskiness of the given input are not just about the number of people in harm's way. They can be about culture, society, and ideology [29].

There are additional underlying cognitive explanations when weighing the risks of known and unknown risks. People place significantly more weight in sticking with the current state of affairs or risk with certain outcomes and have

TABLE 3.8 Loss Aversion and Status Quo Biases [31]

Bias	Explanation
Loss aversion	The potential for a loss is valued higher than a potential for a gain
Certainty effect	Much higher value is place on activities whose outcomes are certain
Status quo bias	Current state is preferred over change
Endowment/Ownership effect	Ownership of an item, department, etc. increases the perception of value to the owner
Regret avoidance	Decisions that have the potential for a person to regret tend to be avoided
Omission effect	Exaggerated preference for inaction

Adapted from Viscusi (1996).

significant tendencies to steer clear of decisions that may create regrets (see Table 3.8).

3.3.2 Perception and Perspectivism

Perception in risk is also an application of the theory of perspectivism. Two aspects of perspectivism are important to the application of risk management. The first concept is there is no singular objective truth, just the observer's perspective of that occurrence. The second is a conceptual framework that an absolute truth may be obtained via an aggregate of different viewpoints or perspectives.

The use of multiple perspectives is seen in real-life applications of risk management through the selection of the individuals involved in the risk management process. This group should generally encompass different functional areas such that different views on a subject can be obtained. The goal is to create a much more robust output from the risk process because, even with the best intentions, a group approach can have issues with herd/groupthink biases.

3.3.3 Risk's Precautionary Principle

The precautionary principle is a risk approach that places the burden of proof on the provider of the goods or services rather than society. It was developed as an approach to manage uncertainty. The precautionary principle has partial grounding in perspectivism as it tries to address issues where there may be no objective truth available and strives to collect multiple viewpoints on a particular risk. This thought process is evident in ICH Q9, which states:

> *Achieving a shared understanding of the application of risk management among diverse stakeholders is difficult because each stakeholder might perceive different potential harms, place a different probability on each harm occurring and attribute different severities to each harm. In relation to pharmaceuticals, although there are*

a variety of stakeholders, including patients and medical practitioners as well as government and industry, the protection of the patient by managing the risk to quality should be considered of prime importance [32].

It is echoed in the introduction and preamble of ISO 14971 *Medical Devices — Application of risk Management to Medical Devices*:

The concepts of risk management are particularly important in relation to medical devices because of the variety of stakeholders including medical practitioners, the organizations providing health care, governments, industry, patients, and members of the public. All stakeholders need to understand that the use of a medical device entails some degree of risk. The acceptability of a risk to a stakeholder is influenced by the components listed above and by the stakeholders' perception of the risk. Each stakeholders' perception of the risk can vary greatly depending on their cultural background, the socio-economic and educational background of the society concerned, the actual and perceived state of health of the patient, and many other factors [33].

The principle's philosophy states that unacceptable risks should be targeted and prevented from occurring before damage occurs, even if knowledge about the risk is unknown. It is closely aligned with the decision theory's Maximin Principle that states that *the best decision maximizes the minimal output of a specific hazard*.

There is much debate regarding the principle and its impact on trade. The disputes revolve around the amount of weight that should be given to moral, political, ethical, and lifestyle objections to technologies or products and the costs incurred to reduce those potential risks. These discussions emerge from a culture or society's approach and perception to risk aversion. These perceptions can create moving frames of trust with members of that society. The trust is gained or lost by industry and regulators as a result of issues with products such as Vioxx and Avandia and the consent decrees with companies such as Johnson and Johnson [34–38].

The precautionary principle is embedded within the European Community and defines the level of acceptable risk in uncertain situations as an inherently political matter [39].

Implementation of the precautionary principle is based on the potential severity of the risk and the response required. These are also known as the *societal trigger level and the societal response* [40]. The trigger and response are defined by the Commission of the European Communities' 2000 document, *Communication from the Commission on the Precautionary Principle*. The trigger level is a decision for "Recourse to the precautionary principle." This trigger requires both an "identification of potentially negative effects resulting from a phenomenon, product, or process" and a "scientific evaluation of the risk which because of the insufficiency of the data, their inconclusive or imprecise nature, makes it impossible to determine with sufficient certainty the risk in question." The societal response is political and temporal where "the appropriate response in a given situation is the result of a political decision, a function of the risk level that is acceptable to the society on which the risk is imposed."

The trigger level is based on the potential societal damage of the specific hazard along with the amount of knowledge that exists of the hazard. The knowledge would be primarily objective. If the knowledge shifts from objective to subjective, the principle would define it as having significant uncertainty. There are different trigger levels for a low severity hazard that has a degree of uncertainty and compared to a high severity hazard with equivalent uncertainty (Commission of the European Communities 2000, 13).

The policy of regulators in using the precautionary principle shifts evidentiary requirements to the suppliers. However, when looking at multiple stakeholders in the process with varied perceptions, the amount of knowledge varies. These stakeholders are not all within the company that provides the product. They include regulators and the public. What does not exist in the literature is the idea of asymmetry. The companies will always have more information about a product's risks than regulators and the public. Society, however, places an expectation on the provider to reduce the residual risk of a product to its lowest possible level (Commission of the European Communities 2000, Breyer 2005).

3.3.4 Risk Regulation's Unintended Consequence: Asymmetry of Risk Knowledge

The burden of proof creates an unintended consequence, which can be described as Asymmetry of Risk Knowledge. There is an inherent amount of product knowledge asymmetry because of the intimacy the manufacture has with the design and production of the product it intends to provide to the market. The manufacturer naturally has more information regarding the product and process than the regulators or the public.

As one of the stakeholders, the manufacturer makes judgments relating to safety of the product including the acceptability of risks, taking into account the generally accepted state of the art, in order to determine the suitability of the product to be placed on the market [41].

As the guidances show, there are many stakeholders with interest in the process although the manufacturer generally reviews and approves the risk.[§] With the manufacturer owning the data used to determine risk as well as providing and justifying the risk–benefit balance for a given product, the standing or value of the other stakeholders' perception of risk may be reduced, but the responsibility to society by the manufacturers increases.

3.4 INTEGRATION INTO ORGANIZATIONAL PROCESSES

For successful integration of risk management into the quality system, the organization must ensure that individuals engaged in risk management activities

---

[§]These cGMP risks are now being reviewed during inspections.

understand the inherent and intrinsic value of risk management, are adequately trained, and are familiar with risk management tools and the overall risk management process.

When a company is implementing a risk management system, a key first step should be an evaluation of the organization's current understanding of risk management. The evaluation or gap analysis should assess the state of the current risk infrastructure, including the following:

- written policies and procedures;
- risk management practices; and
- training and personnel skills.

Integration of risk management into an organization is a multistep process that begins with this gap analysis of the current state compared to the expected state and ends with the deployment of a fully realized risk management process. As a part of the current state gap analysis, the use of an *organization risk maturity level table*, such as the one shown in Table 3.9, can assist in creating an understanding regarding where the current level of understanding or risk management resides within the organization. For example,

- Does the organization have an accepting attitude toward risk management?
- Is the organization simply complying with the requirements by using a tick-the-box method or has it really imbedded risk management into its systems?

Once the gap analysis has been completed, risk management procedures and activities should be implemented based on the gap analysis findings. Organizations should consider piloting aspects of the new procedures such as risk-rating scales, risk acceptance decision charts, and review processes. Full roll out of the system can occur after collecting feedback on the pilot program.

Organizations should consider a multilayer approach to training. This would include a high-level risk management overview training program for the general employee population. This can be followed by a second level of focused policy, procedure, and tool-based training. Special facilitator-level training for select individuals who have been designated as the risk subject matter experts should also be considered. Other key foundations for the successful implementation of a risk management system are as follows:

- top level management support and commitment;
- understand the path and start simply and avoid complexity;
- understand what are the organization's internal and external risks; and
- a continuous cycle of learning and improvement that creates a robust risk management organization culture.

3.4.1 Training

Training may be the most important factor in creating a successful QRM system. All organizations must provide for training on any formal risk management procedure, although the current training rarely goes beyond "read and understand" requirements for the procedures. This type of training may meet the basic requirements, but the effectiveness of the QRM system and thus the pharmaceutical quality system may not be optimized. Using such "read and understand" training provides for a nonstandard approach to understanding the companies' principles on the subject.

The following are key focus areas of risk training for companies:

- risk management regulations, not just the "what" of the regulations, but the "why" as well;
- the risk management tools the company chooses to use;
- standard risk management terminology to help reduce subjectivity and uncertainty;
- risk assessment facilitation to assure the best outcomes from risk management activities;
- effect of heuristics and biases on the risk assessment process to better understand how we perceive risk and assist in minimizing subjectivity in the organizations' risk decision making (see Table 3.10); and

TABLE 3.9 Risk Management Maturity

Risk Maturity Level	Risk Processes	Attitude	Behavior	Skills and Knowledge
Skepticism	No formal processes	"Accidents will happen"	Fear of blame culture	Unconscious incompetence
Awareness	Ad hoc use of stand-alone processes	Suspended belief	Reactive, "fire fighting"	Conscious incompetence
Understanding and application	Tick the box approach	Passive acceptance	Compliance, reliance on registers	Conscious competence
Embedding and integration	Risk management imbedded in the business	Active engagement	Risk-based decision making	Unconscious competence
Robust risk management	Regular review and improvement	Champion	Innovative, confident, and appropriate risk management	Expert

Table adapted from *A Guide to Supply Chain Risk Management for the Pharmaceutical and Medical Device Industries and their Suppliers, 2010*.

TABLE 3.10 Risk Perception Heuristics and Biases Management and Training Strategies

Area (Example, not an exhaustive list)	Strategy
Training	The organization should create, as a part of their overall risk management training program, a module on the impact of heuristics and biases on risk decision making. This training should include hands-on interactive training coupled with readings on the subject.
Team makeup	Risk assessment team should be multidisciplinary. This encourages the risk to be viewed from many lenses. Include subject matter experts on the team as well. This provides for alternative views and constructive dialog of the risks being assessed. Facilitators who have extensive knowledge of not only the tools but also techniques to minimize impact of individual stakeholder bias should be used.
Assessment of risk	Emphasis on the use of scientific knowledge and hard data. Encourage teams to say "we don't know" or "we don't have the data" rather than a subjective assessment with the goal to collect the data to make more informed decisions. Facilitators should be vigilant for signs of bias during assessments and initiate effective changes if required.

Table adapted from *PDA Technical Report Draft* "Implementation of QRM for Pharmaceutical and Biotechnology Manufacturing Operations," November 2011, page 20.

- the importance of using data to determine occurrence rates (increased understanding of probability and statistics).

3.5 CONCLUSION

Fully functioning mature and effective QRM systems are created via an organization's solid foundation and understanding of the regulations, guidances, and expectations set down by the global health authorities and administrations, robust training programs, senior management commitment, and engagement, and a robust review and continuous improvement program.

Organizations need to have a solid foundation in the underlying mechanisms regarding how individuals assess risk and how heuristics and biases influence an individual or group's risk perception and decision making.

When implementing risk management programs, readers are encouraged to think of the "Whys" of risk management as well as the "How to." The principles presented in this chapter will help create the foundation and infrastructure the reader will need to plan and implement a successful risk management program.

REFERENCES

1. EudraLex, The Rules Governing Medicinal Products in the European Union, Volume 4, EUGuidelines to Good Manufacturing Practice, Medicinal Products for Human and Veterinary Use, Part III, Q9 Quality Risk Management, 2011, p. 1. Available at http://ec.europa.eu/health/documents/eudralex/vol-4/index_en.htm. Accessed 2011 Dec 26.
2. EudraLex, The Rules Governing Medicinal Products in the European Union, Volume 4, EUGuidelines to Good Manufacturing Practice, Medicinal Products for Human and Veterinary Use, Part I, Chapter 1, Quality Management, 2008, p. 2. Available at http://ec.europa.eu/health/documents/eudralex/vol-4/index_en.htm. Accessed 2011 Dec 26.
3. EudraLex, The Rules Governing Medicinal Products in the European Union, Volume 4, EUGuidelines to Good Manufacturing Practice, Medicinal Products for Human and Veterinary Use, Part I, Chapter 1 Quality Management, 2008, p. 2. Available at http://ec.europa.eu/health/documents/eudralex/vol-4/index_en.htm. Accessed 2011 Dec 26.
4. EudraLex, The Rules Governing Medicinal Products in the European Union, Volume 4, EUGuidelines to Good Manufacturing Practice, Medicinal Products for Human and Veterinary Use, Part I, Chapter 1 Quality Management, 2008, p. 2. Available at http://ec.europa.eu/health/documents/eudralex/vol-4/index_en.htm. Accessed 2011 Dec 26.
5. Davis, Matthew, Office of Manufacturing Quality, Australian Government, Department of Health and Ageing, Therapeutic Goods Agency (TGA) Quality Risk Management Audit Expectations and Observations, CAPSIG seminar 'Risk Management within Manufacturing Plants', 4 May 2011 Available at http://www.tga.gov.au/newsroom/events-presentations-manuf.htm. Accessed 2011 Dec 26.
6. Oemmerance, Sven, GMP News: Evaluation of the Warning Letters issued by the FDA for the Fiscal Year 2010 with regard to Risk Management, March 9, 2011, http://www.gmp-compliance.org/eca_news_2450_6764,6917,6971.html. Accessed Dec 26 2011.
7. Davis, Matthew TGA, 4 May 2011.
8. Davis, Matthew TGA, 4 May 2011.
9. World Health Organization WHO Guideline on Quality Risk Management, Working document QAS/10.376/Rev.1 Draft for discussion, August 2011, Available at http://www.who.int/medicines/areas/quality_safety/quality_assurance/projects/en/. Accessed 2011 Dec 26.
10. Davis, Matthew, TGA, 4 May 2011.
11. Davis, Matthew, TGA, 4 May 2011.
12. Davis, Matthew, TGA, 4 May 2011.
13. World Health Organization WHO Guideline on Quality Risk Management, Working document QAS/10.376/Rev.1 Draft for discussion, August 2011, pp. 9–10, Available at http://www.who.int/medicines/areas/quality_safety/quality_assurance/projects/en/.

14. MHRA Good Manufacturing Practice (GMP)—Quality Risk Management: Frequently asked questions, Available at :http://www.mhra.gov.uk/Howweregulate/Medicines/Inspectionandstandards/GoodManufacturingPractice/FAQ/QualityRiskManagement/index.htm. Accessed 2011 Dec 26.
15. Q9 Quality Risk Management, 2011, p. 7. Available at http://ec.europa.eu/health/documents/eudralex/vol-4/index_en.htm. Accessed Dec 2011.
16. ISO 31000:2009 Risk management—Principles and guidelines, International Organization for Standardization, 2009 p. 3.
17. ISO 31000:2009 Risk management—Principles and guidelines, 2009, p. 4.
18. MHRA Good Manufacturing Practice (GMP)—Quality Risk Management: Frequently asked questions available at: http://www.mhra.gov.uk/Howweregulate/Medicines/Inspectionandstandards/GoodManufacturingPractice/FAQ/QualityRiskManagement/index.htm. Accessed 2011 Dec 26.
19. Davis, Matthew, TGA, 4 May 2011.
20. Davis, Matthew, TGA, 4 May 2011.
21. WHO Guideline on Quality Risk Management, 2011, pp. 8–11.
22. Kahneman D, Frederick S. Representativeness Revisited: Attribute Substitution in Intuitive Judgment. In: Gilovich T, Griffin D, Kahneman D, editors. Heuristics, and Biases, the Psychology of Intuitive Judgment. New York: Cambridge University; 2002. p. 50–51.
23. Gardner D. The Science of Fear. New York: Penguin Books; 2008. p. 26.
24. Sloman, S, Two systems of reasoning. Heuristics and biases, the psychology of intuitive judgment. Gilovich, T., Griffin, D., Kahneman, D. Eds Cambridge University, New York, 2002, 383–384.
25. Gardner, D. 2008, p. 28.
26. Tversky, A., Kahneman, D, (1974) Judgment under Uncertainty: Heuristics and Biases, Science, New Series, Vol. 185, No. 4157. pp. 1124–1131. Available at http://links.jstor.org/sici?sici=0036-8075%2819740927%293%3A185%3A4157%3C1124%3AJUUHAB%3E2.0.CO%3B2-M. Accessed 2011 Dec 26.
27. Gardner 2008, pp. 41, 47, 71.
28. Slovic P, Finucane M, Peters E, MacGregor DG. The Affect Heuristic. In: Gilovich T, Griffin D, Kahneman D, editors. Heuristics and biases, the psychology of intuitive judgment. New York: Cambridge University; 2002. pp. 397–420.
29. Slovic, P, Perception of Risk, Science, New Series, Vol. 236, No. 4799. pp. 280–285. Available at http://links.jstor.org/sici?sici=0036-8075%2819870417%293%3A236%3A4799%3C280%3APOR%3E2.0.CO%3B2-N, Accessed 2011 Dec 26.
30. Slovic, 1987, pp. 282–283.
31. Viscusi, W., Magat, W., Scharff, R. Asymmetric Assessments in Valuing Pharmaceutical Risks Medical Care, Vol. 34, No. 12, Supplement: Standards for Economic Evaluation of Drugs: Issues and Answers pp. DS34DS47 Available at http://www.jstor.org/stable/3766351, Accessed 2011 Dec 26.
32. Q9 Quality Risk Management, 2011, p. 3. Available at http://ec.europa.eu/health/documents/eudralex/vol-4/index_en.htm. Accessed 2011 Dec 26.
33. ISO 14971 Medical Devices—Application of Risk Management to Medical Devices, International Organization for Standardization 2007, p. v.

34. Breyer S. Breaking the vicious circle : toward effective risk regulation. Cambridge: Harvard University; 1993.
35. Breyer S. Active Liberty. In: Interpreting our Democratic Constitution. New York: Knopft; 2005.
36. Calabbresi, M., After Avandia: Does the FDA Have a Drug Problem, Time. Aug 12, 2010 Available at http://www.time.com/time/magazine/article/0,9171,2010181,00.html, Accessed Dec 26 2011.
37. FDA Press Release, Justice Department takes action against McNeil-PPC Inc. Charged with manufacturing and distributing OTC drugs in violation of federal law, Available at http://www.fda.gov/NewsEvents/Newsroom/PressAnnouncements/ucm246685.htm. Accessed Dec 26 2011.
38. Wilkinson, A., S. Elahi, and E. Eidinow (2003) Section 2 Background and dynamics of the scenarios.Journal of Risk Research. Vol 6. *No*. 4–6:365–401.
39. Commission of the European Communities. Communication from the Commission on the precautionary principle. Brussels 2000. Available at ec.europa.eu/dgs/health_consumer/library/pub/pub07_en.pdf. Accessed 2011 Dec 26.
40. Commission of the European Communities. Communication from the Commission on the precautionary principle. Brussels 200015–16.
41. ISO 14971 Medical Devices—Application of Risk Management to Medical Devices, International Organization for Standardization 2007, p. v.
42. MHRA Good Manufacturing Practice (GMP)—Quality Risk Management: Frequently Asked Questions Available at: http://www.mhra.gov.uk/Howweregulate/Medicines/Inspectionandstandards/GoodManufacturingPractice/FAQ/QualityRiskManagement/index.htm. Accessed Dec 26 2011.

4

STATISTICAL TOPICS AND ANALYSIS IN RISK ASSESSMENT

MIKE LONG

> *As we know,*
> *There are known knowns.*
> *There are things we know we know.*
> *We also know*
> *There are known unknowns.*
> *That is to say*
> *We know there are some things*
> *We do not know.*
> *But there are also unknown unknowns,*
> *The ones we don't know*
> *We don't know.*
> —Donald Rumsfeld Feb. 12, 2002, U.S. Department of Defense news briefing. Wording arranged by Slate Magazine in the article The Poetry of Donald Rumsfeld, *April 3, 2003.*
>
> *The only thing certain is Uncertainty.*
> —Anonymous

4.1 INTRODUCTION

Understanding probability and its role in the field of statistics is critical for a mature level of understating of risk management. This chapter does not make an attempt to provide a complete understanding of the topic, as that content would need its own library wing. It does, however, attempt to provide a brief overview of points to consider and ponder when you attempt to quantify the probability dimension of risk.

4.2 UNCERTAINTY

John Maynard Keynes declared there are times and places were no "probability calculus" would be possible. He called this true uncertainty [1]. Keynes argued that uncertainty could not be defined objectively as there may be no knowledge of a hazard. Others say risk and uncertainty are and the same and attempt to quantify the subjective portion of risk. This is an attempt to create an enumerative vehicle for a subjective expected utility. There is a difference between what is possible, merely probable, uncertain, and true uncertainty. Something is simply uncertain when we have yet to collect data; think of process development here. True uncertainty lies in areas where it is impossible to measure, such as the probability of war between China and the United States, or the obsolescence of an invention such as a microwave oven or an iPad. We simply do not know. As Keynes put it, even "the weather is only moderately uncertain"; a statement made well before the weather technology of today [2].

In practice, risk is on a sliding scale. In some cases, it can be defined in objective terms, in others it is subjective. In applications of risk assessment where uncertainty exists, people try to assess uncertainty within a subjective rating instead of simply stating "I don't know" and collecting data.*

By combining definitions from several sources, uncertainty is defined here as an unquantitated lack of specific knowledge. "Looks like rain," is uncertainty statement.

By comparison, a quantitated lack of specific knowledge is the basis for the complementary fields of probability, statistics, and risk. "A fifty percent chance of rain" is a probability statement.

4.3 LUCK AND PROBABILITY

Luck is probability taken personally.

—*Anonymous*

The duality of this statement illustrates the nature of probability in our society. Millions of people believe in luck but not in probability. A short example illustrates the point. If we flip a quarter six times and get heads each time, most people on the street or in a casino believe that the probability of getting a tail on the seventh flip is greater than if we had gotten three heads and three tails. They would in fact literally bet money on it.

---

*For an expanded discussion, read Nassim Nicholas Taleb's The Black Swan.

To their mind, the coin is due to come up tails. Yet, probability theory confirms that the probability of a seventh flip coming up heads is still 50/50 as it was on the first, second, and all flips. The coin has no memory. This misunderstanding illustrates the need for an understanding of probability as related to risk and risk management. Clearly, patient health and safety cannot be based on ignorance or, worse, luck.

4.4 DEFINITIONS OF PROBABILITY

It is hard to improve on this three-hundred-year-old definition of probability:

For it should be presumed that a particular thing will occur or not occur in the future as many times as it has been observed, in similar circumstances, to have occurred or not occurred in the past [3].

In addition, there are three common definitions or uses of the term probability:

1. The least quantitative is a degree of subjective belief, gut feeling, or intuition. "I believe there is a small probability that he will be promoted this year."
2. The most empirical is the ratio of the number of outcomes to the total number of possible outcomes. It is written as a fraction, percentage, or proportion. "A four has been observed sixteen times in the last one hundred rolls of the die." This is the approach we prefer if we have the data to do the calculation. Probability estimates from past data projected to future probability assumes that the future will continue to look like the past. Of course, this adds to the uncertainty of the exercise.
3. Last, the most quantitative is a theoretical value determined by the limiting frequency of an infinite random series for an assumed model. In theory, a fair coin will come up heads 50% of the time. In practice, it only comes close.

4.5 RULES OF PROBABILITY

Probability is expressed as a number between 0 and 1. It is also expressed as a percentage, from 0–100%. Zero is assigned to an event that is absolutely impossible to occur. One or hundred percent is assigned to an event that is absolutely certain to occur. In some situations, they can be given in smaller categories such as low, medium, high, 0, 1, 2, 3, 4, or 0, 1, 2, 3, 4, 5, 6. Note an odd number of categories starting with zero permits a center category that is half the maximum. Reporting scales are often specific for a given project.

A probability of 0.5 or 50% is assigned to an event that would occur half the time on average in an infinitely long series. It is also called the indifference point. In some literature, an event is said to be "possible" if the probability is 50% or less. It is "probable" is the probability is greater than 50%.

The probability of flipping a nickel and getting a head is in theory one-half, 0.5, or 50%.

$$P \text{ (Head)} = 0.5$$

In actual practice, it could be slightly more or less. The old buffalo nickel was heavy on the side with the buffalo. It is possible that it could be shown to have an effect.

The numerator and denominator in probability calculation need to be carefully defined. The letters a, e, i, o, u, and y are the usual vowels. Thus, the probability of picking a vowel at random from a page of text is 6 out of 26 or 23%. In some cases, however, y is used as a consonant. Thus, there is some uncertainty in the numerator and the final result. Careful assumptions, operational definitions, and specific rules will clarify these situations.

What is the probability of tossing a nickel twice and getting two heads? The probability is assumed to be one-half for the first toss and one-half for the second toss. Thus, 0.5 AND 0.5, where AND means to multiply. The result is one out of four.

$$P \text{ (Head and Head)} = P \text{ (Head)} * P \text{ (Head)} = 0.5 * 0.5 = 0.25 \text{ or } 25\%.$$

We multiply when we see "AND."

This assumes that the two events are independent of each other. That is, they do not influence each other. If the events are not independent, the rule or formula is shown here.

What is the probability of getting a head or a tail on one toss of a dime? The probability is assumed to be one-half for a head or one-half for a tail. Thus, 0.5 OR 0.5, where OR means to add.

$$P \text{ (Head or Tail)} = P \text{ (Head)} + P \text{ (Tail)} = 0.5 + 0.5 = 1.0 \text{ or } 100\%$$

This holds as long as there is no commonality of the two events. We add when we see "OR."

4.6 CAUTIONS

The cumulative effect of these rules can be impressive. For example, conducting multiple t-tests with the same alpha value can result in a very high probability of at least one failure just by random chance. Assume an assay validation report has 47 t-tests, each with a typical alpha, or α value of 0.05. The probability of at least one or more failures is found by the formula $1 - (1 - \alpha)^n$.
So,

$$P = 1 - (1 - \alpha)^n$$

where

$$n = 47 \text{ and } \alpha = 0.05$$

and
$$P = 1-(1-0.05)^{47}$$

So
$$P = 0.91$$

In this case, the probability that at least one or more t-tests will fail by random chance alone is 91%.

4.7 RISK

Every manufacturing process and product has some risk and probability of failure. Zero risk 100% of the time will cost an infinite amount of time and money and is effectively unattainable as some risk (called *residual risk*), although small, is always present even after mitigation efforts are completed.

As explained elsewhere in this book, risk can be the estimated probability (also known as occurrence or likelihood) of an event times a severity estimate.

$$\text{Risk } (R) = P(\text{or } O) \times S$$

In either case, probability is a root concept for risk. Risk cannot be understood or estimated without an understanding of probability.

4.7.1 Venn Diagram

A good way to visualize the dimensions of probability described earlier is to look at the entire population for a given process or product as shown in Figure 4.1. Within that product population space, there are three subsets. Using the language of Section E of ISO 14971, [4] two of the subsets, events, are of interest. One is the area within the process or product space where Hazards have occurred, the gray area or H. The second is the space defined by where Failure and thus harm to the patient can occur, the blue area or F. The third is the intersection of gray and blue or F & H.

So, when looking at assessing the occurrence or likelihood of an event, we have to look at how often a given hazardous, H, situation happens (it may be quite common) as well as how often these "happenings" lead to failure, F, and thus harm to the patient. A basic discussion on conditional and joint probabilities is important in understanding the topic.

4.7.1.1 General Rule for Addition For the two events in Figure 4.1, the probability of a hazard OR the probability of a failure is the sum of the two minus the intersection of the two as in the formula.

$$P \text{ (Hazard or Failure)} = P \text{ (Hazard)} + P \text{ (Failure)} - P \text{ (Hazard and Failure)}$$

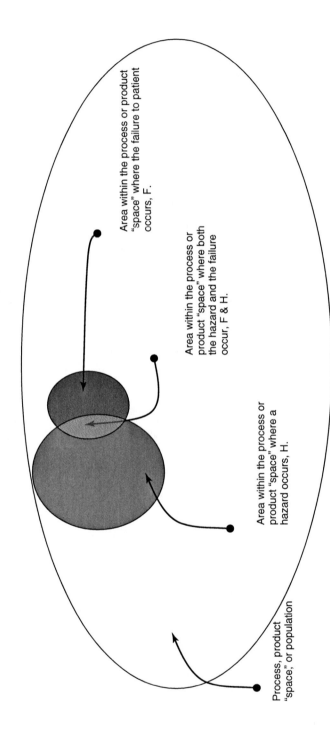

Figure 4.1 Venn diagram visualization of ISO 14971 section E. (*See insert for color representation of the figure.*)

4.7.2 Conditional Probability

A good way to define *conditional probability* is to think about it pictorially as in Figure 4.1. Conditional Probability is defined as

The probability of event Failure occurring given that event Hazard has already occurred, or "F given H".

It is written as:

$$P(\text{Failure}|\text{Hazard}) = P\frac{(\text{Failure}|\text{Hazard})}{P(\text{Hazard})}$$

where

The vertical bar is the symbol for "given."

This traditional definition does not really highlight the fact that we are reducing or sectioning the sample space we wish to assess. Figure 4.2 shows two different sets within the process and clearly shows how the reduction of the sample space looks when compared to Figure 4.1. Set 1 is the process or product population we are trying to describe in the risk assessment. Within this set, two subsets are used to determine the conditional probability. The first subset (i) is simply the magnitude of a hazard that has occurred within the population. The second subset (ii) shows the magnitude of a specific harm that has occurred within subset (i) Visually, Figure 4.2 shows the magnitude of a failure or harm that occurs when the hazard is present.

Another way to say conditional probability could be "Failure within Hazard" rather than "Failure given Hazard."

Thus, the probability is $P(\text{Failure}|\text{Hazard}) = P\dfrac{(\text{Failure}|\text{Hazard})}{P(\text{Hazard})}$

Multiplication Rule or Joint Probability

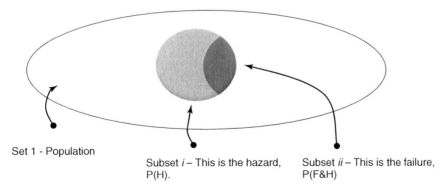

Figure 4.2 Conditional probability described as a subset. (*See insert for color representation of the figure.*)

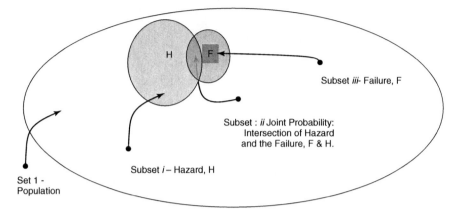

Figure 4.3 Joint probability. *(See insert for color representation of the figure.)*

Joint probability refers to the occurrence of two (or more) events happening and we use the *multiplication rule* shown to define it.

$$P \text{ (Failure \& Hazard)} = P \text{ (Failure|Hazard)} \, P \text{ (Hazard)}$$

It is just a restatement of the equation for conditional probability. Where the *joint probability, P (Failure and Hazard)*, of the failure is equal to the proportion of the number of times the failure has occurred within the hazard, P (Failure|Hazard), multiplied by the proportion of the number of times the hazard, P (Hazard), has occurred within the population.

Conceptually, it is simply the number of joint occurrences of the hazard and the harm in reference to the entire population as shown in Figure 4.3.

4.8 STATISTICAL RISK

In statistical literature, risk is often discussed as a simple probability. Statisticians differentiate between two types. A Type I (one) error or alpha level is the probability of rejecting a hypothesis when it is true. This is the producer's risk. A Type II (two) error or beta level is the probability of accepting a hypothesis when it is false. This is the consumer's risk.

These probabilities can be used to calculate risk for defined models. For a further introduction to probability, see a statistical textbook such as *Introduction to the Practice of Statistics*, Moore and McCabe [5]. For a good (and humorous) overview of basic probability, see *The Cartoon Guide to Statistics*, Gonick and Smith [6].

4.9 RARE EVENTS

> "If anything can go wrong, it will."
>
> —Murphy's Law

RARE EVENTS

A rare event is one that is uncommon or infrequent and unusual or distinctive in some way. Such an event occurs because the scope of possibilities is so much greater than we can imagine. For example, there are thousands of pathways in a computer that can lead to truly weird results.

Rare is also relative. A once in a million per day event for a small village is happening about 300 times a day in the whole United States. Six Sigma programs aim for defects of around four defects per million. In a tablet batch of 1.2 million running at Six Sigma, can we really expect to find those four defective tablets?

It is also helpful to differentiate between extreme values and outliers. An extreme value exceeds the ordinary or the usual. They are situated at the farthest possible point from the center of the data and are still considered to be part of the population. They have a very low probability of occurrence. For example, heavy tablets or capsules or under-filled vials are extreme values. Floods and natural disasters are another example. They are the result of an accumulation of common causes.

On the other hand, outliers are so far from the rest of the population that they are not considered to be part of the expected distribution. They are often the result of nonrandom special causes. Examples include an empty sugar packet, an empty vial, or a double-struck tablet. Manmade disasters are outliers. Outliers are generally rare events.

4.9.1 Calculating POISSON

Poisson is useful in determining the likelihood of small occurrences where there are many chances for the event to occur.

When calculating probabilities of occurrence of hazards such as lack of sterility or the presence of endotoxin, the *law of small numbers* (*or the law of rare events*) is applicable. Poisson distribution describes the law of rare events and is shown mathematically as

$$P(X) = \frac{\lambda^X e^{-\lambda}}{X!}$$

where:
- X = the number of expected occurrences
- e = base of the natural log
- λ = average number of occurrences of the event

For example, if we wanted to determine the probability of having 0 sterility failures in the upcoming year for a manufacturing site that has a historical average of 2, we could use Poisson. In this example, X, the number of expected occurrences, is 0, and λ, average number of occurrences of the event, is 2. So:

$$X = 0$$
$$\lambda = 2$$

Therefore,

$$P(0) = \frac{2^0 e^{-2}}{0!}$$

$$P(0) = \frac{e^{-2}}{1}$$

$$P(0) = 0.1353$$

Using Poisson, we calculate the probability P of having no sterility failures for a site that averages 2 per year to be 0.135, or a 13.5% chance. If we wish to find out the likelihood of exceeding the average of 2 failures per year, we would calculate the probability as such:

$$P(X > 2) = 1 - [P(0) + P(1) + P(2)]$$

where, from the above-mentioned

$$P(0) = 0.1353$$

and

$$P(1) = \frac{2^1 e^{-2}}{1!} = P(1) = \frac{2e^{-2}}{1} = 0.2707$$

and

$$P(2) = \frac{2^2 e^{-2}}{2!} = P(2) = \frac{4e^{-2}}{2} = P(2) = \frac{2e^{-2}}{2} \; 0.2707$$

Therefore,

$$P(X > 2) = 1 - [0.1353 + 0.2707 + 0.2707 = 0.3233]$$

So, the probability for the manufacturing site that averages 2 sterility failures per year to experience greater than 2 failures in the upcoming year would be 0.323 or 32.3%.

4.10 COINCIDENCES

A coincidence is the occurrence of two events that happen at the same time by random chance, but give the appearance of a connection or a cause-and-effect relationship.

> "Given the billions of people and the quadrillions of characteristics, it would be incredible if incredible things didn't happen"
>
> —Anonymous

Coincidences can be rare events. For example, in a four-car accident, all four drivers had the same last name; or in California, a couple had three children all born on May 28. When a coincidence is very rare in an industrial setting, it is tempting to assume a cause-and-effect relationship.

4.11 ESTIMATING PROBABILITIES

Probability estimates and evaluations are based on scientific knowledge and historical data to the extent possible.

Also, the level of effort and rigor in estimating probability should be commensurate with the perceived level of risk. Considerable effort is expected for high risk situations.

Probabilities can be estimated using the three definitions of probability. In the first definition, the assumption is that little information or data is available. The estimate is a best scientific guess based on past experience, wisdom, and intuition.

However, it is human nature to equate publicity with increased risk. For example, as of the week of May 18, 2009, the World Health Organization reported that a total of only 4800 people had contracted swine flu and that 61 people were believed to have died from it. Yet, the massive publicity based on poor estimates of the problem in Mexico prompted many schools to close. The Vice President of the United States told his family not to travel. People were buying face masks and stockpiling food.

Now compare this to auto deaths in the United States. For the first ten months of 2008, there was an average of 3111 deaths each month for auto crashes. This was considered good news since for that same time span in 2007 there was an average of 3450 deaths per month, a decrease of 339 deaths. Notice that the Vice President did not recommend that his family stop using their cars.

Unfortunately, this means that highly publicized events that have happened recently may be perceived to be more likely than events that have happened with less publicity. For example, with many news articles about pharmaceutical supply chain problems, a company may perceive that it is at greater risk for them when in fact they may actually be at greater risk for microbiological contamination.

The second definition assumes some data is available to calculate a probability. This is usually historical data collected in development, validation, and/or production. Having weighed a 1000 tablets from each of 10 sequential lots, we can estimate the probability of finding tablets that will exceed the alert, action, or failure specifications. Of course, we expect the data to be truly representative of the process.

As another example, assume a series of trays of 400 vials each has, on inspection, averaged 40 cracked vials per tray. What is the future estimate of probability of selecting a cracked vial if we pulled one vial purely at random? We calculate the result as 40 out of 400 or 10%, assuming the future will look like the past. So the occurrence rate of the hazard is 10%. So we could assign a probability value of 0.1 for the hazard. But is this the true value? How confident are we of the value? Does this matter for a risk assessment?

For this, we need to calculate a confidence interval around the occurrence rate determined for the potential hazard. For most cases, looking for a 95% confidence interval is perfectly acceptable. This is shown as

$$95\% \text{ CI} = P \pm 1.96 s_p$$

where

S_p = standard error of the estimate
P = occurrence rate sample probability
1.96 = normal Z score for 95% confidence

So

$$95\% \text{ CI} = 1 \pm 1.96 s_p$$

To calculate $S_p{}^\dagger$, we use the following formula:

$$S_p = \sqrt{\frac{P(1-P)}{N}}$$

where

P = the occurrence rate sample probability
N = the number of samples taken

In this case, we know:

$$P = 0.1$$
$$N = 400$$

So,

$$S_p = \sqrt{\frac{0.1(0.9)}{400}} = 0.015$$

Now

$$95\% \text{ CI} = 0.1 + 1.96 s_p = 0.1 \pm 1.96 \times 0.015$$
$$95\% \text{ CI} = 0.1 \pm 0.0294$$

So, we are 95% confident that the true occurrence rate is between

0.071 and 0.129

or

7.1% and 12.9%

$^\dagger S_p$ gives us a standard error based upon the sample size, N, used.

TABLE 4.1 Comparison of Samples Sizes When Estimating the Occurrence Rate for a Hazard

Occurrence Rate Sample P	Sample Size N	Confidence Interval Range [C] at 95%
0.1	400	0.071–0.129
0.1	30	0–0.21

Is this range of any concern when applied to a risk assessment? It depends. What if we took a sample of 30 instead of 400? For $n = 400$, we can assume 1.96 for a 95% CI, but for $n = 30$, we need to use the t distribution with a value of 2.042. We will assume the same occurrence rate $P = 0.1$, or 10% T, and change N from 400–30.
So,

$$S_p = \sqrt{\frac{0.1(0.9)}{30}} = 0.055$$

$$95\% \text{ CI} = 1 \pm 2.042 S_p = 0.1 \pm 2.042 \times 0.055$$

$$95\% \text{ CI} = 0.1 \pm 0.112$$

which changes the 95% CI range to

$$0–0.212$$

or

$$0\%–21.2\%.$$

What we originally thought had a probability of 0.1 could be as little as 0 or as high as 0.212 (if our sample size was 30) as summarized in Table 4.1. Potentially an issue when attempting to assign an occurrence rate.

In the third case, we assume to have a reliable statistical model that can be used to estimate probabilities. The most common models for laboratory data are the normal distribution, the lognormal, and the exponential distribution.

4.12 CONCLUSION

While space does not permit an extended discussion of the common tools mentioned in ICH Q9, most readers have some prior knowledge even if in passing. This chapter was not written to give the reader a full understanding but rather to provide conceptual constructs and points to consider. However, the following references will permit self-study for those needing an introduction or a refresher. An Internet search will find many more.

These are from the excellent NIST web-based statistics book, *Engineering Statistics Handbook*, a highly recommended source.

Control Charts: http://www.itl.nist.gov/div898/handbook/pmc/section3/pmc3.htm

Histograms: http://www.itl.nist.gov/div898/handbook/eda/section3/eda33e.htm

Pareto Plot: http://www.itl.nist.gov/div898/handbook/pri/section5/pri597.htm

Design of Experiments: http://www.itl.nist.gov/div898/handbook/pri/section1/pri11.htm

Process Capability: http://www.itl.nist.gov/div898/handbook/pmc/section1/pmc16.htm

Graphics in general: http://www.itl.nist.gov/div898/handbook/eda/section3/eda33.htm

REFERENCES

1. Keynes, J. M., The Collected Writings of John Maynard Keynes. Volume 29. The General Theory and After: A Supplement, London: The Macmillan Press. 1979.
2. Keynes, J. M. (1937), "The general theory of employment," Quarterly Journal of Economics, vol. 51, pp. 209–233.
3. Bernoulli, Jacob. Ars Conjectandi. Basil: Thurnisiorum, 1713.
4. ISO 14971 Medical Devices—Application of risk Management to Medical Devices, International Organization for Standardization 2007
5. Moore D, McCabe G. Introduction to the Practice of Statistics. New York, NY: W. H. Freeman; 1989.
6. Gonick L, Smith W. The Cartoon Guide to Statistics. New York, NY: Harper Perennial; 1993.

FURTHER READING

Aven T. Foundations of Risk Analysis. New York, NY: John Wiley; 2003.

Bernstein P. Against the Gods. New York, NY: John Wiley; 1996.

Byrd D, Cothern R. Introduction to Risk Analysis. Lantham, MD: Government Institutes; 2005.

Claycamp, Gregg. "Risk, uncertainty, and process analytical technology," The Journal of Process Analytical Technology, vol. 3, No. 2, (2006) 8–12.

DeSain C, Sutton C. Risk Management Basics. Cleveland, OH: Advanstar Communications; 2000.

The Black Swan: The Impact of the Highly Improbable, Nassim Nicholas Taleb.

5

QUALITY BY DESIGN

Bruce S. Davis

5.1 BACKGROUND

The term *quality by design* or QbD is increasingly used in the pharmaceutical industry and normally describes a science- and risk-based approach to developing and manufacturing pharmaceutical products, based on the principles laid out in the following International Conference on Harmonization (ICH) guidances:

- Q8R2: pharmaceutical development [1]
- Q9: risk management [2]
- Q10: pharmaceutical quality system (PQS) [3]
- Q11: development and manufacture of drug substances—chemical entities and biotechnological/biological entities.

These ICH guidelines are voluntary but have set the scene for an alternative way for the pharmaceutical industry to develop and manufacture its products. Each guideline is discrete in its own right, but they should be read together to understand the full impact of this new way of working.

A QbD or "an enhanced approach," a term that ICH Q8R2 uses, is where quality is designed in from the outset, using a science and risk basis, as opposed to a traditional approach, where normally end-product testing used to check quality requirements have been met.

Risk Management Applications in Pharmaceutical and Biopharmaceutical Manufacturing, First Edition. Edited by A. Hamid Mollah, Mike Long, and Harold S. Baseman.
© 2013 John Wiley & Sons, Inc. Published 2013 by John Wiley & Sons, Inc.

ICHQ8R2 states that a QbD approach would include the following elements [1]:

"*A systematic evaluation, understanding and refining of the formulation and manufacturing process, including*

- *identifying, through e.g., prior knowledge, experimentation, and risk assessment, the material attributes and process parameters that can have an effect on product critical quality attributes (CQAs);*
- *determining the functional relationships that link material attributes and process parameters to product CQAs.*"

Taking a science- and risk-based approach to developing a product is not a new concept. Indeed, companies normally apply such methodologies during product development, but historically, the output has focused on regulatory compliance, which may have meant that science- and risk-based information gained at the time was not captured as rigorously as it could have been.

It would, however, be incorrect to suggest that industry was not approaching product and process development with the right intentions. Companies have traditionally had a strong focus on the needs of the patient and have developed their products with this in mind. For example, risk-based approaches for new plants have historically used such tools as hazard and operability (HAZOP) studies, where risks are evaluated to assure adequacy of engineering designs, particularly for operational safety, and have become established processes.

The move toward a strengthened science- and risk-based approach was galvanized by the publication by the Food and Drug Agency's (FDA's) "GMPs for the 21st Century" [4] and their "PAT Guidance" [5]. Another significant step that helped provide an enhanced environment between regulators and industry was the FDA's invitation for companies to join their pilot program. Companies applying were encouraged to discuss, with the FDA, science- and risk-based approaches to new product development proposals, before any formal submission was made. This pilot program was also extended to include biotechnology products.

The European Union (EU) also encourages a science- and risk-based approach, as illustrated by European Medicines Agency (EMA) (formerly EMEA) in instigating their process analytical technology (PAT) team [12]. This is made up of experts from different parts of Europe, including inspectors and assessors with wide experience of submissions and inspections. They have encouraged contact by companies and part of their role, quoting from their mandate, includes the following statement, "perform review and assessment of "mock" submission of applications using PAT and QbD principles" [5].

The PAT team has built up considerable experience of QbD and, similar to FDA, has apparently approved a number of submissions that include QbD principles.

5.2 DEVELOPMENT OF PRODUCTS USING A QBD APPROACH

There are several stages in developing a new product using a QbD methodology. These stages are applicable in principle whether the products are small or large molecules.

The principle steps based on Q8R2 start with patient safety and efficacy requirements, to ensure the final drug product meets quality requirements. In order to explain the QbD approach, it uses the following terms:

- quality target product profile (QTPP);
- critical quality attribute (CQA);
- risk assessment: linking process parameters and material attributes to drug product CQA;
- design space;
- control strategy;
- product life cycle management and continual improvement.

These are diagrammatically shown in Figure 5.1. It should be noted that although the flow goes broadly from left to right, there can be many iterations on the way.

The key output of this flow is to develop and put in place a control strategy to assure product quality and to meet patient safety and efficacy requirements over the life cycle of the product.

Risk assessment is a key part of this overall process and is normally carried out at each stage, to ensure, as knowledge is gained, that potential risks are understood, mitigated, and controlled.

ICH Q9 provides the basis for risk management and uses the following diagram as guidance (Fig. 5.2).

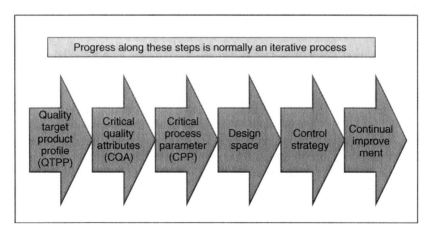

Figure 5.1 Diagrammatic flow of key steps for QbD. (*See insert for color representation of the figure.*)

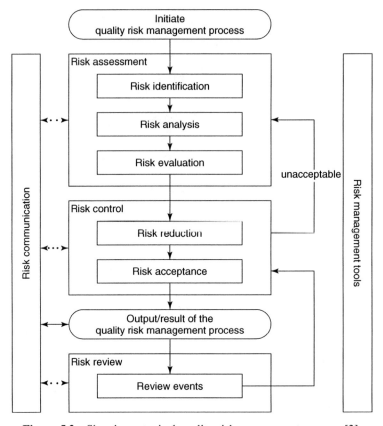

Figure 5.2 Showing a typical quality risk management process [2].

ICH Q9 also suggests a list of the key risk assessment and statistical tools, including, for example:

- failure mode effects analysis (FMEA);
- failure mode, effects, and criticality analysis (FMECA);
- fault tree analysis (FTA);
- HAZOP;
- preliminary hazard analysis (PHA);
- risk ranking and filtering;
- design of experiments (DOE); and
- process capability analysis.

A typical diagrammatic example of the output of an initial risk assessment is shown in Figure 5.3. This output uses risk ranking to categorize potential impact on a patient (normally based on severity) into high (red), medium (yellow), and low (green).

DEVELOPMENT OF PRODUCTS USING A QBD APPROACH

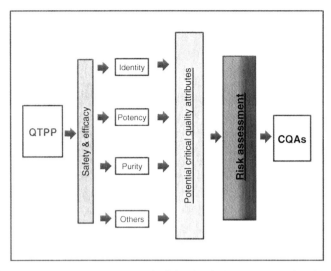

Figure 5.3 Example to illustrate the principles behind the output of a risk assessment and its potential impact on the patient. (*See insert for color representation of the figure.*)

One of the major benefits of diagrams such as Figure 5.3 is that they paint a clear picture of where potential risks might be and are useful, particularly at an early stage, to establish which areas should be subject to further investigation and analysis.

A useful risk assessment approach in relation to QbD is FMEA or FMECA and the output from one of these is illustrated in Figure 5.4.

	Safety	Efficacy	Quality
Potential CQA 1			
Potential CQA 2		High impact	
Potential CQA 3			
Potential CQA 4	High impact		
Potential CQA 5			High impact

Low impact   Medium impact   High impact

Figure 5.4 To show an example of the output of an initial risk assessments, e.g. based on prior knowledge. (*See insert for color representation of the figure.*)

5.3 MAIN STEPS FOR A QUALITY BY DESIGN APPROACH FOR A NEW PRODUCT

Each of the steps as diagrammatically illustrated in Figure 5.1 is now expanded as follows:

The *quality target product profile* (QTPP) is described in ICH Q8R2 [1] as "*a prospective summary of the quality characteristics of a drug product that ideally will be achieved to ensure the desired quality, taking into account safety and efficacy of the drug product.*"

This is where fundamental information about the purpose of a new drug is outlined. A QTPP, as a minimum, may typically include Drug Product quality criteria information for description, potency, dose, impurity limits, microbiological limits, etc; and set specifications or limits to assure quality targets will be met for patient safety and efficacy.

The QTPP is the starting point for QbD and it is important to note that this provides the link from patient requirements to the drug product. The QbD logic must start at the drug product design and then work back to drug substance and include the manufacturing requirements for all important unit operations—for both drug product and drug substance—that may influence the final drug product (Fig. 5.5).

The CQAs are described in ICH Q8R2 [1] as "a *physical, chemical, biological, or microbiological property or characteristic that should be within an appropriate limit, range, or distribution to ensure the desired product quality.*"

	Unit operation 1	Unit operation 2	Unit operation 3	Unit operation 4	Raw material attribute 1	Raw material attribute 2
Drug product CQA 1						
Drug product CQA 2						
Drug product CQA 3						
Drug product CQA 4						
Drug product CQA5						
Drug product CQA 6						

| Low risk | Need to be controlled to keep risk low | High risk |

Figure 5.5 Illustration to show the output of a risk assessment, linking manufacturing unit operations, raw material attributes and CQAs links. (*See insert for color representation of the figure.*)

MAIN STEPS FOR A QUALITY BY DESIGN APPROACH FOR A NEW PRODUCT   95

These are the attributes that impact patient safety and efficacy and therefore it is important they are clearly established for the drug product. An experienced company would normally use their prior knowledge to make an initial assessment of the potential CQAs and then carry out a risk assessment to establish whether these potential CQAs are critical to patient safety and efficacy, and product quality. This prior knowledge could comprise clinical and preclinical data, development records from similar compounds, manufacturing experience, other published data, etc.

Many attributes may be well known—for example, for a small molecule oral solid dosage product, dissolution can be considered as related to efficacy, whereas degradant (impurity) level would be related to patient safety. Alternatively, for a large molecule product, where there are usually many potential CQAs, viral load would be an example of a patient safety attribute.

Normally, but not always, a specification is established for a CQA.

The intent of the early risk assessment is to establish whether experimental work, such as DOE or other multivariate techniques, should be employed to understand which unit operations' parameters and material attributes may impact the CQAs.

For example, it may be known that blending in an oral solid dosage product, lyophilization for a parenteral product, or micronizing for an inhalation product are important unit operations that could affect particular product CQAs, but there may not be sufficient data to assess the risk, and so a company will carry out experimental work to investigate this. The attributes of raw materials (such as nonactive excipients or drug substance) can impact CQAs and therefore these should be included in the risk assessment.

On the basis of their experience, a company may have an initial idea of the kind of manufacturing process envisaged and therefore the type and number of unit operations expected, and choose to tailor their approach, for business reasons, to align with, for example, their manufacturing capability within the company. Its business strategy may favor one type of unit operation over another, such as choosing roller compaction over wet granulation for granulating of a tablet product.

A *critical process parameter (CPP)* is defined in ICH Q8R2 [1] as *"a process parameter whose variability has an impact on a CQA and therefore should be monitored or controlled to ensure the process produces desired quality."* It is therefore important to establish which process parameters for a particular unit operation impact the CQAs and the degree of this impact. An FMEA or FMECA risk assessment approach is commonly used to establish which process parameters and material attributes are critical. An example of the output from a risk assessment is shown in Figure 5.6. This illustrates both the initial risk assessment and final risk assessment, where controls have been put in place to bring the potential risks to an acceptable level, as shown by the dotted line.

Design Space is defined in Q8R2 [1] as, *"the multidimensional combination and interaction of input variables (e.g., material attributes) and process parameters that have been demonstrated to provide assurance of quality."*

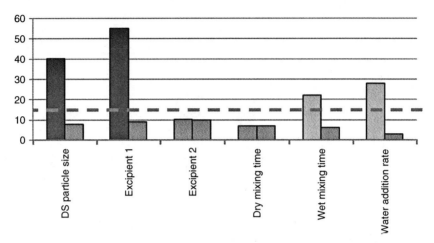

Figure 5.6 Diagrammatic part-outputs of initial and final a risk assessments to show the impact various unit operations may have on a CQA. (Dotted line shows level of acceptable risk). (*See insert for color representation of the figure.*)

It is important to note that design space is optional and normally multidimensional. Many attributes and parameters will have some interrelation with each other and therefore the design space should take into account the basis of these interactions and their boundaries. Having said this, it is not necessary to investigate the limit of every boundary as it may be expensive in time and resources to carry out this work. Where, through prior knowledge, it is known that a particular unit operation has a wide design space, then, after assessing the appropriate risk to product quality, perhaps minimal or no experimental work might be carried out.

It is also important to note that it is not essential to demonstrate the edge of failure of design space, unless of course there is a need to operate very close to this edge. Companies should assure themselves that the most important boundaries of design space have been assessed to understand where further data may be needed.

ICH Q8R2 [1] gives some examples of design space and their Figure 5.2c is shown below (Fig. 5.7):

Establishing the *control strategy* is probably the most important step of all the QbD steps. Control strategy is defined in Q10 [3] as, "*a planned set of controls, derived from current product and process understanding that ensures process performance and product quality. The controls can include parameters and attributes related to drug substance and drug product materials and components, facility and equipment-operating conditions, in-process controls, finished product specifications, and the associated methods and frequency of monitoring and control.*"

This wide definition illustrates the importance of putting in place all the controls that may be necessary at each stage of manufacture to assure the final drug product CQAs are met. The term control in this case may include a wide

MAIN STEPS FOR A QUALITY BY DESIGN APPROACH FOR A NEW PRODUCT

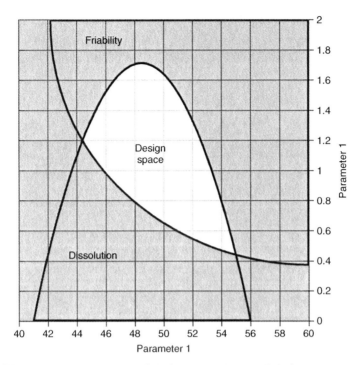

Figure 5.7 To show an example of design space (proposed design space, comprising the overlap region of ranges for friability and or dissolution) Ref ICH Q8R2 [1].

variety of controls, such as automated process controls, manual controls, standard operating procedures (SOPs), etc., to assure product quality. Some product controls may be highly product specific. For example, it may be necessary to control particle size distribution of an aerosol product to assure patient efficacy and this, in turn, may highlight the need to control CPPs at micronizing and blending unit operations. The control strategy should provide a clear rationale as to how the product CQAs are assured at all stages of manufacture.

Hand in hand with the product control strategy is the need to consider the requirements of ICH Q10, [3] the PQS. The PQS is defined in ICH Q10 [3] as, "*a management system to direct and control a pharmaceutical company with regard to quality.*" The product-specific controls and broader GMP controls that apply to more than one product (such as the adequacy of changing facilities for operators to dress into pharmaceutical clothing) should be integrated with the PQS and at the same time permit business requirements, such as operational efficiency, general safety, health and environmental considerations, financial, etc., to be met.

Finally, in regard to the QbD flow diagram in Figure 5.1, there is *Continual Improvement*. This is to ensure that as knowledge is gained over the life cycle of the product, such knowledge can be used to seek improvements to both products and processes. For example, it may be that practical manufacturing experience brings to light new efficiencies that can be applied in the development of other products. Alternatively, it may be that new technologies, developed at research and development (R&D), can later be applied in manufacturing.

The important part is to understand that learning is a continuum and that such learning should be applied across the whole product life cycle. It also highlights the need for R&D and manufacturing to work together.

5.4 EXAMPLES OF QBD APPROACHES

At the time of writing, some examples of QbD approaches for small molecule and large molecule products are available, as follows:

- the EFPIA Mock P2 paper issued 2006 [6];
- the ACE tablet study, issued 2009 [8];
- Mock QOS P2: "Sakura" tablet issued 2008 [7];
- A-Mab biotechnology study issued 2009 [9].

Reference should also be made to International Society of Pharmaceutical Engineers' (ISPE's) Product Quality Lifecycle Implementation (PQLI®) [10] published documents, particularly their Part 1–Concepts and Principles and Part 2–Illustrative Example, and Parenteral Development Associations (PDAs) Paradigm Change in Manufacturing Operations (PCMO*^{SM}) [11] initiative, as well as QbD training offered by other reputable organizations. The FDA has also issued case studies for immediate release and modified release tablets.

5.5 CONCLUSION

The implications of ICH developing Q8R2, Q9, Q10 and Q11 guidances have yet to be fully felt. It is important to understand there is no single way of implementing QbD in practice.

At the time of writing, the major pharmaceutical companies generally are starting to adopt these QbD principles for developing new products and smaller companies and generic and consumer health companies are also taking an interest, but overall adoption has been relatively slow. Some of this may be because of concerns that additional time is needed for this new approach and hence product launch may potentially be delayed or concerns about complexity and the need for learning new techniques in DOE or multivariate statistical methods or maybe how quickly the regulators themselves will become trained in these new techniques. Regardless of this, QbD does require the use of some new skills and approaches and it will be necessary for both industry and regulators to become familiar with applying these.

However, despite the potential concerns, it is important to understand QbD is about enabling improved product and process understanding and that its approach will certainly lead to greater clarity about what is needed for achieving product quality. This clarity will enable activities to be prioritized and to concentrate on what is most critical. Over time, this should lead to efficiency improvements, such as more effective R&D processes, improved manufacturing process capability, reduction in product and process failures, and enabling of cost-effective process validation. And these will bring business benefits.

But, most importantly, QbD will help assure high quality products that patients expect and deserve.

GLOSSARY AND DEFINITIONS

QbD: a systematic approach to development that begins with predefined objectives and emphasizes product and process understanding and process control, based on sound science and quality risk management [1].

Quality attribute: a physical, chemical, or microbiological property or characteristic that directly or indirectly relates to predefined product quality (safety, identity, strength, purity, and marketability of the product).

CQA: a physical, chemical, biological or microbiological property or characteristic that should be within an appropriate limit, range, or distribution to ensure product quality [1].

Process parameter: a process variable (e.g., temperature, compression force) that can be assigned values to be used as control levels or operating limits.

CPP: a process parameter whose variability has an impact on a critical quality attribute and therefore should be monitored or controlled to ensure the process produces desired quality [1].

Design space: the multidimensional combination and interaction of input variables (e.g., material attributes) and process parameters that have been demonstrated to provide assurance of quality [1].

Control strategy: a planned set of controls derived from current product and process understanding that ensures process performance and product quality. The controls can include parameters and attributes related to drug substance and drug product materials and components, facility and equipment-operating conditions, in-process controls, finished product specifications, and the associated methods and frequency of monitoring or control [1] and [3].

Life cycle: all phases in the life of a product from the initial development through marketing until the product's discontinuation [1].

Quality target product profile: a prospective summary of the quality characteristics of a drug product that ideally will be achieved to ensure the desired quality, taking into account the safety and efficacy of the drug product [1].

Quality risk management: a systematic process for the assessment, control, communication, and review of risks to the quality of the drug (medicinal) product across the product life cycle [2].

Risk assessment: a systematic process of organizing information to support a risk decision to be made within a risk management process. It consists of the identification of hazards and the analysis and evaluation of risks associated with exposure to those hazards [2].

PAT: a system for designing, analyzing, and controlling manufacturing through timely measurements (i.e., during processing) of critical quality and performance attributes of raw and in-process materials and processes with the goal of assuring final product quality [1].

REFERENCES

1. International Conference on Harmonization of Technical Requirements for Registration of Pharmaceuticals for Human Use, ICH Harmonized Tripartite Guideline, Pharmaceutical Development, Q8 R2, August 2009.
2. International Conference on Harmonization of Technical Requirements for Registration of Pharmaceuticals for Human Use, ICH Harmonized Tripartite Guideline, Quality Risk Management, Q9 Step 4, 9 November 2005.
3. International Conference on Harmonization of Technical Requirements for Registration of Pharmaceuticals for Human Use, ICH Harmonized Tripartite Guideline, Q10, Pharmaceutical Quality System, Step 4, 4 June 2008.
4. Pharmaceutical CGMPs for the 21st Century: A Risk-Based Approach; A Science and Risk-Based Approach to Product Quality Regulation Incorporating an Integrated Quality Systems Approach, United States Department of Health and Human Services, U.S. Food and Drug Administration, 21 August 2002, http://www.fda.gov/Drugs/DevelopmentApprovalProcess/Manufacturing/QuestionsandAnswerson CurrentGoodManufacturingPracticescGMPforDrugs/ucm137175.htm.
5. PAT—A Framework for Innovative Pharmaceutical Development, Manufacturing, and Quality Assurance, United States Department of Health and Human Services, U.S. Food and Drug Administration, Center for Drug Evaluation and Research (CDER), Center for Veterinary Medicine (CVM), Office of Regulatory Affairs (ORA), Pharmaceutical CGMPs, September 2004. http://www.fda.gov/downloads/Drugs/GuidanceComplianceRegulatoryInformation/Guidances/UCM070305.pdf.
6. European Federation of Pharmaceutical Industries and Associations (EFPIA) Mock P2, http://www.efpia.eu/Content/Default.asp?PageID=559&DocID=2933.
7. Sakura Tablet—Japan, http://www.nihs.go.jp/drug/section3/English%20Mock%20QOS%20P2%20R.pdf.
8. Ace tablet, http://www.ispe.org/acetablets.pdf.
9. A-MaB Biotech, http://www.casss.org/associations/9165/files/A-Mab_Case_Study_Version_2-1.pdf.
10. ISPE Product Quality Lifecycle Implementation (PQLI®) Good Practice Guide: Overview of Product Design, Development and Realization: A Science and Risk-Based Approach to Implementation, http://www.ispe.org/ispepqliguides/overview ofproductdesign.
11. PDA Paradigm Change in Manufacturing Operations (PCMO)*[SM], http://www.pda.org/pcmo/dossier.
12. European Medicines Agency PAT team, http://www.ema.europa.eu/ema/index.jsp?curl=pages/regulation/general/process_analytical_technology.jsp&murl=menus/regulations/regulations.jsp&mid=WC0b01ac058006e00e&jsenabled=true.

6

PROCESS DEVELOPMENT AND CLINICAL PRODUCT MANUFACTURING

The Use of Risk Management Tools in Development of Investigational Medicinal Products

KAREN S. GINSBURY

6.1 QUALITY VISION OF PHARMACEUTICAL DEVELOPMENT

One definition of a product is any thing/item that can be offered to a market in order to satisfy a potential want or need. For a clinical product, intended for use in human trials, the want or need is related to sickness/disease/pathology and therefore specifically to a *patient's* want or need. Therefore, the first quality attribute of any clinical product is that it must be developed to potentially satisfy some patient's want or need.

A product starts out as an idea or concept, which is carried forward, hopefully to realization, by following a set of processes. It is noteworthy that for pharmaceutical products the success rate for commercial product realization is low and the process is long. However, clinical trial materials are also a form of realized product and they are brought to the patient much earlier in the overall product life cycle. Product realization can be defined as

> "The sum total of all the processes that are used to bring a product into being" and involves starting with raw materials and working them into the defined finished product.

> In order to satisfy a patient want or need, an idea must be transformed, using defined processes so as to bring a product into being.

For pharmaceutical products under development, there are some basic rules that must be followed during transformation of an idea into a product in order to ensure patient protection.

1. *Products must be safe*—Clinical products can never be guaranteed as safe with respect to their pharmacological profile as this is the main purpose of performing the trials. The risks in controlled clinical trials are mitigated as much as possible by prior studies (e.g., in animals). Clinical products, however, must be safe with respect to the conditions under which they were manufactured and controlled. This is especially true regarding the quantity of active substance(s) that are claimed to be in them and the absence of unexpected and potentially harmful substances/contaminants other than those that are unavoidably part of the drug substance itself. In such cases, the maximum permitted levels must be closely controlled. This rule arises first from a moral obligation not to harm a patient participating in a study, but equally from the economic reality which determines that no company will realize a product if a patient is harmed as a result of manufacturing errors. This can result from inaccurate formulation or analysis of their product as well as through lack of sterility or the presence of particulate contamination.
2. *Products should be effective*—Ideally, clinical products must be effective, yet in many, possibly most cases, it will not necessarily be so. One of the main reasons for conducting a clinical study is to determine whether or not the proposed product is effective. If meaningful results are to be obtained from a study, the amount of active substance in the trial product must be defined and there should be a means of accurately analyzing and determining this value before the product is sent to a trial site. It is the responsibility of the sponsor to ensure that the product distributed for human use in a clinical trial contains the active substance(s) in the amount(s) stated on the product label. This requires reliable and controlled manufacturing and analytical methods even in early-phase trials.
3. *Product must meet a patient's want or need*

These rules can be summarized as requiring the product to be suitable for its purpose—that is, fit for use in the clinical trial.

In the European Directive 2003/94/EC, a definition is provided for pharmaceutical quality assurance:

QUALITY VISION OF PHARMACEUTICAL DEVELOPMENT

> "Pharmaceutical quality assurance" means the total sum of the organized arrangements made with the object of ensuring that medicinal products or investigational medicinal products are of the quality required for their intended use.

As the patient may not know how to precisely define his or her wants or needs and because society as a whole requires that the patient be protected, all pharmaceutical products intended for human use are controlled. In fact, the ICH quality vision encompassed in July 2003 as a prelude to developing the Q8, 9, and 10 trilogy of product and process life cycle guidance described just such a system:

> "The goal is to develop a harmonized pharmaceutical quality system applicable across the lifecycle of the product emphasizing an integrated approach to quality risk management and science."

This concept is a natural continuation from the ICH Q6A [1] guidance, which states:

> The quality of drug substances and drug products is determined by their design, development, in-process controls, GMP controls, process validation and by specifications applied to them throughout development and manufacture.

The goal is to ensure the quality of the product destined for a clinical trial by identifying product characteristics—critical quality attributes (CQAs) [2] and their critical process parameters (CPPs)—that need to be controlled to manage variability in the manufacturing and quality control sampling and testing process.

Note: During development it is not usually possible to fully identify or define critical as opposed to key attributes and parameters, so care should be taken to set relevant specifications that are refined as knowledge is gained through development and manufacture, that is, throughout the product life cycle. This chapter uses the designations (C)PP and (C)QA to indicate that the criticality is under ongoing investigation, verification, and refinement along with the control strategy.

ICH Q10 [3] describes the tools or enablers of product realization as knowledge management and risk management. Knowledge is the systematic translation of scientifically collected data into useful instructions by analysis and drawing of valid conclusions. The integration of science and risk management should allow for successful product development as is shown later.

Regional GMPs have considered the need for controls in clinical product manufacturing for quite some time.

Medicinal products must be designed and developed in a way that takes into account the requirements of good manufacturing practice (GMP) [4].

Any drug product intended for human use must be manufactured in accordance with the GMP or it is considered adulterated [5].

6.2 BASIS FOR DESIGN—TARGET PRODUCT PROFILE AND PRELIMINARY HAZARD ANALYSIS

In order to achieve product realization, a basis for design is needed to facilitate the transformation from idea to product. The basis for design of a pharmaceutical product is the quality target product profile (QTPP) defined in ICH Q8R1 [6] as

> "A prospective (but dynamic) [7] summary of the quality characteristics of a drug product that ideally will be achieved to ensure the desired quality, taking into account safety and efficacy of the drug product."

The QTPP forms the basis of design for the development of the product and the process and should be one of the very first documents to be placed in a product specification or design history file [8]. The first version of a QTPP for a new product might look something such as the one shown in Figure 6.1 and is also a good starting point for the initial risk assessment.

Quality attribute	Target
Target population	Immunocompromised
Route of administration	Oral
Dosage form	Tablet
Strength	0.6 mg
Packaging	Securitainer, plastic cap with polythene liner and paper overseal, dessicant
Stability	3 years at room temperature
Pharmacokinetics	Immediate release enabling tmax in 2 hours or less
Appearance	White to pale yellow, round with break mark
Assay	90–110% (Stability); 95–103% (Release)
Impurities	Individual impurities: NMT 0.1% Total Impurities: NMT 0.5%
Content uniformity	Meets USP
Dissolution	NLT 70% of labeled amount dissolved in 30 min : (500 ml water; USP apparatus II (paddles); 50 rpm)
Microbiology	NMT 100 CFU / tab total count NMT 10 CFU / tab yeasts and molds

Figure 6.1 Example of a preliminary quality target product profile.

Figure 6.1 shows a QTPP for a solid oral dosage form. A QTPP should be the first controlled document developed when designing a new product. All disciplines involved in ultimately bringing the product to market—engineering, maintenance, production (the process owner), quality, R&D, purchasing, sales and marketing, as well as any other stakeholders—should sign off on the document. Sales and marketing requirements can be particularly important to allow for designing of any unique features such as package configurations for tablets that may directly impact product stability. The QTPP is a controlled document that is assigned a version number and date. The document will be periodically updated as development progresses and some of the ideal requirements are found to be unachievable except at the expense of others. A team meeting should be called to make a decision regarding which requirement takes precedence and this usually requires senior management input. For example, marketing wants the tablets packaged in a securitainer, with a three-year shelf life at room temperature. After studying the stability profile, R&D offers several possibilities: three years at room temperature in a blister pack; 18 months at room temperature in a securitainer; and three years refrigerated in a securitainer. Marketing will probably have the overriding vote regarding the final presentation, but the preliminary hazard analysis, which considered product packaged in a securitainer with a shelf life of three years at room temperature, will now need to be reviewed in light of the new decision. Some of the previously identified risks and their controls may have changed as indicated in Tables 6.1 and 6.2.

The QTPP will also be the basis for the first draft of the finished product specification. Consideration should be given to the fact that drug substance or active pharmaceutical ingredient (API) development often progresses in parallel with the finished product and analytical methods development. As a result, there is an overlap of activity between the different project development teams, which can result in increased uncertainty and therefore increased risk. For example, API synthesis experiments with different solvents to reduce certain impurities while development of analytical methods is under way. QC invests effort in developing methods to identify the impurities from solvent A, while process development eliminates them and introduces others. At the same time, work moves forward on formulation and development of analytical methods for drug products (often different to the API ones taking into account excipients). A preliminary hazard

TABLE 6.1   Initial Hazard Analysis

Item #	Identified Risk for Original QTPP	Control	Monitoring	Comment
1.	Moisture absorbed by tablets during storage	Insert dessicator in each container	Stability data	
2.	Wrong number of tablets in container	Check weight	Calibration Periodic manual count	

TABLE 6.2 Revised Hazard Analysis with Stability Data Input

Item #	Identified Risk for Revised QTPP	Control	Monitoring	Comment
1.	Moisture absorbed by tablets—risk eliminated by blister pack	Eliminated	Not applicable	
2.	Wrong number of tablets	Eliminated	Not applicable	
3.	Absence of tablet/presence of broken tablet	Vision camera	Challenge by deliberately passing blister with missing or broken tablet	Part of line clearance procedure/set up

analysis at an early stage followed by formal risk communication to the parallel project groups can mitigate the possibility of misunderstandings and divergent rather than interleaved development pathways.

As a result of miscommunication, the start of human trials might be delayed. This, in turn, can result in financial hardships that bankrupt the company and cause the potential therapy to be lost. Consider a preclinical study (toxicological safety studies in animals), where there is no well-characterized reference standard available. The drug substance assay is performed against a reference material that is simply an early batch synthesized and put aside for this purpose. It is unlikely that there will be any reference material for impurities available at this stage. The major risk is of sending material for testing that is inadequately characterized with respect to its impurity profile and inadvertently described as purer than it really is. This could lead to subsequent production of a "dirtier" batch for first-in-human trial, that is, one where additional impurities are identified, in which case further toxicology studies would be needed before the human study could proceed. However, this risk could be mitigated by a simple strategy whereby material from the toxicology batch is placed at $(-)70°C$ storage. The retained material can be reanalyzed as the analytical method develops, and if it can be shown that the "new" impurities were also present in the toxicology batch, then the initial studies would be valid and the first-in-human trial can proceed as planned.

To summarize, a preliminary hazard analysis identifies those risks that should be mitigated in the early stages of product development. Identification of the risks allows a systematic approach to risk mitigation as well as communication of the risks to stakeholders. This approach can reduce misunderstandings and delays that arise from less formal techniques.

Once the QTPP is established, a process flow diagram can be prepared for drug substance, and usually a little later, for product manufacture and packaging.

The process flows will assist in performance of an initial risk assessment that might be no more than a brainstorming session using, for example, an Ishikawa

DESIGN—TARGET PRODUCT PROFILE AND PRELIMINARY HAZARD ANALYSIS

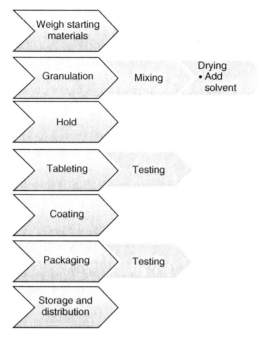

Figure 6.2 Process flow for manufacture of tablets. (*See insert for color representation of the figure.*)

or fishbone diagram to break down the process into smaller steps and identify risks associated with each part of the process. The purpose of this assessment is to prepare a first draft of the product and process control strategy by identifying processing and product-related risks that are readily apparent and require controls to be implemented in the manufacturing process. This will ensure that the finished product will meet its critical and other quality attributes.

Figures 6.2 and 6.3 show, respectively, a block diagram for tablet manufacture and an Ishikawa (fishbone) diagram breaking the process into the 6M's: Man, Machines, Measurement, Materials, Methods, and Management. These diagrams identify risk factors that need to be considered and possibly controlled in designing the manufacturing and control process.

A multi-disciplinary team, which could include personnel from production operation, maintenance, Human Resources (HR), R&D, Quality, Analytical and Microbiological, engineering, purchasing, and marketing, should participate in the brainstorming session. The more the disciplines represented, the more effective the session. The idea is to identify as many potential hazards as possible, working in a systematic manner through the process flow diagram. It is not appropriate for the Quality Unit to sit in their office and develop a risk assessment on their own! For example, including a representative from HR brings objectivity and a completely different skill set to the brainstorming session. Issues such as process flows and ergonomics from

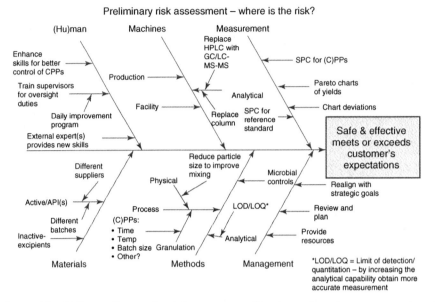

Figure 6.3 Ishikawa diagram for risk assessment. (*See insert for color representation of the figure.*)

the human engineering perspective will be considered and designed into the process.

The Ishikawa diagram in Figure 6.3 provides an example of potential hazards.

The diagram shows many items that are clearly potential risks. For example, weighing could fail because personnel are careless about discarding single-use scoops —or because they think they will save the company money and reuse the same scoop—potentially contaminating another material. Therefore, the human element should consider the possibility of cross contamination as a high risk and allow for developing controls that might include a high level of training, testing, and ongoing oversight and verification regarding personnel understanding of this issue. Weighing could also fail because of a mix-up of materials, so process flow and placement of items in the weighing area as well as labeling would all be important control points under "machines" (facility layout) and "methods" but would also need emphasis under "man" because once the controls are implemented it is critical to their ongoing functioning that personnel understand why they are there and how to implement them correctly.

Mixing can fail for numerous reasons, but if the particle size of the active ingredient is substantially different from batch to batch, many (C)PPs might be affected and the mixing will no longer be homogeneous. If purchasing understands this is a critical control point, they will ensure that the specification is incorporated into the quality contract with the API manufacturer, not merely as one more bullet point but as a nonnegotiable, go/no go parameter and they might even ensure a margin of safety by requesting that batches which are borderline are not supplied even if within the defined profile.

Taking another look at the diagram, it seems as if some of the points raised during the brainstorming are not really hazards, but process control parameters. However, this is the point of the session. Failure to identify individual process parameters at this stage (when their criticality is often only vaguely understood) as being potentially capable of harming a quality attribute will result in the lack of investigation of that potential during the subsequent development process. Should the parameter prove to be critical or necessary for process control, failure to investigate will also result in lack of a control strategy and, ultimately, rejected product batches.

A word of caution—risk assessment includes tools for expanding thinking followed by focusing on the perceived risks to identify control strategies. Brainstorming is intended to expand thinking and participants should be encouraged to put down apparently "zany" ideas. At the time of focusing, the group will decide together if there is a real risk behind the thought, but stifling any idea too early can result in an incomplete or less effective assessment.

6.3 PRODUCT CONTROL STRATEGY

In the QTPP shown in Figure 6.1, the target population is immunocompromised. The attribute of microbial limits, normally less important, becomes a critical quality attribute and the limit is considerably more stringent than for most tablets. This is a very important point and persons involved in development and manufacturing operations need to be made aware of this CQA in order to ensure the development of an appropriate microbial control strategy. Making personnel aware of specific requirements and (C)QAs and translating them into a control strategy is known as *risk communication* and is one of the most neglected areas of risk management. You may never assume that someone is aware of a specific restriction or that the facility and processes currently in operation are sufficiently forgiving to allow for successful manufacture of the new product. Most tablets have a microbial specification of not more than 1000 CFU/g such that a minimally controlled production environment, laboratory coats, hairnets, gloves, face masks, as well as routine cleaning of equipment will suffice to ensure that finished product batches meet specifications. For the product described in Figure 6.1, potential microbial risks are far greater and therefore the control strategy will need to consider upgrading the facility classification, providing additional gowning, disinfection of equipment and facility (sampling, weighing, and production), more stringent microbial control limits for starting materials, etc. In order to be sure that the controls which are designed into the process are then functioning correctly, monitoring will be required. The monitoring will also be designed into the process control strategy and might include, but not necessarily be limited to, frequent microbial monitoring of air, surfaces, and possibly personnel (which is not usual for nonsterile production); regular oversight (quality assurance) of disinfection procedures to ensure procedures are being carefully followed; oversight of personnel gowning to ensure procedures are carefully adhered to; gowning operations performed in the correct order; etc.

Once the initial Ishikawa analysis has been performed, the hazards are arranged according to the group's assessment of their potential impact based on prior knowledge and experience. As this is a subjective assessment, a democratic approach can be appropriate where majority rules. The facilitator should be careful to avoid letting a particular person override the majority because they have a powerful personality—this will introduce bias and detract from the usefulness of the exercise. For example, QC may have experience in analyzing a similar formulation and explain that they had ongoing problems with the specificity of the method because of one particular excipient. Without that piece of the puzzle, the "analytical method" might have been marked as low risk. Now it can remain low risk, if the excipient in question is eliminated. However, if the excipient is selected because other considerations take priority, the analytical team will immediately have a high risk item for failure of the method and will need to develop an appropriate mitigation strategy.

Table 6.3 shows how the risks identified in the Ishikawa diagram might have been prioritized by the group and the rationale for the same. Table 6.3 could serve as a preliminary risk assessment with proposed controls for the development of the control strategy. This should be formalized as a controlled document, assigned a version number and date, and signed off by key players involved in performing the assessment.

6.4 USE OF DESIGN OF EXPERIMENT (DOE) TO ELIMINATE AND STUDY (C)PPS AT LABORATORY SCALE

Where a large number of factors, which could potentially impact the success of the process, are identified, design of experiment allows a systematic method to reduce the number of tests needed to study these factors as well as to study potential interactions between parameters. When combined with prior knowledge, it provides a superb tool for exploring and learning how process parameters need to be controlled in order to achieve the (C)QAs. The outcome of small-scale experiments will always require close scrutiny and documented verification in a formal protocol when initial full-scale batches are produced. For elucidation of these concepts and the requirement to capture the information subsequently generated in a formal development report, refer to FDA's Process Validation Guidance [9].

For example, consider the synthesis of an API where, during the synthesis step, there are numerous parameters that might influence the process such as the following:

- concentration of reactants;
- pressure;

TABLE 6.3 Initial Risk Prioritization and Rationale

Risk # (Priority)	Identified Risk	Proposed Controls	Rationale
1.	Microbial contamination	Facility classification, equipment design and disinfection; personnel: gowning and training; materials specifications and handling procedures	Microbial contamination risk increased because target population immunocompromised
2.	Analytical methods	Select a more sensitive method to ensure accurate quantitiation and identification	Excipient known to interact and affect specificity of method
3.	Stability	Switch from securitainer to blister pack	Product sensitive to moisture
4.	API sensitive to moisture	Use (existing) glove box with humidity control and production facility with humidity microenvironment	Small quantities but moisture sensitive
5.	Cross-contamination	Existing controls will suffice	This product poses no special risk—toxicity profile known, low potency

- temperature;
- mixing speed; and
- reaction time.

If these were placed in a matrix with a potential high range, low range, and mid range (levels) for each of the parameters, the matrix would look like Table 6.4.

In a full factorial design, that is, where each of the six factors is tested at two levels (low and high) leaving out the mid range, there would be 2^6 experiments required = 128 experiments needed. Design of experiment software (commercially available and inexpensive) allows a partial factorial design to be used where, with prior knowledge (e.g., reactant 2 is present in excess, such that this cannot be a factor affecting the success of the process), fewer than half of the experiments are needed and information can be gained not only for the

TABLE 6.4 Matrix for DOE of API Process

Parameter	Low Range	Mid Range	High Range	Units
Reactant 1	300	350	400	Kg
Reactant 2	500	550	600	Kg
Pressure	N/A	1	1.2	Atmospheres
Mixing speed	60	80	120	Rpm
Temperature	Room temperature	60	120	°C
Reaction time	60	120	240	Minutes

proven acceptable range (PAR) for each individual variable but also regarding potential interactions between different factors. This information is then translated into batch manufacturing instructions that allow assignment of appropriate, valid ranges to each parameter and successful manufacture of each batch as long as each parameter is within its allowed range throughout the process. This will ensure that the (C)QAs are met at the end of the process.

NOTE: Referring to the QTPP, one of the CQAs would include stability data at the end of the product's shelf life, which is not included in a product specification but is of course essential.

A similar matrix can be designed after using an Ishikawa to brainstorm possible modes of failure of an analytical method under development (refer to Figure 6.4 and Table 6.5).

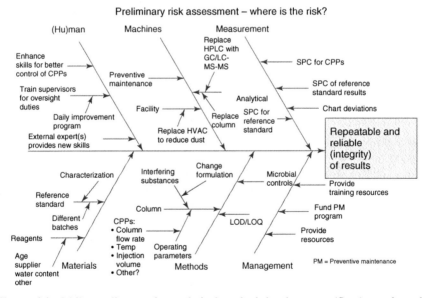

Figure 6.4 Ishikawa diagram for analytical method development. (*See insert for color representation of the figure.*)

TABLE 6.5 DOE Matrix for Analytical Method Development—Column Considerations

Parameter	Low Range	Mid Range	High Range	Units
Flow rate	A	B	C	
Injection volume	A	B	C	
Temperature	A	B	C	
Height	A	B	C	
Diameter	A	B	C	
Packaging volume	A	B	C	

Five variables at two levels would result in 64 experiments just for the column and probably not all the parameters. Some of the parameters can be eliminated on the basis of prior knowledge of similar methods, but in order to perform methodical development, which will allow trouble-free analytical testing, the use of a partial factorial design is needed to select the appropriate experiments.

6.5 PRECLINICAL STUDIES AND RISK MANAGEMENT FOR FIRST-IN-HUMAN TRIALS

Those active in the area of clinical trials and investigational medicinal products will remember the phase 1 trial of TGN1412 from March 2006 that took place at Northwick Park Hospital Research Center. Six patients, participating in a phase 1 trial to determine the safety of the product, took the drug, while two were given placebo. The drug initiated a life-threatening cytokine storm in all six patients, causing horrendous side effects because of an unanticipated biological effect. Fortunately, the research unit was adjacent to the main hospital and all participants received first-class treatment and apparently suffered no long-term harm. Nevertheless, regulators and industry alike had numerous questions regarding procedures for first-in-human trials and risk management. The outcome was a guidance document from the European Medicines Agency (EMA) on risk management [10] or, specifically, strategies to identify and mitigate risk for first-in-human trials. The guidance ties in with preclinical studies and prior knowledge, in particular, requiring the sponsor to demonstrate the relevance of the animal model.

While not directly related to the risks identified in the aforementioned trial, from the perspective of the drug product manufacturer, one critical risk control is to ensure the formulation and analytical methods used to determine the assay and impurity profile of the preclinical material are comparable to those used to manufacture products for use in humans. Without this assurance, the animal studies may not provide an accurate picture of the toxicological profile of the drug product.

6.6 PHASE 1 THROUGH PHASE 3 CLINICAL TRIALS

Risk management for early-stage trials tends to focus primarily on safety. At this stage, the toxicity profile of the product may not yet be fully understood, increasing the risks related to exposure of employees (research, manufacturing and analytical) and of course potentially for the patient. While cleaning equipment, usually the pilot plant, the inherent risks need to be considered, especially because analytical methods may not be sufficiently developed to provide a full impurity profile. Controls should focus on worst-case scenarios where cleaning is required to remove all traces down to a minimal acceptable level, often using total organic carbon (TOC) testing and assuming that any measurement is ascribed to the product, whereas in fact some will be because of detergent residues or excipients.

Other factors of concern in early-phase trials could include, but are not limited to, the following:

- level of supplier qualification for active and inactive substances;
- skills and familiarity with the process: may be the first time personnel have produced at this scale, in which case risk mitigation might involve a "dry run"—production of a "learning" batch that will not be used in humans;
- how to document production and how to provide written instructions;
- how much qualification of analytical methods; and
- how much stability data and determination of retest date/expiry or use by date.

Table 6.6 shows an example of a preliminary risk assessment performed in one company before phase 1 manufacture. At the time the company was only handling a single API in a novel delivery platform. Since the API was well characterized, the risks were significantly reduced as compared to a new chemical entity or unknown molecule. Nevertheless, risk assessment is lengthy as will be the case for any systematic assessment. This was a start-up company, so some of the risks may be different (not necessarily greater or lesser) than for a development department within an existing commercial company.

As can be seen in Table 6.6, a very detailed risk assessment has been conducted, but without adopting a formal risk assessment tool of the type mentioned in ICH Q9. There is no scoring system, not even low/medium/high. There is, however, an objective assessment of the risks involved, made by a multidisciplinary team using a process flow diagram to systematically review the manufacturing and control process for the batch to be produced for the phase 1 clinical trial. Mitigation measures are identified and can therefore be communicated to the personnel responsible for manufacturing and controlling the batch—both as formal, written instructions and "in frontal training" as well as "on-the job" training.

Table 6.7 shows a full FMEA table for just a few of the points in the original table, revised for phase 3 production. In this case, the company has introduced additional APIs so that there are new risks related to cleaning and

TABLE 6.6 Preliminary Risk Assessment for Phase 1 Clinical Trial

#	Failure Mode	Cause of Failure	Effect	Controls	Comments	Controls in place Yes/No
1.	Facility unsuitable	HVAC system and associated controls	Product contamination	• Work only with solid, oral dosage forms; no sterile products	Severity of failure of air handling system significantly reduced as no aseptic processing and less likelihood of contamination. Risk acceptable	YES
2.	Facility unsuitable	Other potent compounds in facility	Cross-contamination	• Company does not work with potent compounds at this time, including for R&D purposes • GMP manufacturing facility separated from R&D activities	Risk does not exist at present and change control procedure in place to capture any change of the status quo (e.g., R&D request to work with a potent compound or one of unknown toxicity)	YES
3.	Facility unsuitable	API to be used in trial is potent or unknown toxicity profile	Unsafe product or Operator safety hazards	• Company is working with one, characterized API ONLY at this time	Risk does not exist and change control procedure is in place to allow for assessment if a new trial with a different API is planned	YES

(*continued*)

TABLE 6.6 (*Continued*)

#	Failure Mode	Cause of Failure	Effect	Controls	Comments	Controls in place Yes/No
4.	Lack of adequate work areas	Mix-up/ contamination or cross-contamination	Product unsafe	• Facility designed to allow adequate space for operations, ease of cleaning and maintenance • Facility operated as a suite i.e., only the active (API) to be used in the clinical trial can be exposed in the facility from the time of initiation of the manufacture of the clinical trial batch. This includes sampling and weighing of starting materials • Rigorous, documented cleaning and line clearance before the start of the above-mentioned campaign	Risk mitigated and acceptable	YES
5.	Shared facility unsuitable	Air handling system lets R&D material carry into drug product for human use	Cross-contamination	• Separate facility for activities for human use • Restricted access (coded) • Sealed rooms in GMP facility • Fan Filter Units with HEPA filtration on incoming air in GMP facility	R&D facility can continue uninhibited GMP facility is protected provided controls are enforced and monitored Risk mitigated once facility is commissioned and SOPs are in place.	YES

			• Over pressures • Double door airlock • Gowning changes	Risk acceptable	
6.	Shared facility unsuitable	Animal house located in same building	Contamination of drug product	• Door separating animal facility from general corridor *in addition to* double door airlock and overpressures, HEPA filtration etc. (see earlier text) for GMP production facility as well as negative pressure in animal facility • Different staff—personnel working in GMP facility forbidden to do any animal work on same day they work in production • Doors self-closing • Access controls enforced at all times • Shared equipment e.g., dishwasher replaced with dedicated equipment to mitigate need for R&D personnel to access GMP area	Risk mitigated and acceptable. Risk needs to be monitored (pressures)	YES

(continued)

TABLE 6.6 (Continued)

#	Failure Mode	Cause of Failure	Effect	Controls	Comments	Controls in place Yes/No
7.	Shared equipment	Failure to clean properly or use of equipment for R&D and QC glassware/biological glassware together with GMP items	Cross-contamination/contamination	• Provision of separate equipment such that equipment (dishwasher) and all manufacturing items are only for GMP use – there is no longer any need for R&D or biology/QC to use them as they have their own equipment • At present only one API used in GMP facility so that cleaning is a minor concern • Line clearance procedure before manufacture for human use	Risk mitigated	YES
8.	Equipment dirty	Poor design	Cross-contamination/contamination	• Item of primary concern is the lyophilizer which is not inherently sanitary design. Risk mitigated by use of sealed, single-use Lyoguard trays for product drying. Together with a visually clean criteria after cleaning the inside of the chamber	Risk mitigated bearing in mind that no potent compounds are used at company at present	YES

9.	Equipment dirty	Failure of cleaning procedures	Contamination/cross-contamination	• Develop detailed cleaning procedures and TOC for verification at product changeover. At present there is no changeover because single API used	Risk acceptable—detailed documentation of line clearance and equipment cleaning before GMP production	YES
10.	Equipment malfunctions	Equipment defective	Faulty product	• Qualify critical equipment/items • Calibrate critical instrumentation • Procedures for malfunction handling including assessment of potential product impact	Risk acceptable	YES
11.	Materials unsuitable	Animal sourced materials	TSE/BSE	• Manufacturers identified (not just suppliers) so that supply chain is transparent • Certificates of suitability obtained	Risk acceptable	YES

(continued)

TABLE 6.6 (*Continued*)

#	Failure Mode	Cause of Failure	Effect	Controls	Comments	Controls in place Yes/No
12.	Materials unsuitable	Incoming material fails to meet specification or specification not sufficiently robust	Contaminated product	• Questionnaires received from each manufacturer along with detailed specifications • Reputable suppliers/sources used—located in countries with recognized regulatory systems in place • COA received with every batch of starting material	Residual risk remains—on-site audit would be preferable to questionnaire especially for API but not at this stage FDA phase 1 guidance [11] allows identity testing only; full COA testing to be performed on API batch used in the clinical trial in lieu of site audit. In addition, COAs for each of the materials (active and inactive) to be used will be reviewed and retained.	Risk mitigated and residual risk accepted
13.	Materials unsuitable	Lack of traceability	Unsafe product	• SOP for materials receipt including receiving goods logbook recording: receipt date, quantity in shipment, supplier's name, material lot number, storage conditions, and corresponding expiration date	Risk acceptable	YES

			• Batch manufacturing record with requirements to document: manufacturer, manufacturer's unique lot number and manufacturer's catalog number and expiry date for each material used in the batch		
14.	Lack of adequately controlled procedures	Failure of any of the above-mentioned controls	Unsafe product	• Develop a quality system with a list of SOPs that must be in place before start of GMP production • Develop detailed batch manufacturing instructions and quality control records to capture activities (e.g., manufacturing operations, in-process sampling) as they are performed and record any changes/deviations • Train, coach, supervise, and test personnel	Risk mitigated/ acceptable	YES

(continued)

TABLE 6.6 (Continued)

#	Failure Mode	Cause of Failure	Effect	Controls	Comments	Controls in place Yes/No
15.	Lack of adequately controlled procedures	Labeling mix-up	Mislabeled product	• Very small quantities handled per batch • GMP facility operates as a dedicated suite during the campaign • Documented reconciliation of labels (from issuance through use and destruction)	Risk mitigated/acceptable	YES
16.	Lack of adequately controlled procedures	Distribution of product inadequately controlled	Poor quality product	• Product to be used for initial trial is not temperature sensitive and packaging will be adequate to ensure quality through delivery to trial sites within a 25 km radius (one, local site only)	Risk mitigated/acceptable	YES
17.	Lack of adequately controlled procedures	Failure to retain batch sample/place batch on stability	Inability to conduct investigations	• SOP for sample retention • Stability protocol at least for duration of the trial and qualified test methods shown to be stability indicating	Risk acceptable	YES

TABLE 6.7 FMEA for Pre-Phase 3 Manufacture and Control

# Process Step	Failure Mode	Failure Causes	Failure Effects	Occurrence [1–5]	Detection [1–5]	Severity [1–5]	Risk Priority Number (RPN)	Risk Acceptable Yes/No	Actions to Reduce Occurrence of Failure
9. All	Equipment dirty	Failure of cleaning procedures	Contamination/cross-contamination	5	3	4	60	No	Disposable equipment for sampling and weighing, cleaning procedures; line clearance procedure upgraded. NOTE: company has now introduced additional APIs so risk has greatly increased
12. Incoming material	Materials unsuitable	Material OOS	Contaminated product	2	2	4	16	No	On-site audit Quality agreement with supplier Full COA testing with validated methods
16. Shipping to site	Lack of adequately controlled procedures	Distribution of product inadequately controlled	Poor quality product	1	5	3	15	No	Use calibrated data loggers to monitor shipping Qualify shipping process
21. QC	Lack of adequately controlled procedures	QC procedures unreliable	COA results lack integrity	3	3	4	36	No	Validate methods according to ICH

cross-contamination. The table has additional columns relating to failure causes and, of course, assignment of risk scores using a scale of 1–5 for likelihood of occurrence, and likelihood of detection and severity, respectively. A risk priority number (RPN) is calculated by multiplying the occurrence by detection by severity and then prioritizing the risks by the size of the number obtained. The item numbering is retained from Table 6.6 so that you can check back and see how the controls have been ramped up and those that were acceptable for a phase 1 study are no longer accepted for a phase III (one step before commercialization study).

To conclude this section, it is important to emphasize that there is no single risk management tool that is appropriate for assessing the risks involved in investigational products. The specific risks are product and process related and are closely tied in with the quality system already existing within a particular company. Incremental application of GMPs ties in with increased risks associated with larger populations using the drug in late-stage trials, although the safety of a few in early-phase trials cannot be overlooked. More importantly, a company's knowledge base increases as it moves through the phases of clinical study, such that some risks designated as "high" in phase 1 are substantially reduced by phase 3 (e.g., toxicity profile may now be well understood and found to be benign). The common thread throughout the development process is risk identification, development, and implementation of appropriate risk controls, risk communication (most commonly overlooked or rushed), risk monitoring, and then event review—deviations and unexpected events that might send you back for a revision of your initial assessment and implementation of new controls. Likewise, all these need to be documented with a brief rationale supporting the decision.

6.7 PRODUCT SPECIFICATION FILE, DEVELOPMENT HISTORY, AND TECHNOLOGY TRANSFER

The product specification file is defined in Annex 13 of the EU GMPs as "A reference file containing, or referring to files containing, all the information necessary to draft the detailed written instructions on processing, packaging, quality control testing, batch release and shipping of an investigational medicinal product."

The annex requires the following information at a minimum to be available or referenced in the product specification file:

- specifications and analytical methods for starting materials, packaging materials;
- intermediate, bulk, and finished products;
- manufacturing methods;
- in-process testing and methods;
- approved label copy;

- relevant clinical trial protocols and randomization codes, as appropriate;
- relevant technical agreements with contract givers, as appropriate;
- stability data; and
- storage and shipment conditions.

It is important to retain company memory of the development process. The development history should be captured in a development report or the equivalent of the medical devices quality system requirement (QSR)s Design History File (see Reference 8). This report should provide an audit trail of manufacturing instructions and analytical methods development including changes and showing comparability from preclinical through commercialization. This should be a life cycle document that is updated regularly, including after commercialization, as additional knowledge of product and process is gained. FDA's Process Validation Guidance describes the first stage of validation as process design. If the history file/development report is carefully constructed with reference to small-scale experiments, pilot batches, and scale-up through technology transfer, it may meet the requirements for the first stage of validation. To that end, it would be appropriate to write a protocol for data collection and analysis toward the later stages of product and process development. The following points could be added to the list of items to be identified in the product specification file:

- quality target product profile;
- list of (critical) process parameters;
- list of (critical) quality attributes;
- product control strategy;
- drug product development report;
- analytical method development report;
- review of comparability of batches from preclinical through human trials for comparability;
- risk assessments;
- risk communication documentation; and
- risk review (including corrective and preventive actions (CAPA) and feedback).

Each of the documents in the product specification file will be controlled and earlier versions will be archived to allow traceability of the development process and transparency regarding the changes implemented.

The file should provide a compilation of all the necessary documentation for successful technology transfer and initial commercial scale manufacture. This documentation must contain the science and knowledge as well as risk management to enable the repeated and reliable manufacture of batches while controlling (C)PPs in order to ensure that each of the (C)QAs is achieved. A vital part of the process is risk communication to concerned personnel to highlight the successes and "failures" of the development process so that they may learn from the collective company experience.

This is not to say that there will not be surprises during technology transfer—there will be and those will be "risk events" that need to be assessed, analyzed, and used as feedback to refine the risk assessments and issue new versions of the batch manufacturing instructions and/or analytical control procedures or shipping or labeling procedures. Continual improvement is the final and overriding element in any quality system, so that there is no ultimate state of "quality" but a constant feedback of knowledge gained to decrease uncertainty and reduce risk in feedforward.

6.8 SUMMARY

To summarize, the use of risk management tools in process development and manufacture of clinical trial material allows for increased process understanding. By identifying (critical) control points and providing clearly defined risk mitigation measures, the product control strategy becomes a systematic means of ensuring product and process consistency. By clearly documenting risks and the measures implemented to reduce them, risk communication is greatly enhanced. Personnel involved in the manufacture and quality control of the product can be provided with clear, written instructions accompanied by focused emphasis on identified risks. Those areas where there is increased uncertainty because the risks are not yet clearly understood can also be identified and strategies including increased sampling and testing may allow increasing the likelihood of detection of any failures.

Documentation of the risk assessments and their outcomes as well as periodic updating when new information comes to light ensure life cycle management and maintenance of the product control strategy and a safe product for participants in the trials.

REFERENCES

1. ICH Q6A: Specifications: Test Procedures and Acceptance Criteria for New Drug Substances and New Drug Products: Chemical Substances.
2. Chapters 8, 12, and 14 describe and provide examples of CPPs and CQAs.
3. ICH Q10, Pharmaceutical Quality System, June 2008.
4. Eudralex Volume 4, EU GMPs, Chapter 1, Quality Management.
5. US FD&C Act.
6. ICH Q8(R1), Pharmaceutical Development, November 2008.
7. Author's parentheses—this version appeared in the draft Q8 R1 guide, but was eliminated from the final version. Nevertheless, it is relevant because the QTPP must be updated as development progresses.

8. Product specification file is defined in Annex 13 of the EU GMPs: Manufacture of Investigational Medicinal Products. The concept of design controls and a design history file is described in the Quality System Regulation for Medical Devices, 21CFR part 820.
9. FDA Guidance for Industry Process Validation: General Principles and Practices, January 2011.
10. Guideline on strategies to identify and mitigate risks for first-in-human clinical trials with investigational medicinal products, EMEA, July 2007.
11. FDA Guidance for Industry CGMP for Phase 1 Investigational Drugs, July 2008.

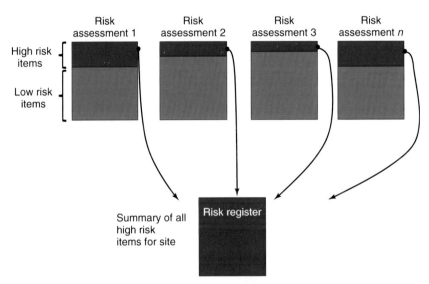

Figure 3.1 Risk register.

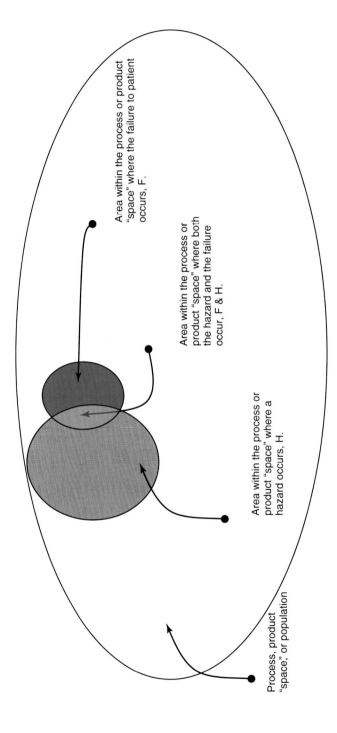

Figure 4.1 Venn diagram visualization of ISO 14971 section E.

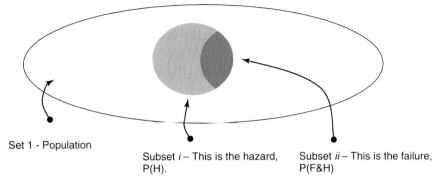

Figure 4.2 Conditional probability described as a subset.

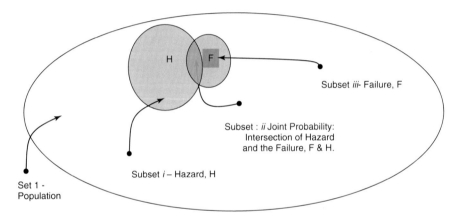

Figure 4.3 Joint probability.

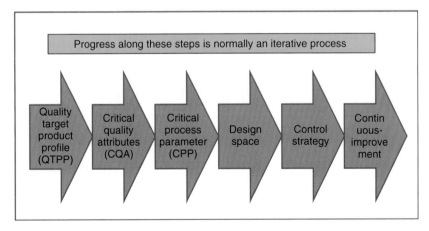

Figure 5.1 Diagrammatic flow of key steps for QbD.

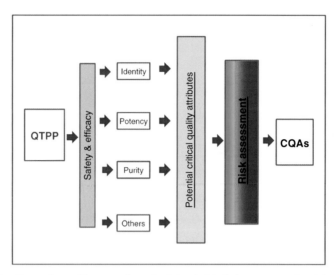

Figure 5.3 Example to illustrate the principles behind the output of a risk assessment and its potential impact on the patient.

	Safety	Efficacy	Quality
Potential CQA 1			
Potential CQA 2			
Potential CQA 3			
Potential CQA 4			
Potential CQA 5			

Low impact　Medium impact　High impact

Figure 5.4 To show an example of the output of an initial risk assessments, e.g. based on prior knowledge.

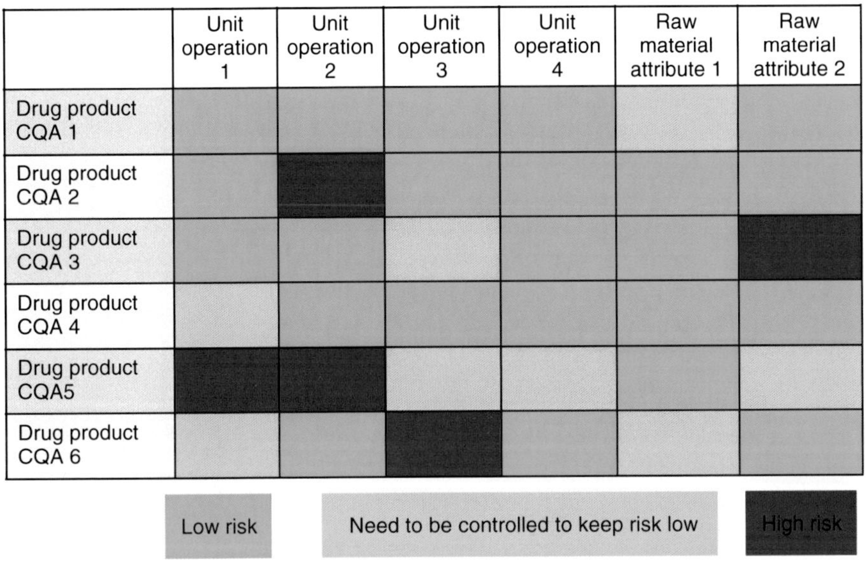

	Unit operation 1	Unit operation 2	Unit operation 3	Unit operation 4	Raw material attribute 1	Raw material attribute 2
Drug product CQA 1						
Drug product CQA 2						
Drug product CQA 3						
Drug product CQA 4						
Drug product CQA5						
Drug product CQA 6						

Low risk | Need to be controlled to keep risk low | High risk

Figure 5.5 Illustration to show the output of a risk assessment, linking manufacturing unit operations, raw material attributes and CQAs links.

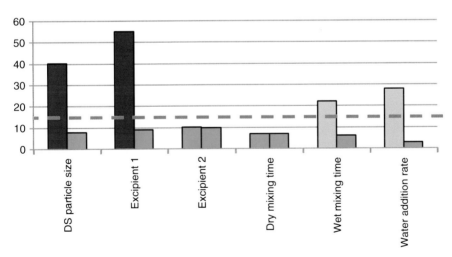

Figure 5.6 Diagrammatic part-outputs of initial and final a risk assessments to show the impact various unit operations may have on a CQA. (Dotted line shows level of acceptable risk).

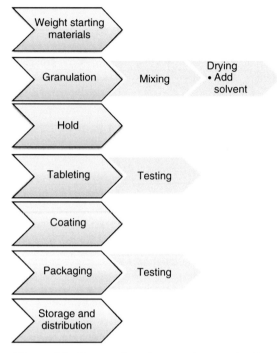

Figure 6.2 Process flow for manufacture of tablets.

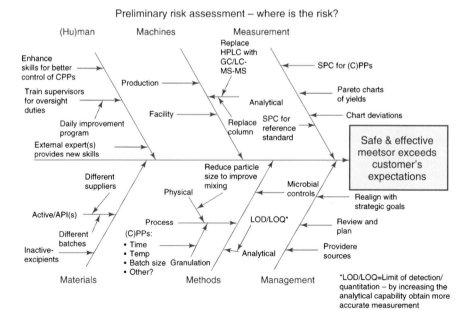

Figure 6.3 Ishikawa diagram for risk assessment.

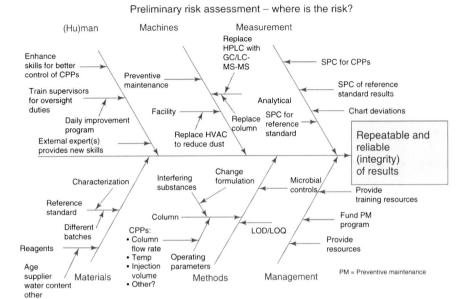

Figure 6.4 Ishikawa diagram for analytical method development.

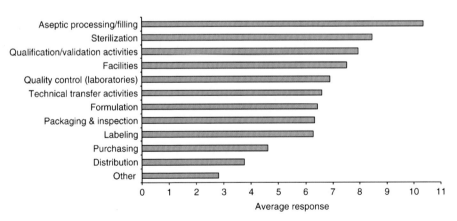

Figure 7.3 2006 Parenteral Drug Association survey on the functional area that has the most need for risk assessments.

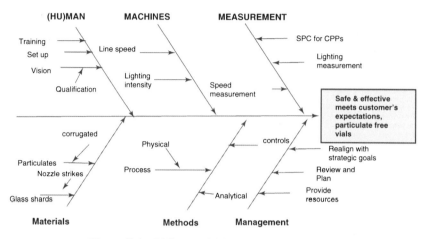

Figure 7.6 Ishikawa Fishbone analysis example.

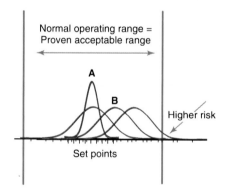

Figure 8.3 When PAR is unknown and considered equal to NOR.

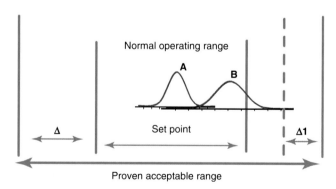

Figure 8.4 Determining level of risk-based on the spread between PAR and NOR, variability of the operating parameter, and the location of the set-point.

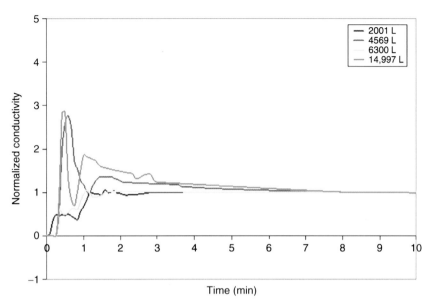

Figure 8.6 Mixing in a 15,000 L tank.

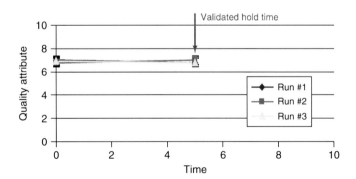

Figure 8.8 Traditional hold time validation (performing three experiments).

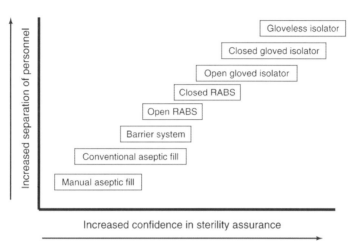

Figure 9.1 Aseptic processing continuum.

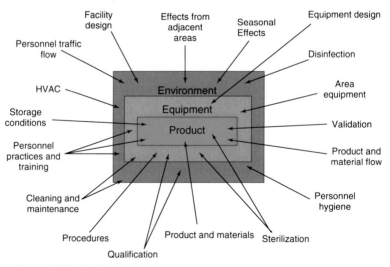

Adapted from Leonard Mestrandrea.

Figure 9.2 Factors influencing aseptic processing.

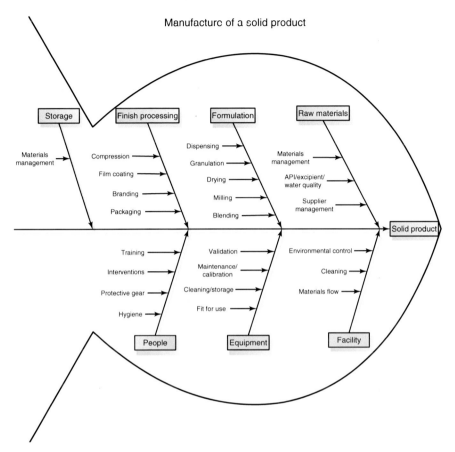

Figure 11.1 Example of a cause and effect diagram showing the manufacture of a solid product.

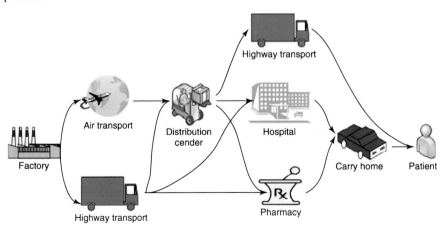

Figure 12.8 Typical distribution chain.

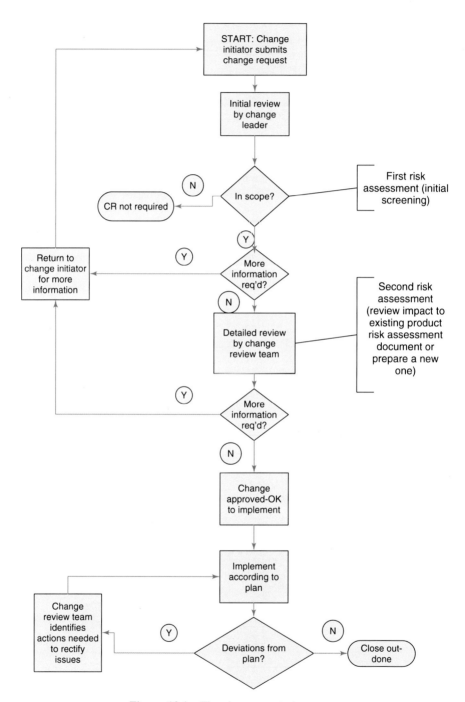

Figure 13.1 The change control process.

7

POINTS TO CONSIDER FOR COMMISSIONING AND QUALIFICATION OF MANUFACTURING FACILITIES AND EQUIPMENT

Harold S. Baseman and Michael Bogan

7.1 INTRODUCTION AND BACKGROUND

This chapter presents *points to consider* on the use of quality risk management and assessments to help the reader make decisions needed to plan, develop, and conduct more effective qualification efforts. The chapter is not meant to be an all-inclusive discussion on how to perform installation and operational qualification. Nor is the objective of this chapter to present an exclusive and complete guide to the use of risk assessment techniques for commissioning and qualification. Instead, the chapter presents some of the areas where risk assessment can effectively be used to help develop and implement a sound, efficient qualification program.

In order to assure that drug manufacturing processes perform as expected and result in products meeting quality specifications, equipment and systems supporting those processes must work in a reliable manner. Commissioning and qualification efforts should be designed to ensure that these systems are designed, installed, and operating properly.

The management of risk to product quality can be used to help companies effectively plan, prioritize, and perform facility and equipment qualification studies. Risk assessment techniques can provide the information needed to make decisions related to the following:

- which systems and components need to be qualified;
- the level to which this qualification needs to be performed;
- prioritization of the efforts needed to qualify these systems;
- determination of the appropriateness of information sources such as commissioning, FAT (factory acceptance tests), and vendor-provided test data and the extent to which this information needs to be independently confirmed;
- development of the acceptance criteria for qualification studies;
- addressing failures, discrepancies, and deviations uncovered during the execution of the qualification studies; and
- the extent, frequency, and criteria for repeating qualification studies.

Risk assessments should be used to help make sound and logical qualification-related decisions. *Risk assessments should not make the decision. They should provide the information needed to make the decision.* The key to an effective qualification program is to understand the relationship between the function of the facility/equipment and product quality. The risk to product quality and patient safety should be considered when making decisions related to the planning and performance of qualification studies.

7.1.1 Qualification, Risk, and Regulatory Expectations

Governmental regulatory agencies require companies marketing healthcare products to provide an assurance that the products are safe and effective. The US FDA (Food and Drug Administration) states in Part 21 CFR 211.100 of the Current Good Manufacturing Practice regulations that companies must have process controls designed to assure the products possess the defined critical attributes of strength, quality, identity, and purity [1]. Therefore, companies that manufacture and distribute regulated healthcare products have an obligation to meet these requirements and to provide safe products that meet the claims of effectiveness. Assurance of product quality and process reliability is a regulatory requirement as well as a sound business practice.

Reinforcing this requirement, in January 2011, the FDA published a revision of its 1987 Guidance for Industry on process validation (PV): General Principles and Practices [2]. The guidance presents a three-stage approach to the validation of processes used to manufacture pharmaceutical and biopharmaceutical products.

- Stage 1 – Process design: The commercial process is defined during this stage based on knowledge gained through development and scale-up activities.

INTRODUCTION AND BACKGROUND

- Stage 2 Process qualification (PQ): During this stage, the process design is confirmed as being capable of reproducible commercial manufacturing.
- Stage 3 – Continued process verification: Ongoing assurance is gained during routine production that the process remains in a state of control.

Under Stage 2 a. Design of a Facility and Qualification of Utilities and Equipment, the FDA states that "Proper design of a manufacturing facility is required under part 211, subpart C, of the CGMP regulations on Buildings and Facilities. It is essential that activities performed to assure proper facility design and commissioning precede PPQ (process performance qualification). Here, the term qualification refers to activities undertaken to demonstrate that utilities and equipment are suitable for their intended use and perform properly. These activities necessarily precede manufacturing products at the commercial scale [3]." The linking of qualification to the GMPs indicates that the FDA considers the function and the qualification of equipment and facility to have an impact on product quality. Therefore, decisions made to plan and perform qualification should be based on risk to product quality.

Figure 7.1 illustrates a three-stage approach to validation similar to the one discussed in the FDA PV guidance. Qualification steps, which are set between process design and commercial manufacture, demonstrate and provide confidence that the process or system design will result in reliable and consistent process performance and, therefore, product quality.

The FDA PV guidance does not specifically endorse a method for facility and equipment qualification. However, it is important to show that the equipment,

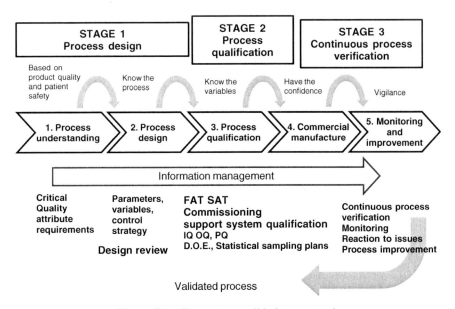

Figure 7.1 Three-stage validation approach.

systems, and components involved in the manufacturing process are suitable for use and function properly, reliably, consistently, and predictably. The method for doing so is to be determined by the manufacturing company. While the FDA PV guidance does not contain specific sections on risk management and qualification, there is an expectation that product quality and, therefore, qualification-related decisions are made using information based on risk to product quality. The FDA expects firms to develop programs for assuring that facilities and equipment involved with critical process steps be qualified and that the rationale for the qualification program should be based on risk to product quality.

While the FDA position on the use of formal risk assessments is not mandatory (at the time of this writing), regulators do expect companies to use a logical approach, taking into consideration risk to product quality, when making decisions related to regulated product manufacture and release. Therefore, the use of risk assessment and risk management in decision making is considered important, useful, and expected.

FDA citations reinforce the expectation of using risk assessment results as decision-making criteria in facility design. In a 2006 warning letter, the FDA noted: "Regardless of how often any product... is manufactured, because of the potential risk of cross-contamination, a risk assessment is necessary to determine whether you need separate and defined areas for manufacturing potent and nonpotent products." And further states "..., your firm provides no risk assessment to determine the hazard classification of your products [4]." It is interesting to note that the firm did not appear to get the warning letter because it failed to do a risk assessment; rather, the warning indicated a lack of clear decision-making criteria linked to patient safety. In other words, performing formal risk assessment may not be a requirement, but making sound decisions based on maintaining product quality is and risk assessments help to achieve that.

The drug industry has presented its views on risk-based qualification. To help industry clarify the relationship between proper design and the necessity of risk-based qualification, the ASTM published its *Standard Guide for Specification, Design, and Verification of Pharmaceutical and Biopharmaceutical Manufacturing Systems and Equipment* (E2500-07) in July 2007. The document presents an industry standard guidance, written to help people understand the relationship between risk to product quality, process design, and having equipment and facilities that work in a reliable manner. The ASTM E2500 Standard Guide states:

> "*Product and process information, as it relates to product quality and patient safety, should be used as the basis for making science and risk-based decisions that ensure that the manufacturing systems are designed and verified to be fit for their intended use. Further, the Guide reinforces the need for sound engineering and design as the key to effective facility and equipment qualification—good engineering practice (GEP) should underpin and support the specification, design, and verification activities by stating that quality by design concepts should be applied to ensure that critical aspects are designed into systems during the specification and design process* [5]."

INTRODUCTION AND BACKGROUND 133

7.1.2 Risk-Based Qualification and Assurance of Quality

Proving assurance is to make certain, provide confidence, and remove doubt or to be free from doubt. In the context of our industry, assurance can be achieved by observation or prediction. Observation or verification is the confirmation by examination and provision of objective evidence that specified requirements have been fulfilled. One is assured that something exists because one can see it, observe it, and test it. If one cannot observe something, then one must predict that it will happen or has happened. Confidence in the prediction of an outcome based on observation or evaluation of sampling is a key element of pharmaceutical PV. Validation and qualification are ways to provide assurance of product quality through assurance of process performance validation, or qualification is confirmation by examination and objective evidence that the particular requirements for a specific intended use can be consistently fulfilled. In other words, validation is a combination of what can be verified and our confidence that even with reduced verification, that condition will still exist. Put another way, validation or qualification is prediction of outcome we cannot fully observe based on conditions that we can observe.

As long as the outcome, in this case product quality, can be observed, assurance is achieved through observation or inspection. However, if the outcome cannot be observed completely or always, then relying solely on observation or inspection would not be effective. For that reason, there needs to be a way to assure the reliability of the performance of the process in order to assure the quality of the outcome. For the process to be reliable and predictable, the systems and equipment that support and perform the process must be reliable and predictable. The systems and equipment must be fit for use and must perform in a manner that will consistently result in the desired outcome. The desired outcome is a process that results in a product of a specified quality and purity, a product that is safe and effective.

Qualification confirms that systems operate in a manner that adequately supports the process. Qualification provides information and observable criteria, including the design, installation, and operation of a system that supports a process, which will be needed to predict the outcome of the process. A quality-risk-based approach to qualification focuses efforts on those aspects and functions that adversely affect product quality. There may be and there are sound reasons to commission and qualify systems whose functions do not affect product quality. This chapter is not meant to dissuade those efforts. However, this chapter focuses on the efforts required to qualify those functions and conditions affecting product quality.

7.1.3 Role of Quality and the Quality Unit in Risk-Based Qualification

The ASTM E2500-07 Standard Guide states that the "*acceptance criteria of critical aspects (that is, critical to product quality and patient safety) should be approved by the quality unit.*" This is consistent with requirements and expectations of many regulatory agencies including the U.S. FDA. The FDA presents requirements in Subparts C and D of 21 CFR Part 211, the cGMPs for assuring proper design and function—the link to product quality and the requirement stated in other sections, including 211.22, which assign the responsibility for control of quality to the quality unit. These recommendations and requirements reinforce the need for quality oversight of those efforts that qualify facility and equipment design and function, which have the potential to adversely affect product quality.

Following this logic, facilities and equipment that support critical process steps and, therefore, whose failure would have an adverse affect on product quality must be qualified and that qualification and acceptance criteria must be approved by the company's quality unit. Having said that, it would then be logical that the function of those facility and equipment systems that do not support critical process steps or the failure of which would not adversely affect product quality do not have to be qualified. It is still prudent to have confidence that these systems are installed and operating to specification and expectation. This can be accomplished through commissioning efforts. In this case, these commissioning efforts or their acceptance criteria would not have to be approved by the quality unit.

In performing risk assessments, one should assemble as diverse a team as practical. Team members should include relevant stakeholders who can contribute to the assessment. It may not be necessary to include all stakeholders, if they do not have such information or input. The quality unit, laboratory, engineering, manufacturing, process development, technical support, validation, and others may have information and input that are beneficial to the assessment effort and therefore should be considered for inclusion on the team. The risk assessment is an exercise to determine and mitigate risk elements. While it is not specially designed to be a communication effort, it can still effectively work as such. In that context, having quality involved in the early stages should allow for a better understanding of the rationale on which decisions are based. The use of a diverse team provides valuable input and also reduces the potential for biased conclusions.

The quality unit has the responsibility to assure product quality and therefore to approve the testing approach, acceptance criteria, and conclusions of the qualification study, including data received from sources directly represented in the qualification. As such, they will have a role in the transfer of qualified systems to commercial manufacture and should be included in the planning and final approval of qualification efforts.

7.1.4 Role of Commissioning in Risk-Based Qualification

The relationship between engineering, construction, commissioning, and qualification phases of a new system or equipment project are illustrated in Figure 7.2. Engineering represents the phase where design requirements are determined and systems and equipment are designed. Construction represents the phase where

INTRODUCTION AND BACKGROUND

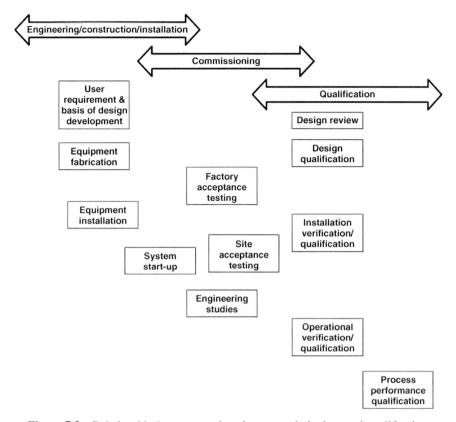

Figure 7.2 Relationship between engineering, commissioning, and qualification.

the systems and equipment are fabricated and installed. Commissioning represents the phase which establishes that the systems and equipment are designed and functioning as intended. Qualification represents the phase which confirms or proves that the systems are suitable for and capable of functioning correctly. There is overlap between phases and activities within each phase. This overlap provides the opportunity to share and leverage information. The degree of overlap and which activities fall into each phase will in large part depend on the company's experience and confidence in the people performing these activities and the systems in place for controlling these activities.

Healthcare product companies have an obligation to manufacture and distribute products that are safe and effective. There is a regulatory requirement that these companies assure that procedures and processes are in place to do so consistently [6]. To meet this requirement, manufacturing processes affecting product quality must be designed to control variables to the extent that the outcome is predictable and consistent. The processes must be validated to prove that these control measures are effective. A validated process relies on mechanical systems that are reliable and suitable. The systems must be well designed, fabricated, and installed. They must be qualified to assure that their function meets expectations. To be able to claim this, the systems should be commissioned to verify that they are working properly and fit for use. The certainty of function of a system can be obtained through design assessment (design qualification, design review), testing (data acquisition), and the accumulation and evaluation of design, installation, and operation-related information. The extent and effort of qualification should be commensurate with the level of risk to product quality.

Qualification provides documented evidence that items, systems, and processes are suitable for their intended purpose. These things must function in compliance with regulations. There should be an understanding of the correlation between the function and its impact on the quality, safety, efficacy, strength, or identity of the product. The confirmation through a formal documentation procedure proves that something is capable of its intended performance. Commissioning and FAT, which are similar to qualification, can ensure acceptable qualification results and if performed and documented properly may be used to supplement qualification efforts.

Systems supporting product-quality-related processes should be commissioned and must be qualified. Because companies possess limited resources, it is prudent to prioritize the qualification of those systems based on their effect on product quality. All systems affecting product quality should be qualified; however, qualification efforts and related acceptance criteria should be commensurate with the level of risk associated with their respective process step or equipment function.

Before a system can be qualified it must be in good working order. Systems that are not in good working order are unreliable and may pose a risk to product quality through malfunction. Commissioning confirms that facilities, equipment, and systems are functioning properly. The ISPE defines commissioning "a well-planned, documented, and managed engineering approach to the start-up and turnover of facilities, systems, and equipment to the end-user, that results in a safe and functional environment that meets established design requirements and stakeholder expectations [7]."

Commissioning may involve the start-up of equipment and getting the equipment to function correctly as expected. Once it has been started up and working properly, qualification approaches can be used to confirm or assure that the system functions the same way consistently, reliably, and predictably.

The relative effect on product quality of a system's function can be obtained through analysis of relative risk and impact of that function on product quality. This can be accomplished through a risk assessment or through an impact assessment. The consideration of relative risk and the use of risk assessment and risk management are essential to making decisions related to the commissioning and qualification of facilities and equipment used for the manufacture of pharmaceutical products.

INTRODUCTION AND BACKGROUND

7.1.5 Qualification and Risk Management

There are key questions to consider when developing a risk-based qualification approach.

1. Which items should be qualified?
2. What is the correlation between system function and product quality or patient safety?
3. What are the acceptance criteria or information needed to qualify those items?
4. What is the source of that information?
5. How will information or results be evaluated?

Quality risk management (QRM) can be a method for providing information needed to help answer those questions. QRM can also be used to prioritize and focus commensurate efforts on commissioning and qualifying those things, systems, and processes that, if they fail, could have an adverse affect on product quality, specifically to the point of adversely affecting patient safety. Patient safety is directly related to product quality. However, these terms are not synonymous. System failures will result in loss of product quality, but if that failure is detected or that product rejected before reaching the patient it will not cause harm. On the other hand, disruption of product supply as a result of system failure may adversely affect patient safety without affecting quality of product.

The pharmaceutical and biopharmaceutical industry has recognized the value of QRM in the planning and performance of qualification activities. Figure 7.3 shows the results from a 2006 PDA survey of 129 companies. The response to the question: "From a theoretical risk management perspective, what functional area has the most need for risk assessment" indicated that companies felt that

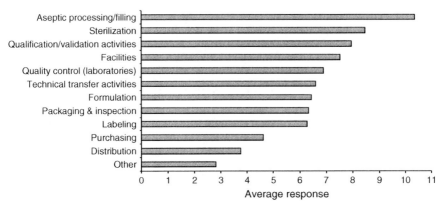

Figure 7.3 2006 Parenteral Drug Association survey on the functional area that has the most need for risk assessments. (*See insert for color representation of the figure.*)

qualification- and validation-related activities were among the most beneficial uses for QRM and risk assessment [8].

The evaluation of the risk to quality should be based on scientific knowledge and ultimately link to the protection of the patient. The level of effort, formality, and documentation of the QRM process should be commensurate with the level of risk. These principles should be applied to specification, design, and verification of manufacturing systems. The scope and extent of QRM for specification are as follows.

The ASTM Standard Guide states that "... risk management should underpin the specification, design, and verification process, and be applied appropriately at each stage." The evaluation of the risk to quality should be based on scientific design, and verification activities and documentation should be based on the risk to product quality and patient safety [9].

In May 2007, the ISPE published a draft revision of its Baseline® Pharmaceutical Engineering Guide for New and Renovated Facilities entitled Volume 5 – Installation and Verification – "A Revision to the Commissioning & Qualification Baseline Guide" Version 0.2.1. The draft revision noted that "In 2005, the international regulatory community published the final versions of ICH Q8, *Pharmaceutical Development*, and ICH Q9, *QRM*. In 2006, the U.S. FDA published *Quality Systems Approach to Pharmaceutical cGMP Regulations* and the international regulatory community is working on ICH Q10, *Quality Systems*. These documents emphasize a science-based process understanding and the use of risk management principles to focus the quality management system including the design and delivery of facilities [10]."

Science and process understanding are the starting point for the equipment, system, automation, and facility design. Science and process understanding are the bases by which deviations and changes are evaluated. Risk management should underpin the specification, design, and verification process, and be applied appropriately at each stage. The scope and extent of specification, design, and verification activities and documentation should be based on the risk to product quality and public health.

The primary principles of QRM as identified in ICH Q9 reinforce the concept that manufacturing- and qualification-related decisions should be based on the relative risk to product quality.

- The evaluation of the risk to product quality should be based on scientific knowledge and ultimately link to the protection of the patient.
- The level of effort, formality, and documentation of the QRM process should be commensurate with the level of risk [11].

QRM involves risk assessment. Risk assessments ask three sets of questions.

1. What can go wrong?
   What are the hazards or failures associated with the process and the equipment which supports that process?

2. How bad is it if it does go wrong?
   What is the impact of the failure on product quality?
3. What is the likelihood that it will go wrong?
   What is the probability that the failure will occur to an extent where it will adversely affect product quality?

The next sections of the chapter present some thoughts on the development of an approach and plan for the qualification of facilities and equipment using risk as a logical means for making qualification-related decisions.

7.2 RISK-BASED QUALIFICATION PLANNING

Commissioning, qualification, and risk-assessment-related activities should be documented in formal procedures, policies, and directives. Evaluations, justification, test methodology, acceptance criteria, and results should be documented in properly prepared and approved plans, protocols, and reports.

A key element in risk-based qualification is the prioritization of tasks and the allocation of resources commensurate with relative risk to product quality. Poor prioritization and allocation can result in project delays and the potential to overlook or miss activities essential to the qualification effort. A sound project plan is an effective tool for a successful risk-based qualification project.

The starting point for risk-based qualification planning should be the development of a plan that defines the *who, what, when, where, and how* of the project. A typical project plan will address areas such as project description, roles and responsibilities, project controls, schedule development strategies, information leveraging and management, and overall integrated approach.

7.2.1 Project Description

The project description provides information on which the reader or reviewer can rely on to ascertain if the approach will result in the qualification of the systems and equipment. It should present an explanation of the project and may include location, scope, facility or process function, and notable directives or policies. It should also establish that decisions related to acceptance criteria will be based on risk to product quality and patient safety.

Understanding the interdependence of processes related to risk to product quality is important. For example, a cooling water system used to cool a bulk product may not be considered to have an effect on the product if there is an effective monitoring system that indicates water and product temperature. However, for that to be the case that monitoring system must be qualified. A less reliable monitoring system might redirect efforts to qualifying the cooling water system.

To begin the planning process, one should assemble the key players that will be involved in the project and outline the project description and business objectives along with the critical success factors. Some items such as a defined organizational structure, list of assumptions and dependencies, and business risks would be critical aspects of the plan. Work streams should be outlined on the basis of the project's principal activities. These should align with the existing business units as it is helpful to organize specific work streams with multidisciplinary resources, focused on certain project activities.

Development of milestones should be identified. This is an important project area that is sometimes overlooked. The title of the milestone should be defined, as well as the extent of the deliverable. Key items such as mechanical completion, care, custody, and control should be clearly defined, especially where there will be a transfer of task ownership during commissioning and qualification. For example, consider the impact to the schedule and budget if resources were planned and mobilized to start commissioning and qualification activities only to find that the definition of mechanical completion does not support the systems being in an operational state.

Many aspects of the project contain interrelated activities. These activities may need to be managed using an automated business tool or interactive data base, so other teams can more easily sort through, utilize, and leverage the information as needed. For example, changes in a particular system during construction and commissioning may require additional assessment related to the design, and this may change the approach to qualification of that system.

The plan should define project roles and responsibilities, identifying those responsible for project controls, scheduling, overall project strategy, document or study review and approval, change control, document preparation, study execution, punch list resolution, failure investigation, and other critical project planning and execution activities. In order to clearly understand and outline roles and responsibilities, process maps can be utilized. These process maps are helpful, because they can easily identify main process steps as well as any changes in ownership with the process steps outlined. As the project plan progresses, the process map matures and helps outline even more detailed deliverables, durations of activities, department specific activities such as review and approvals, critical inputs into building an effective schedule. A signature matrix should present documents that require review and approvals. Each department, along with the individuals responsible for those activities, is listed. Each department should develop service-level agreements with all suppliers, vendors, and contractors.

Critical commitments such as review and approval of documentation should be captured in the service-level agreement. By doing this, expectations are more clearly defined rather than suggested or assumed. This document is particularly useful, because people, commitments, and expectations over the term of the project change as resources may be reassigned or added to the project.

7.2.2 Project Controls

A sound project control system can help ensure the qualification project is executed efficiently. A poorly controlled qualification project may be prone to errors, which can affect the schedule and budget, but may also distract the efforts of execution personnel and could therefore lead to errors by omission and possible project quality issues.

Project controls should include the means to assure test functions and studies are executed and completed in the proper manner and sequence and deviations are addressed before release of the system for PV or commercial use. Systems that represent higher risk to product quality, through impact of failure, probability or occurrence, or control strategy should have more robust control measures that assure satisfactory completion.

It would be difficult to manage any project without specific project controls. These project controls not only address budget concerns but also the ability to easily and effectively communicate the progress and status of activities that impact the project. The development and use of dashboards or project summary by the project team also requires the development of performance metrics (e.g., earned value curves, etc.). The dashboard is a one-page presentation that contains all of the information needed to senior management to understand the current progress of the various project work streams.

7.2.3 Project Performance Metrics

A good qualification program will include methods for monitoring the effort to assure that the plan is being properly executed. The monitoring program should be designed to uncover problems early, when it is less costly to make modifications to program design, schedule, and execution. This involves the development of project performance metrics for the determination of performance based on project quality, schedule, budget, safety, and effectiveness or compliance. Project control systems should be in place to track, analyze, and report all relevant project metrics. Clear project expectations may be one way to mitigate the risk of inaccurate information and errors.

Performance metrics should be developed to easily help identify areas of concern by listing certain high level project activities. For example, many projects have a multitude of documents that require drafting, review, approval, and execution. The performance metrics should outline progress based on planned activities completed versus actual activities completed within a specific timeline. It is helpful to add an outline of what is expected by the next update. Any metrics that identify areas of concern may require corrective actions summarized at the end of the dashboard.

7.2.4 Schedules

Schedules are an important tool employed in the management of any project. It is important that agreed on durations and specific deliverables outlined from each perspective work stream be included in the schedule. Selecting the appropriate level of schedule detail is essential to the development of a useful schedule. Information should not have too high or too low of a level of detail, such that each work stream cannot effectively manage the respective deliverable. Each work stream should be scheduled to sufficient level of detail to manage the day-to-day tasks and have available data to generate performance metrics. These work stream schedules should then be fed into a higher level project schedule. The higher level schedule would then be the driving force during multidepartmental work stream meetings typically held on a daily basis in a "war room" setting.

Other strategies should be defined within the project plan so that all associated work streams understand how any preceding work may impact their ability to meet their commitments to the project. Defining the scope of your risk approach as listed in the project charter or plan to a level where the assumptions, dependencies, and constraints is a key to project success.

It is important to define project deliverables and understand the associated process step dependencies. For example, clean room sanitization qualification may include sampling of surfaces before cleaning, while clean room environment qualification may include monitoring of the room after cleaning. Performing the environmental study before the sanitization study could result in an ineffective qualification and possible project delay. Another example might be the sampling of clean steam condensate from the distribution loop before qualifying the clean steam generator. In this case, the source of a failure would be difficult to discern, as it could be from failure of the generator or failure of the distribution loop.

Other examples of schedule dependencies include the sealing of walls, blocking views of mechanical installations, or the installation of insulation on piping before sloop verification. In these cases, improper scheduling could result in failure to adequately inspect the system and might result in risk to product quality.

Projects that fall behind schedule or exceed the budget may be subject to resource or sequence compromises. These compromises could pose a risk to product quality if the project is not executed properly. More care and control may need to be placed on projects that have fallen behind schedule or budget to assure that "corners are not cut" to the point where product quality is at risk.

7.2.5 Program-Level Alignment

Once the definition of the scope of work is complete, a list of all procedures and policies to manage the project activities from one phase to another should be developed. Each phase may be owned by a few different functional groups, which would provide a detailed view of related activities. Usually, these functional groups are related to the different phases of a project and therefore assigned accordingly. Provided here is a list of typical functional areas, which is followed by the activities, policies, and processes that will be the mechanism that provides the information required to satisfy your qualification program.

It is important to know the "big picture." Understanding the strategy being employed by a preceding activity is critical to developing an overall project execution program. Adjustments may be made to work streams to mitigate any possible risks. Table 7.1 presents a list of interrelated activities, all of which are programs and deliverables managed by various functional groups that can have an

RISK-BASED QUALIFICATION PLANNING

impact on project execution. They should take into account the interrelationship of facility design, process requirements, product quality, regulatory expectations, and system capability. Much can be learned from ineffective practices of the past. This is helpful if system and component impact assessments are to be used to leverage construction or engineering documentation to gain project execution efficiency.

Integrated strategies for construction turnover, commissioning, qualification, and equipment release should be developed and linked to product quality risk. Construction of critical utilities that can affect product quality, such as HVAC, purified water, clean steam, and transfer line piping, can be commissioned and qualified on an individual equipment or system basis. Critical process systems, such as bioreactors, buffer tanks, media tanks, and purification equipment, can be commissioned and qualified on an individual or process system basis. Commissioning and qualification of the process systems require prior completion of the commissioning and qualification of the utility and support equipment systems. This will mitigate the risk of changes to the utility systems, which can affect the performance of the process systems. Consideration should also be given to other nonproduct impact ancillary activities such as floor sealing, wall painting, and sealing of walls and ceilings. These activities may affect the ability to perform commissioning checks, because of limits to plant and equipment accessibility. These activities may also affect utility or process system performance. For example, solvents from floor sealing or painting, dirt from construction, and damage from installation adjacent to critical systems and lines may adversely affect performance and product quality or at the least require additional cleaning and repair steps.

7.2.6 System Boundaries

The commissioning and qualification plan should define boundaries between functional areas for the listed programs. Boundaries are an important element in protocol development as well as ongoing operations. The manner in which these boundaries are defined and managed is critical to an effective project plan. They will help organize tasks to ensure that critical systems are properly qualified and not missed.

It is important that facility construction is managed in a disciplined manner. Commissioning and qualification should be planned and executed on a system level. Any system boundary may and will most likely contain components of other systems with their boundaries. Systems boundaries are included in a risk management strategy for traceability of routine operations. Once the initial project is completed, validation maintenance or the procedures necessary to assure the system remains in the state in which it was qualified are implemented. These procedures include preventive and reparative maintenance, change control, software security, audits, and risk assessment reviews. It is important to keep in mind that throughout the project, people may move in and out of positions responsible for maintenance of the program. If not properly managed, maintaining the

TABLE 7.1 Typical Construction/Qualification Project Phases and Activities

Engineering	Construction	Commissioning	Qualification
• System and component impact assessment	• System/construction impact assessment	• Equipment and automation FAT/SAT	• Installation verification or qualification
• Vendor quality audit plan development	• Quality vendor audit	• Turn over package review	• Operational verification or qualification
• Design qualification plan development	• Design qualification	• Instrument check	• Process performance qualification
• Construction quality plan development	• Construction quality plan execution	• Functional check	• cGMP change control
• Engineering change plan development	• Engineering change management	• System shake down	• Resolution of deviations
• LOTO (lock out tag out) plan development	• LOTO procedure execution	• System performance testing	
• Equipment inspection plan development	• Equipment and automation FAT	• Punch list mitigation	
• Mechanical completion plan development	• Mechanical plan execution	• Calibration and maintenance plan execution	
• Systems boundaries defined	• Commissioning plan development	• SOP development	
• Automation requirement and design	• Qualification plan development	• Training plan development and execution	
• Schedule development	• Calibration and maintenance plan development	• Resource plan development	
	• Construction inspections	• Engineering change control	
	• Automation development testing		

traceability of changes for future assessments could become a challenge. A well-planned, organized project will reduce the potential for systems to be overlooked or inadequately commissioned and qualified.

Once boundaries are defined and maintained on the P&IDs (piping and instrumentation diagrams), they should be managed as living documents and controlled as part of the construction and/or engineering change control program. As many changes will and do occur during construction, the change control programs must accurately indicate the system and boundary in which the change occurs. Changes during design and construction that are missed and inadequately addressed may affect system performance and result in repeated testing and project delay.

It is not an efficient use of time and resources to continually modify CAD (computer-aided design) red line drawings. A controlled program should be employed to manage drawing red lines "on a stick." Engineers should be able to identify if the red line drawings reflect the most current approved changes or if any approved changes are awaiting CAD modifications. Changes or modifications to drawings that are not captured in change control may result in construction issues or in commissioning and qualification problems. These may also result in designs that are not in compliance with the approved basis of design, user requirements, or regulatory expectations.

Project definitions such as mechanical completion or any other milestones that indicate the completion of one operation and transfer of responsibilities and deliverables to another functional area should be clearly defined. The definition of mechanical completion usually indicates the finishing of a level or element of construction and identifies that a system is ready for commissioning. However, this does not necessarily mean that the system is functionally operable to perform the tasks linked to this activity. In some instances, it may not even indicate that a system is completely installed. Depending on the interpretations, mechanical completion may mean different things during the course of a project. The definitions of these types of activities should be clear and not change to new definitions simply to claim completion of a milestone. Issues related to the definition of project completions may include the following:

- failure to identify and control changes to critical systems;
- warranty and system ownership disputes;
- incomplete or inaccurate transfer of information needed for commissioning and qualification; and
- delays in start-up, commissioning, and qualification.

Likewise, activities such as construction change control and engineering change control have a direct impact on your risk-based strategy, particularly if you are planning on leveraging documentation. It is important to understand which changes occurred to a particular system within a defined system boundary, so that material requests and approvals of changes have been documented and are traceable. In some cases, it may be beneficial to employ quality approval of those changes in the event the documentation will be leveraged into your qualification.

7.2.7 Leveraging Risk-Based Qualification Information

Information and data used to determine if a system or item is qualified may come from the qualification study itself or from other sources. Regardless of the source, it is essential that this information be accurate and complete. Inaccurate or incomplete information can lead to incorrect decisions related to the qualification of the system. This can result in false failures and unnecessary efforts, which presents a risk to the efficient execution of the qualification project. It can also lead to acceptance of systems that have not been properly qualified, which would present a risk to product quality and perhaps patient safety.

Information may be obtained from sources other than the qualification study, including design engineers, equipment vendors, and construction or mechanical contractors. It is often beneficial to use or leverage this information to support the qualification study or reduce efforts.

For best results, the equipment inspection plan and a vendor audit plan should be aligned with the project plan. Information and data obtained from engineering and construction personnel must be accurate and should support the qualification effort. In other words, there should be a link to product quality. It should be reliable and follow an appropriate level of good documentation practices (GDPs). This would be a requirement for quality unit acceptance as information to be leveraged for qualification. The decision to leverage information and the criteria for making these decisions should be based on risk to product quality.

Information about the installation and operation of a critical system or piece of equipment can come from several different sources, including design drawings, submittals, fabrication and construction records, FAT, site acceptance testing (SAT), commissioning and start-up records, the historical performance data from similar systems, and formal qualification efforts. These efforts contain the information that can be used or leveraged for the qualification program. The source of this information can be from a multitude of companies and disciplines such as A&E firms; design engineers; vendors; contractors; construction workers; and mechanical, maintenance, operations, validation, and qualification contractors.

The challenge of relying on information from sources other than those specifically responsible for qualification of the facility and the equipment is that the objectives of those groups may not be aligned with the objectives of the qualification effort. The key to successful leveraging is the alignment of objectives of all parties involved. The objective of vendors and construction personnel may be different from that of the qualification effort. For those responsible for building the facility, the objective would be completion of the construction project. Those responsible for qualification of the facility will have the objective of providing assurance that the equipment and system function in a manner that adequately supports the critical process steps. This is not to say that the vendor or contractor is not trustworthy or would deliberately provide inaccurate information. To the contrary, vendor and contractor personnel are often very aware of and sensitive to the need for accurate information. However, they may not fully understand the information needed, when it is needed, in what form the information is needed, or even why. In addition, their time and resource commitment may be limited to completion of their construction objective. Therefore, it is an objective of a risk-based qualification program to ensure that those providing the leveraged information understand and appreciate the need for accuracy and reliability.

Having said this, there remain significant benefits for using vendors and construction contractors to provide qualification-related information.

1. Vendors and construction personnel may have the best understanding of the fabrication, construction, and installation process and, as such, are subject matter experts (SMEs) in those efforts.
2. Using this information will reduce the need for additional qualification efforts to obtain the information, thus eliminating redundant efforts and the risk of transcription errors.
3. Vendor and construction staff represent a significant source of personnel early in the project and can be a highly leveraged and valuable resource.

The leveraging of information from sources other than the qualification team should be a risk-based exercise, considering the risk of inaccurate information to the success of the project. The decision and justification to use vendor documentation to support the verification of critical aspects of the manufacturing element, including the intended use of the manufacturing system, should be documented and approved by SMEs including the quality unit. Therefore, deciding on the source of qualification information is a matter of risk management in and of itself. The factors to be considered to mitigate information inaccuracy include the following:

- experience of the vendor/source of information;
- experience with the vendor/source of information;
- criticality or function of system being built or installed;
- availability of redundant information sources or checks and balances; and
- robustness of the vendor quality system.

When planning the leveraged use of information it is prudent to consider the risk that inaccurate information would have on product quality, as well as qualification project success. Steps may be taken to mitigate such risks. For systems with high criticality or potential effects on product quality, it may be prudent to employ additional qualification testing or information gathering regardless of vendor capability. Companies may also consider audit of information obtained from the vendor or contractor. If errors are found, then additional testing may be necessary.

Risk assessments can be used to provide information to help determine the extent of additional confirmation required for acceptance of vendor-provided information. Useful information can be developed during the design, fabrication, installation, commissioning, and vendor testing phases of a project. There may be less need to perform additional efforts to obtain similar information during qualification studies and tests if this information is accurate and in a form suitable for the support of the qualification program.

The key to effective leveraging of information is establishing a partnership between vendors, contractors, and the qualification team. The way to assure a good partnership is by aligning the objectives of all parties. The objectives of parties performing project phases may differ. The role of the vendors building the equipment is to provide equipment as per specification. The objective of the vendor testing the equipment is to prove to the customer that the business transaction has been completed by proving that the equipment has been built to specification, thus motivating the customer to complete his end of the transaction and pay the vendor. In providing this proof, the vendor may also be proving that the equipment works as specified and therefore is ready for qualification. Understanding these differences in objectives and taking steps to better align objectives will help ensure the efficient leveraging of vendor-provided information.

People commissioning the equipment are responsible for getting the system in good working order and getting the system to work. Their objective is to confirm that the system is working to the expectation of the user. In doing so, the commissioning team will obtain information that confirms the system is working properly. This is an important element in determining that the system will work reliably and predictably to the level of performance needed to support the process. Those qualifying the equipment are responsible for proving or providing assurance that the system will work in the manner needed to support the process to the level needed to produce safe and effective products. The objectives are proof and confidence.

To accomplish this, the qualification team may choose to rely on vendor-provided information or commissioning data or may choose to repeat, audit, or confirm the accuracy of this information entirely or in part. The decision may be based on the following risk-based criteria.

1. Source reliability
   - Is the source of information reliable?
   - Is it free of error, credible, accurate, and useful?
   - What has been your experience with the source of information on this project and other projects?
   - Do you have any experience with this source of information? Are you certain of the reliability of their performance?
2. Source capability
   - Is the source knowledgeable in qualification-related efforts?
   - Does the source of information have quality systems and training in place to assure accuracy?
   - Do they have experience with providing this type of information in a qualification setting?
3. Impact of the information

RISK-BASED QUALIFICATION PLANNING 149

- What is the impact of inaccurate information on product quality and patient safety?
- Are there additional safeguards in place to catch mistakes or are the systems to which the information is used not of direct or significant quality impact?

The less reliable the source, the more the qualification person should mitigate the risk or impact of inaccurate information. Mitigation might take the form of the following:

- not using the information;
- confirming the information through audit of systems and test results; or
- providing assistance to help the source provide more reliable information through detailed specification of information needed or training.

Sometimes the accuracy of information is based on the level of understanding between the source and the qualification person. The meaning of nomenclature and terminology may differ from company to company, industry to industry, or discipline to discipline.

To illustrate the point, consider the example of a simple steam line purge. The qualification protocol called for a purge of the steam lines. The mechanical contractor agreed to perform this step. After it was completed, the qualification group tested the cleanliness of the line by inspecting the effluent for clarity and lack of visible particles. They found an unacceptable level of particles, specks, and debris. The test failed and the project was delayed. On investigation, it was discovered that the person writing the protocol defined purge as running steam through the line in a sufficient quantity to blow out any debris, thus cleaning the line of residue. The contractor did in fact perform the purge, but to him a purge of a steam line meant replacing the air in the line with steam. To accomplish what the protocol writer had in mind was a simple task, but it was not a purge, it was a steam cleaning or flushing operation. The result was a test failure. This mistake presented a risk to the execution of the project. The risk of this type of failure could have been mitigated by a common understanding of terminology; perhaps through training, better specifications and test protocol descriptions, or using a contractor more experienced in GMP facility testing.

Another example illustrates a potential risk to product quality as a result of accepting information which would not be reconfirmed later and occurred shortly after on a different project. In this example, the commissioning effort included verification of process line slope. The slope was needed to provide proper drainage and allow for cleaning and reduce the possibility of pooled material and contamination. As the mechanical contractor needed to verify line slope at the completion of their work and possessed the expertise to do so, it was decided to use that verification in the qualification protocol and not require an additional qualification team measurement. The line sloops were verified by the mechanical contractors properly as soon as the system installation was completed.

The sloop verification was performed well and accurately documented as part of the system close out. Several weeks later, the HVAC contractors moved into the same space as the process lines. In order to move around in the confined space they were constantly bumping up against, leaning on, and in a couple of instances standing on the process lines. This activity could have and in this case did change the sloop of some of the lines. The result could have been process failure during cleaning or contamination control. Mitigating this risk of failure could have been accomplished by more schedule coordination, training, HVAC contractor selection, notices, or subsequent and final sloop verification as part of qualification at the end of the project.

Utilization of information is a risk management exercise in and of itself. A key element of a successful qualification program is being able to efficiently leverage information from different sources. However, the information must be accurate. The key to accurate information is making sure the prerequisite support programs are in place. These include vendor quality systems and training, construction, and engineering change. Definitions must be clear and consistent. Procedures for maintaining qualified systems must be in place. Can the quality unit allow the acceptance of vendor-generated information? Do they have enough confidence in the accuracy of the information? Is the information accurate and is it useful?

The ASTM E2500-07 Standard Guide suggests that "... if inadequacies are found in the vendor quality system, technical capability, or application of GEP, then the regulated company may choose to mitigate potential risks by applying specific, targeted, additional verification checks or other controls rather than repeating vendor activities and replicating vendor documentation [12]."

If inaccurate information is found, then how does that affect all of the other leveraged information? This, again, may depend on the level and cause of the inaccuracy. To mitigate the risk of inaccurate information, the following steps have to be taken.

- Audit vendors.
- Monitor vendor performance.
- Provide vendor personnel training.
- Communicate expectations to vendor personnel.
- Choose vendors with relevant experience.
- Provide quality unit review of critical studies.

Having said this, companies using vendor-provided information should be careful not to overreact to inaccuracies or errors, as well as overrely on accuracy of the information. Information from vendors, including FAT and SAT, should be useful, as well as accurate. The information needs to fit the objective of the qualification. To assure that relevant qualification information can be gleaned from FAT and SAT, prepare and communicate qualification test functions, studies, and acceptance criteria to the vendor as soon in the project as practical. The requirement for this information as well as expected documentation quality should be included as part of the purchase agreement to avoid misunderstandings later in the project.

7.2.8 Other Types of Information Leveraging

Aside from leveraging of information from vendor and contractor sources, companies may find it useful to leverage information from similar equipment testing and qualification, previous qualification testing, and historical performance of this equipment or of similar equipment.

Information from these sources may be beneficial in supporting the qualification effort, in deciding if a system is qualified, or in determining if particular functions or conditions pose a risk to product quality.

When deciding whether or to what extent to use this information, consider the following:

- the credibility of the source;
- its accuracy of information;
- the controls in place when it was generated;
- any changes that may have occurred since it was generated; and
- the relevance it has to the equipment, system, or function being qualified.

7.2.9 Information Management as a Tool for Risk Management

A challenge to the use of accurate information comes from mistakes or omissions in recording and transferring information between phases of the commissioning and qualification of the project, just as inaccurate information and transmission mistakes without further controls can result in a risk to project execution efficiency as well as a risk to product quality. An interactive project information management database is helpful for the efficient capture and transfer of information from design through engineering, construction, commissioning, and qualification, and ultimately transfers that information to operations for inclusion in maintenance and operational procedures. The project information management system should be designed to assure accurate transfer of information, including changes to the appropriate project teams on a timely basis. Seamless transfer of information should result in a more efficient use of resources, but will also result in less data-transcription-related errors.

Automated project information management systems can reduce the risk of inaccurate or incomplete information because of errors in transmission and transcription, as well as, and therefore, reduce deviations. Project information management systems can be used to capture and transfer changes, thereby reducing risk of errors, omissions, and deviations.

Project information management systems are typically automated, often web-based applications that can facilitate architectural/engineering/construction project management by providing a single repository of project-related documentation and data. Web-based systems can provide access to project data from remote locations and can alleviate the issues stemming from global project team collaboration. The systems should include security administration to ensure that users have appropriate access to data. Project information management systems can include online repositories for engineering data that allow users to maintain engineered system information and provide access to the most current and relevant project data. They can also include a project library, allowing users to catalog and classify project change and project documentation as part of the project life cycle. Systems may allow users to define workflows for project library items and include e-mail notifications to ensure all project team members are kept updated on outstanding and upcoming deliverables; and may include integrated project punch listing, request for information tracking, project estimating, detailed cost control, and forecasting.

From a commissioning and qualification perspective, when used as the single repository of project-related documentation and data, a project information management system allows users to accurately capture and transfer the information needed to qualify and operate systems. This mitigates the risk of corruption of data that can occur when it passes through multiple levels of handling and transfer from one user to another.

7.3 IMPLEMENTATION

7.3.1 Risk-Based Qualification Plan

Regulatory guidance recommends the preparation of qualification plans as a means to perform an effective qualification program. The FDA states in the 2011 process validation guidance that [13]:

Qualification of utilities and equipment can be covered under individual plans or as part of an overall project plan. The plan should consider the requirements of use and can incorporate risk management to prioritize certain activities and to identify a level of effort in both the performance and documentation of qualification activities. The plan should identify the following items:

1. the studies or tests to use;
2. the criteria appropriate to assess outcomes;
3. the timing of qualification activities;
4. the responsibilities of relevant departments and the quality unit; and
5. the procedures for documenting and approving the qualification.

The project plan should also include the firm's requirements for the evaluation of changes. Qualification activities should be documented and summarized in a report with conclusions that address criteria in the plan. The quality control unit must review and approve the qualification plan and report (21CFR part 211.22).

IMPLEMENTATION

However, compliance with regulatory expectations should not be the only reason to plan and perform qualification programs. The overall objective of a qualification program should be to create a robust and reliable process which assures that the process is capable of manufacturing a product that meets the company's requirements for product quality. Companies can make costly mistakes when planning and conducting commissioning and qualification of a facility by qualifying beyond what is necessary to mitigate risk to product quality or by not adequately qualifying systems that may pose a risk to product quality. They may choose to include systems and efforts that are redundant and unnecessary. Limited time and resources may then result in the potential to overlook or not adequately qualify items that should be included in the qualification. An approach based on relative risk to product quality can help a company prepare an optimal plan that will be effective, efficient, and compliant.

7.3.2 Define which Items to Qualify

To understand the role of risk assessment in qualification, it is helpful to note that risk and uncertainty are directly related. The less certain something is, the more risky it is. Qualification can reduce uncertainty by providing assurance of system function reliability and appropriateness of design in relation to process outcome. As such, qualification is a means of reducing risk. However, effective qualification approaches rely on an understanding of the correlation between system function and process outcome. Without that correlation, it is difficult to design an effective means to test and challenge the function. The correlation between the process step failure and product quality should be defined. Table 7.2 presents a simple correlation analysis between process steps and the quality attributes they affect.

Table 7.2 illustrates the correlation between process steps, associated equipment, and product quality attributes. In this example, failure of the mixing process

TABLE 7.2 Correlation between Process Step and Quality Attribute

What is the link?

	Process step/operation		Product quality/patient safety
1.	Mixing	➢	Efficacy, strength
2.	Cleaning & sterilization	➢	Purity, safety
3.	Labeling & coding	➢	Identity
4.	Stability testing	➢	Efficacy, strength, safety
5.	Lighting	➢	Strength, identity, purity

or failure of the equipment performing the mixing process would likely have an adverse effect on the strength and therefore effectiveness of the product, as it could cause nonuniformity of active ingredient or solubility issues. This failure could be a result of improper mixer design, impeller sizing, positioning, or speed variation. Likewise, failure of the cleaning or sterilization systems could be the result of clean-in-place pumps malfunction, improper line sloping, and so on. On the basis of this analysis, both mixing and cleaning are process steps that should be qualified. Further, the equipment and systems installed to control these steps should be qualified.

It is essential to understand the relationship between product quality—the process steps that can adversely affect quality—and the facility conditions and equipment that affect or perform those process steps. Item 5 is facility lighting. Can lighting affect product quality? It depends on what the lighting is used for. If it is used for illuminating a process floor—perhaps, but the effect would not be direct. What if the lighting was used to illuminate an inspection box where filled vials are inspected by visual observation? What if the product is light sensitive?

7.3.3 Linking Product Quality Risk to Qualification Testing

Deciding on items and systems to test as part of risked-based qualification follows a logical and sequential approach. The approach can help identify the elements of a process that can adversely affect product quality and choose test methods to confirm that those elements are properly functioning and controlled. Figure 7.4 presents a simple example using an aseptic process. The sequence starts with identifying the critical quality attributes (CQAs) of the product. The CQAs are those elements or characteristics of the product that define it for commercial use. Columns 1 and 2 list the CQA categories and the attributes that define the product as safe, effective, pure, and identifiable. Column 3 lists the corresponding process steps that can have an adverse effect on those CQAs. Column 4 lists the systems that support those steps. Column 5 lists the function that the system performs. Once this is understood, the next step, as presented in column 6, would be to determine test functions and acceptance criteria for demonstrating that those functions occur on a consistent and reliable basis. These tests may also include documentation of proper design and installation. In this way, a link has been created from the qualification test back to a particular CQA. This assures that qualification tests are performed on systems where there may be an affect or risk to the product.

7.3.4 Impact Assessment: Deciding What to Qualify

One method for determining what to qualify is to use a system impact assessment. An impact assessment is a form of risk assessment, where the impact of the function of a system, piece of equipment, or facility condition on product quality is determined. It is a method of determining whether a system or system component requires qualification or to what extent it should be qualified. This

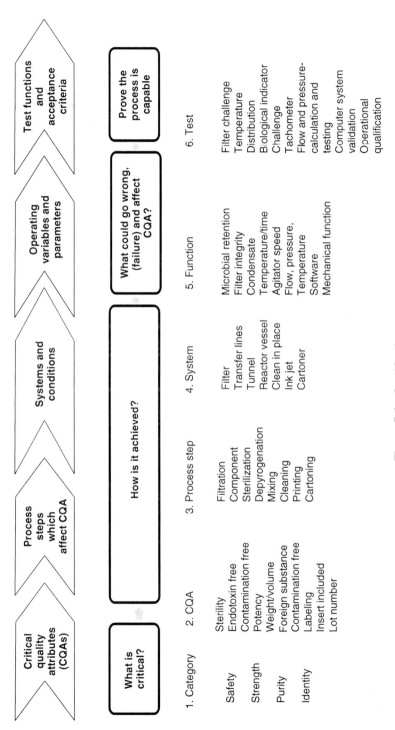

Figure 7.4 Qualification sequence.

involves listing all of the systems and components in the process and then determining if each item has an impact on product quality. To be able to assess the impact of a system or component to product quality, one must understand the process, the variables inherent in the process, the source of process variation, and the effect of variation.

The system or component is assessed as to the effect its failure may have on the quality of the product. If the failure of the system or component affects the quality of the product, then it should be qualified. A relative impact assessment would determine to what level the failure would affect quality. The greater the effect on product quality, the more important the qualification. As the concern is ultimately patient safety, detection of process failure or loss of product quality attribute plays a role in mitigating the risk. If the failure or its effect can be detected and removed before reaching the patient, then the risk to patient safety is eliminated. The risk to product quality may still remain and would result in the loss of the product. If detection is used as a mitigating factor, then the detection method should be qualified.

The impact on quality of a system or component will vary according to the process step it is supporting. For illustrative purposes, let us continue to use lighting as an example. Lighting to illuminate an area may or may not require qualification, depending on the effect on product quality. While lighting in an inspection booth where visible defects are inspected would likely require qualification, lighting to strictly illuminate the packaging area floor may not. Lighting in a stability chamber, where the product may be light sensitive, may require qualification, while the lighting in the laboratory room may not.

Systems with functions that affect product quality are sometimes referred to as *direct impact systems, quality impact systems, or GMP impact systems*. However, some systems may affect the ability to manufacture the product, but not the quality of the product. In other words, failure of these systems may shut down operations or limit yields. For example, a vacuum system that transports waste might not affect product quality and therefore not require qualification, but a vacuum system that removes dust from a tableting operation might have an effect on the quality of the product and would require qualification. Another such system might be plant steam. If plant steam is lost, then production may not be possible. However, the quality of the product is not affected. In this case, the company could choose not to qualify the plant steam system, and rely on commissioning to assure reliable operation.

Although it may be sound business judgment to qualify or commission these systems in some way to ensure proper functionality and capability, it may not be necessary to qualify these systems to assure product quality and patient safety. However, the failure of a clean steam system, which produces steam for the autoclave and does contact filler parts, may affect product quality if it results in poor quality steam. A notable exception might be assurance of continued product supply. If patient safety is an objective, the loss of product because of the malfunction or failure of "nonimpact systems" while not affecting product quality might still result in harm. For the purpose of this chapter, product supply

IMPLEMENTATION 157

issues are not addressed. However, it is worth mentioning that it would be logical to consider inclusion of these systems in the qualification program.

In any case, an impact assessment could provide information that will guide decisions related to the design of the qualification study, including the extent of qualification. Equipment systems could be categorized as having direct impact, indirect impact, or no impact on product quality. To that end, basic impact assessment guidance questions might include the following:

- Does the system or component contact the product or a material that contacts the product?
- Does the system or component control or maintain a critical environment?
- Does the system or component control or measure a parameter or condition that can affect product quality?
- Will the failure of the system or component affect product quality?

7.3.5 System- and Component-Level Impact Assessments

System- and component-level impact assessments are critical to the risk management strategy. Not all systems require qualification and not all components require qualification. The basis of this program requires an accurate system and component list. This can be a challenge because complete system lists are sometimes difficult to find early in the project and accurately maintain.

The initial system list may be a result of previous engineering activities and thus be documented in an engineering design review report. Equipment and components may be obtained by review of approved P&IDs, design-related equipment lists, or process descriptions. There should be a system in place to accurately identify when the P&IDs are updated with additions or deletions of systems. Components of systems may also be obtained by review of material requisitions, which is also labor intensive.

Subsequent to the initial documentation of the systems and components, these lists need to be managed as living documents and therefore controlled as part of the construction and/or engineering change control program. Changes may and likely will occur during construction. The change control programs must accurately contain information necessary to complete impact assessments, be easily located, and contain a trigger to revise impact assessments.

7.3.6 System Impact Assessments

A good system impact assessment approach starts by defining different levels or classifications of impact. These might include GMP critical (direct impact), cGMP noncritical (indirect impact), and no impact. The approach should include a series of questions that will help determine the systems classification. This is best completed during a risk assessment review session with stakeholders having information that could contribute to the decision to classify the system.

Stakeholders who are interested but do not have information to contribute may still be included as a means to communicate the results of the assessment. Again, effective impact assessment relies on a sound understanding on the process, process variables, and the effect of equipment and systems on the cause and control of those variables.

An example of specific impact assessment questions might include the following:

1. Does the system come in direct physical contact with the product (e.g., excipients, ingredients, solvents, or processing materials or aids) at, or after, a critical process step?
2. Does the system directly affect product quality (e.g., does the system clean, sanitize, and sterilize or does the system maintain a classified environment)?
3. Do the system measurements directly influence or impact the product characterization or impurity profile? Are they control critical process parameters (CPPs)?
4. Does the system preserve product status? Is the system used to store and preserve the status (e.g., preserve temperature, humidity, etc.) of interim and final products or product sample?
5. Does the system generate data used or evaluated to support product disposition and regulatory filings?

7.3.7 Component Impact Assessments

Component impact assessments are similar to system impact assessments, but are focused on the elements that make up the system. Component impact questions might include the following:

1. Is the component used to demonstrate compliance with production records requirements?
2. Does the normal operation or control of the component have a direct effect on product quality?
3. Does failure or alarm of the component have a direct effect on product quality?
4. Is information from this component recorded as part of the production record, lot release data, or other cGMP-related data?
5. Does the component have direct contact with the drug product, its excipients, or device contact surfaces?
6. Does the component control critical process elements that may affect product quality, without independent verification of the control system performance?
7. Is the component used to create or preserve a critical system status?

Once the impact assessments are completed, the information can be used to develop the specific commissioning and qualification plans. Systems or components that have a direct impact will likely require qualification and quality unit oversight/acceptance criteria approval. Systems or components that have an indirect impact may require qualification depending on the level of impact. Systems or components that have no impact on product quality may not require qualification. All systems may require commissioning to assure proper function. However, the commissioning of direct and indirect impact systems may require additional levels of control and quality unit review, if that information is to be leveraged as part of the qualification effort.

It is important to note that the information obtained from the impact assessments may also be considered when entered into the maintenance program and change control/change management procedure. The impact items have on product quality should be considered when addressing the response to process, system, or procedural changes. Noncritical components on a critical system would not need to undergo qualification. A simple work order might suffice. However, if a new piece of equipment or component was added to a system or removed, then the assessment would need to be revised. This is a much more time-effective activity than qualification.

Table 7.3 presents an example of an impact assessment. The answers to the following questions will determine if a system is direct, indirect, or no impact. An affirmative answer to questions #1 through #6 generally indicates that the system is direct impact. An affirmative answer to only question #7 generally indicates that the system is indirect impact. If the answer to each question is negative, then the system may be deemed no impact.

TABLE 7.3 System Assessment

	Direct Impact System Assessment Criteria	Yes/No
1	Does the system have direct contact with the product?	No
2	Does the system produce or have direct contact with raw material, excipient, or solvent that comes in contact with the product?	No
3	Is the system used in cleaning or sanitizing production equipment or systems?	No
4	Does the system preserve product status (physical, chemical or microbiological integrity)?	Yes
5	Does the system produce data that is used to accept or reject product, medical devices, or stability data that will be used to make future quality or regulatory decisions?	Yes
6	Is the system a process control system that may affect product quality and there is no system for independent verification of control system performance in place?	No
7	Does the system supply a utility or function to a direct impact system or otherwise affect the performance of a direct impact system?	Yes

For the purpose of this example, consider the lighting in a semi-manual inspection booth, where inspectors look for particulates and defects in vials of a product.

On the basis of this assessment, the system, the inspection booth, and the lighting in the booth would be deemed critical according to affirmative answers to #5 and #6. Utilities that support the booth, such as electrical power, would be deemed as having indirect impact. So the booth and the lighting should be qualified. The decision to qualify the electrical power source would depend on the amount of variation in the power source, the effect on lighting intensity that such a variation might have, and the effect of intensity variation on product quality or the ability of the inspector to see and reject defective units.

7.3.8 To What Extent Would it be Qualified?

Once a system or component has been classified according to level of impact, a decision should be taken as to how to assure the correct function of that system. This can be done by asking the question—what could go wrong and what would cause such a situation or failure? Tools such as fault tree analysis, partially illustrated in Figure 7.5, and fishbone analysis, partially illustrated in Figure 7.6, can be helpful in identifying those process steps or conditions that pose a risk to product quality. These tools are designed to break the process into smaller steps where risk can more effectively be addressed and reduced.

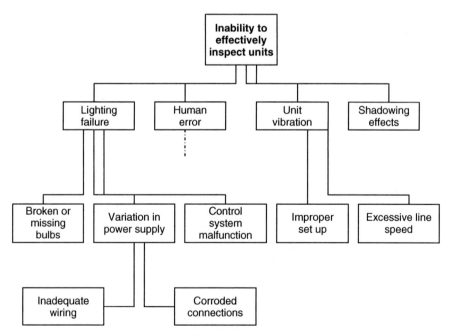

Figure 7.5 Fault tree analysis example.

IMPLEMENTATION

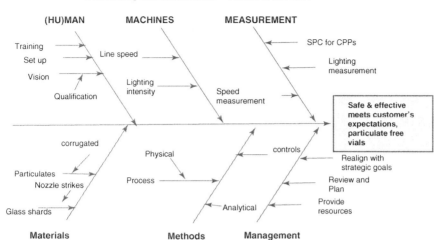

Figure 7.6 Ishikawa Fishbone analysis example. (*See insert for color representation of the figure.*)

7.3.9 Criticality Analysis

Another method of impact assessment might ask questions related to: How important is the function of the equipment or system to the quality of the product? Let us look at this from the perspective of the light example.

1. What is the impact on product quality if this process step—visual inspection—fails?
2. How important is the system or component—lighting—to the success of the process step?
3. How robust is the process? How much variance can the process take and still be effective—what are the operating parameters?
4. How effective or reliable are the control systems in place to control variation and prevent failure?
5. How effective are the ways to detect a failure before it adversely affects the quality of the product or patient safety?
6. What is the control strategy or risk mitigation/risk reduction strategy for assuring that process results remain in specification?

For the purpose of this exercise—one could use a simple HIGH, MEDIUM, and LOW rating system. HIGH means very important, severe, definitely could happen. LOW means not very important, not of much concern, doubtful it would happen; MEDIUM is everything in between. Question (6) requires a more extensive answer.

TABLE 7.4 System Risk Determination

	Question	Answer or Consideration	Risk Ranking
A	What is the impact on product quality if the visual inspection process step fails?	Particulates could be overlooked and therefore cause vascular damage and clotting	HIGH
B	How important is the system or component lighting to the success of the visual inspection process step?	This is a manual system, dependent on lighting conditions	HIGH
C	How robust is the visual inspection process? How much lighting variance can the process take and still be effective?	There needs to be adequate lighting, but that covers a relatively broad range	MEDIUM
D	How effective or reliable are the systems in place to control the process within acceptable levels of variation and prevent failure?	Lights are not adjustable and there are no meters, but the procedure requires a check that all lights/bulbs are functioning and turned on	MEDIUM
E	How effective are the ways to detect a visual inspection failure before it adversely affects the quality of the product or patient safety?	Statistical sampling and testing of finished product is performed before shipment	MEDIUM
F	What is the mitigation strategy for reducing the risk and assuring that process results remain in specification?	Use of rheostats and light meters to set the level of lighting and design of system with adequate lighting sources	LOW

Table 7.4 presents the questions, considerations, and answers. Where high risk potential is identified in any of the questions, it would be prudent to seek ways to mitigate or reduce that risk through process design changes. Risks that remain higher than desired may require additional qualification and validation efforts to provide assurance of adequate process control.

In this case, qualifying the lighting may be of moderate but not high importance, because the process is relatively robust and the likelihood of failure of the system going undetected (i.e., bulb out) is low. The qualification would involve confirmation that rheostats and meters are operating correctly and that lighting system, bulbs and background materials, have been installed as per specification. With these systems confirmed, the PQ might involve running sets of defects through the inspection system and confirming trained inspector performance given

IMPLEMENTATION 163

TABLE 7.5 Example of an FMEA Template

Process Step/Unit Operation	Failure	SEV	Cause	OCC	Current Control	DET	RPN	Risk Accepted

system function, speed, lighting levels, etc. The qualification would verify that the system will function in a consistent manner.

7.3.10 Risk Ranking Level of Detail

Risk assessment and risk ranking tools provide the information needed to help make an informed risk-based decision. However, they should not necessarily make the decision. The focus should be on evaluating the outcome of the risk assessment, rather than evaluating strictly the risk ranking value.

Consider how an FMEA (failure mode effects analysis) model is used to rank risk. Table 7.5 presents a typical FMEA spreadsheet template for ranking risk. The level of numeration detail or granularity of the ranking choices, shown in Table 7.6, is the number or quantity of choices one has in ranking risk elements, including severity, occurrence, and detection. Some FMEA models utilize ranking value choices ranging from 1 to 10. This would represent a relatively high detail or granularity method. Some models use less numbers or qualitative choices, such as high, medium, and low as illustrated in Table 7.8. This represents a relatively low detail or granularity method.

Where quantitative data is available, such as number or rate of defects, production yields, or failure rates, higher granularity methods should be objective and useful. However, where quantitative data is not readily available and qualitative ranking is used, such as quality of, confidence in, or experience of a vendor; a lower granularity method may be more useful and effective. In these cases, lower granularity, such as high, medium, and low may reduce the potential for bias and ranking value "debate."

In the FMEA assessment example presented in Tables 7.6 through 7.8, the difference between a "marginal" and a "critical" risk is one point, 500 versus 501. Taking into account the descriptions presented in Table 7.7, this may mean that the difference between an acceptable and unacceptable risk is one point. In this case, the assessment team may be "tempted" to select individual scores in order to achieve a lower risk ranking. For instance, the team might spend much of the time debating whether an occurrence value, is a four, a five, or a six.

If, on the other hand, the choices were high, medium, and low, as presented in Table 7.8, then occurrence might be often (high) or seldom (low), or anything in between (medium). The team would ask questions such as: Does the event occur often? If so, then the value is high. Does it occur seldom? If so, then the value is low. Otherwise it is medium. This might lead to productive discussion over

TABLE 7.6 Example of a 10-Point Quantitative FMEA Risk Analysis Calculation, Where: Risk = Risk Priority Number = Sev × Occ × Det

- Severity (SEV): Classify the severity or importance of the effect.
  - Very low to no impact 1
  - Unimportant failure 2–3
  - Failure of medium importance, may cause customer dissatisfaction 4–6
  - Critical failure, may cause harm to patient 7–8

  Extremely critical failure, will likely cause significant harm to patient 9–10
- Occurrence (OCC): Estimate the probability of occurrence of the failure.
  - Very low probability 1
  - Failure might happen, but very seldom 2–3
  - Failure happens from time to time 4–6
  - Failure happens frequently 7–8

  High probability that failure will happen 9–10
- Detection (DET): Evaluate the probability of the failure detection.
  - Failure detection is ensured 1
  - High probability of failure detection 2–3
  - Failure detection not certain 4–6
  - Low probability of failure detection 7–8

  Failure detection is highly improbable 9–10

TABLE 7.7 Examples of Risk Category Descriptions and Risk Priority Rankings for a 10-Point Quantitative FMEA

Category	Description
Category I Catastrophic	Serious and/or unexpected product adverse experiences or serious bodily injury
Category II Critical	Unexpected product adverse experiences, severe injury, or inconvenience
Category III Marginal	Minor injury or inconvenience, or possible product adverse experience
Category IV Minor	Not serious enough to cause injury or inconvenience or other product adverse experiences

Risk Priority Number	Category Description	Category
1–125	Very low process risk	Category IV
126–250	Low process risk	Category IV
251–500	Moderate process risk	Category III
501–750	High to very high process risk	Category II
751–850	Very high process risk	Category II
851–1000	Extreme-critical process risk	Category I

TABLE 7.8 Example of a Three Point Qualitative Risk Priority Rank Determination [14]

Ranking	Risk Factors		
	Severity	Occurrence	Detection
HIGH	Impact of the unwanted event is severe	Occurrence is often	Process failure will almost certainly escape detection
MEDIUM	Impact of the unwanted event is moderate	Occurrence is periodic	Controls may detect the existence of a process failure
LOW	Impact of the unwanted event is low	Occurrence is seldom	Process failure is obvious and readily detected

DETECTION*

OCCURRENCE	LOW	MEDIUM	HIGH
HIGH	This cause is likely to occur, but when it does it will be detected. If we are certain it will be detected it is low risk, but if we are not certain then it should be a Medium Risk	This cause is likely to occur and the detection is not certain. It is a High Risk	This cause is likely to occur and is not likely to be detected. It has a High Risk
MEDIUM	This cause could occur, but if it did it would be detected. Depending on the frequency of occurrence and the confidence in the detection, it is a Low or a Medium Risk.	This cause could occur and it could be detected. Depending on our confidence in the detection its risk would be Medium or High	The cause may occur and it will not be detected. The Risk is High
LOW	This cause is not likely to occur and if it does it will be detected. This is a Low Risk	The cause is not likely to occur and if it did it may be detected. Depending on the frequency of occurrence and confidence in detection methods, it would be Low or Medium Risk	The cause is not likely to occur, but if it did occur it would probably not be detected. The Risk is Medium

Note: This example involves aseptic process risk—where severity is a constant HIGH and therefore not factored. If severity is a variable it should be factored on an additional analysis.

the relative risk of the occurrence, than discussion over what the risk "number" should be.

While these examples are purely hypothetical and do not represent specific manufacturing situations, they present descriptions that are not uncommon in the industry and could indicate potential ranking determination considerations.

7.3.11 Test Function and Acceptance Criteria

Qualification test functions should be developed to demonstrate the capability of equipment as it relates to process performance and product quality. It may not be necessary to test equipment and systems capability beyond the range expected during the process. However, the tests should be based on sound scientific and engineering criteria and match with those functions that affect product quality.

The FDA process validation guidance suggests that qualification of utilities and equipment verify that the utility system and equipment operate in accordance with the process requirements in all anticipated operating ranges [15]. This should include challenging the equipment or system functions while under load comparable to that expected during routine production.

The acceptance criteria should be useful, attainable, and verifiable. Useful means that the criteria represent a challenge to the system. It should not be so easily obtained that it does not provide confidence of the continued reliability of the system operation. An example might be qualifying a clean room, which must meet Class 100 or less than 100 particles of 0.5 microns or larger per cubic foot per minute sample. Modern, well-designed clean rooms should be able to perform below the 100 limit. Setting criteria at 100 may be technically correct, but may not provide an indication if the system is performing properly.

Attainable means that the criteria are not beyond the range of performance and therefore too difficult to obtain. An example might be setting automated inspection machine qualification limits at zero. This certainly is the optimal performance level, but obtaining zero may not be realistic using current technology or may result in unacceptable levels of "false" defects rejected. A more realistic level of defect rejection, above zero, may be attainable and represent an acceptable risk level to patient safety.

Verifiable means that the criteria can be quantified or objectively confirmed. An example of unverifiable criteria might be requiring that the fill line run well or the vessel be "free of foreign material residue." These terms lack quantifiable limits. They would be subjective, open to interpretation, and difficult to prove. Likewise, setting criteria that all product transfer line surfaces be visually inspected and shown to be free-of-pitting might be an acceptable criteria, but the test function might be impossible to perform. In this case, the test function to satisfy the free-of-pitting criterion might be a confirmation of material of construction combined with an inspection of a representative sampling of surfaces.

Risk to patient safety and product quality should be considered in determining the test function and acceptance criteria. Let us say the transfer lines will carry a relatively noncorrosive material, which will go through further downstream purification steps. The test function and acceptance criteria might be confirmation of material along with weld inspection. However, what if the material is more corrosive, with no further purification? Then, the test function might include more extensive confirmation, perhaps metallurgic analysis along with a higher level of inspection.

7.3.12 Qualification Plan Implementation Notes

7.3.12.1 Subject Matter Experts   SMEs are responsible for the preparation of respective qualification documents and plans. The ASTM E2500-07 Standard Guide defines SMEs as *"individuals with specific expertise and responsibility in a particular area or field"* ... who *"should take the lead role in the verification (commissioning and qualification) of manufacturing systems as appropriate within their area of expertise and responsibility ... includ(ing) planning and defining verification strategies and acceptance criteria* ... [16]" As such, it is important that the SME has a good understanding of the individual and collective risk and impact of systems on process performance and product quality. SMEs should participate in risk and impact assessments, and use product quality risk as a decision-making criteria.

7.3.13 Risk-Based Design Review

Merely testing the system for proper function is often not sufficient to confirm reliable and suitable function. An essential part of a risk-based qualification approach includes review of the facility, equipment, or system design. In a risk-based review, the reviewer should focus on those characteristics or aspects of the design that can affect product quality. The review should confirm that user requirements and quality specifications have been addressed in the design of the system. The basic user requirement may require that a process transfer line be cleanable and drainable. Therefore, the design should include smooth, nonreactive materials of construction, welds, slope, and sanitary fittings. This can be determined during the design review. Later, tests can be incorporated in the qualification or leveraged from prior information to confirm these aspects of the installation.

A design review should familiarize the reviewer to the layout of the system, including those characteristics that could affect the qualification test and information gathering. Attention should be on installation of systems that may be covered by walls or otherwise may not be easily viewed after construction, e.g., utility piping, transfer lines, duct work, wiring, slope of insulated line, etc. The incorrect installation of these components may pose a risk to product quality and should be confirmed. Confirmation may need to be performed before construction completion or the qualification team may need to leverage or rely on construction installation information rather than direct observation.

A design review should also confirm or identify sampling ports and places for line transporting fluids, steam, gases, and other materials. A qualification test may require direct sampling of materials in lines or as confirmation of cleanliness; there, the review should confirm that these places or ports are included.

7.3.14 Risk of Using Information before Identification of Critical Process Parameters

One of the potential pitfalls of using early leveraged commissioning information is not knowing the critical process parameters and operating conditions needed to successfully perform the process. At times, the commissioning must be performed before the process has been fully developed and finalized. If the process parameters and conditions are not fully understood, then it may be possible that the tests conducted during commissioning will not adequately challenge or demonstrate the operating ranges. To mitigate that risk, one can perform the following:

- Test at the full capability or at an expanded operating range of the system.
- Utilize smaller scale process development studies or engineering studies to determine most likely operating parameters and ranges.
- Repeat testing during qualification, if CPPs fall out of ranges tested during commissioning.

7.3.15 Installation Qualification

The installation qualification confirms that the facility and the equipment have been constructed and installed to user specification and regulatory requirements. To do this, the installation qualification relies on the inclusion and alignment of engineering and commissioning documentation. However, very often qualification risk management strategies are outlined without consideration of the preceding engineering activities. The result, if not managed properly, can undermine the entire process late in the project where cost and schedule are adversely impacted. For example, it is important to capture the numbering strategy used to uniquely identify each protocol so that the equipment itself, as well as the type of protocol, can be archived and easily retrieved for future reference. Manufacturing plants must have efficient and effective means to retrieve historical information.

There may be a benefit to creating and using commissioning and support function templates with test details, instructions, acceptance criteria, and executable attachments. Different functional areas may have the ability or need to change template wording to suit specific applications. However, these changes should not change the original intent of the tests or the acceptance criteria to an extent that could affect the use or leveraging of the information. Consideration should also be given to developing templates that are conducive to retesting.

Organizations vary on their level of comfort or tolerance to risk when developing and using commissioning templates. One suggestion is to outline the execution attachments. These outlines may later be used to create the detailed commissioning or engineering documents. The installation qualification execution attachment can then list what will be attached and referenced back to the commissioning documents.

For example, the P&ID section will list the drawings associated with that particular system and then document the location of the commission execution attachment that lists the P&IDs that were walked down and required red lines. P&IDs listed in the qualification document, but missing from the commissioning documentation would then require a deviation to resolve the discrepancy. Likewise, any P&ID that is missing from the qualification document will also require a deviation to resolve the discrepancy.

Another suggestion would be to utilize the installation verification portion of the protocol. This section should list each system and component assessed as critical with the associated criticality assessment and will list what will be attached or reference back to the commissioning documents. All applicable sections of the protocol should be presented in an approved template so that all tests included meet the expectations of your organization. Similar critical components should have the same level of qualification or attributes inspected and qualified. The component-level assessments discussed earlier can be used to find similarities in common systems and assure a consistent qualification approach.

It is possible to streamline the installation qualification process through use of commissioning and information. However, the qualification process works best when it is aligned with the commissioning process. Discrepancies need to be resolved, reviewed, and approved before close of the protocol. These discrepancies are resolved using an approved deviation process, which can have the ability to consume time and adversely impact time and cost savings. One way these discrepancies can be avoided is by ensuring that they are adequately addressed in the punch listing process, which should be included as an aspect of construction and/or engineering change control. A method should be employed to track the deviations by deviation types. Metrics can be established to track the occurrences. This can provide the ability to see what types of deviations and how many of each type occurred during execution.

This also provides the opportunity to focus on the "low-lying fruit" and develop mitigation plans that eliminated those deviations. Rather than focusing on the most painful deviations that seem to draw attention and resources from the project team, consider focusing on the deviations that had the largest numbers and relatively fastest corrective actions. These deviations usually can be quickly and easily mitigated by minor adjustments in strategies, policies, or procedures resulting in a dramatic gain of efficiencies.

7.3.16 Operational Qualification

Operational qualification provides assurance that the facility and equipment function in a reliable manner to the level of performance needed to support the process. This involves the development of test functions, studies, and acceptance criteria designed to challenge the system through its intended operating ranges. The functions to be tested and the ranges of operation should be based on support of those process steps that can affect product quality. These tests can be based on information uncovered during the impact and risk assessments.

In many cases, information can be gleaned from prequalification testing. This includes construction testing, FAT and SAT, as well as commissioning tests. It is important that this information is accurate, from a reliable source, follows GDP, and is aligned with the qualification requirements. As mentioned earlier, the qualification team should consider risk to product quality as a criterion for deciding what prequalification information to use in lieu of or in combination with qualification tests. A simple risk formula analysis can be used to help determine the use of the information.

$$Risk = Severity \times Occurrence \times Detection$$

Risk is a factor of severity or impact of a failure, the probability of the failure occurring, and the likelihood that the failure will be detected and corrected before it can affect product quality. For equipment functions that have a relatively low impact on product quality, are not likely to fail, and have other controls in place to detect the results of such a failure, accepting prequalification information may be an efficient and effective means of obtaining qualification information. However, higher impact functions or those with no or little means of detection represent higher risk and may require additional testing. The functions that may be prone to failure should be considered for redesign as a means of mitigation. Again, the capability of and confidence in vendors and construction personnel will also be a strong factor in considering the use of prequalification information. In many cases, the vendor and construction personnel have such a strong expertise or utility in the testing as to overcome process failure concerns. Quality unit review and approval is recommended for any product-quality-risk-related test information and results.

7.3.17 Computer System Validation

Many of the points presented in this chapter on general system and equipment qualification are applicable to computer system qualification. Systems should be designed to user requirements and specification. System and instrument installation, wiring, and component verification should be based on risk to product quality, as should functional and operational checks, alarms, and monitoring results and displays.

Some automated systems perform multiple functions, with some affecting product quality and some not. It may not be possible to qualify those functions that impact quality without also testing those functions that do not—especially if the functions are interrelated. Caution should be used to assure that non-impact functions do not interfere with impact functions. Security of the entire system—including access restriction, change control, and audit trail may be interrelated. This may especially be relevant to open systems that have connections to larger systems or to the Internet.

A clear understanding of the computer system function, including a review of software logic, would mitigate the risk of inadvertent or unexpected interference between applications.

IMPLEMENTATION   171

Qualification efforts, including software validation, verification, and testing should be based on the level of confidence in the system as well as the use of the system and complexity (FDA guidance). Consider the effect of software modifications on product quality. Modifications to critical function modules may be controlled, but modifications to noncritical functions or applications may not be—although they may indirectly have an effect on product quality. Computer system functions or software requiring additional efforts may include systems that:

- control, monitor, record operations that can adversely affect product quality or patient safety;
- collect, manipulate, or report data used to make decisions that may have an adverse affect on product quality or patient safety;
- are relatively new or for which the company does not have previous experience; and
- affect complex operation that could have an adverse affect on product quality or patient.

Other computer system qualification points to consider are as follows:

- Risk to product quality may include interaction with nonvalidated systems, including Internet access.
- Systems should provide access security and audit trail for changes.
- Automated control systems should be qualified before the qualification of the mechanical systems they control, monitor, or operate.
- Companies may use information from automated monitoring systems to show that operating or environmental systems are operating properly if those automated systems have been qualified.
- Be cautious of changes made by the vendor after FAT or other commissioning testing.
- Be cautious of damage to electronic circuitry during shipment and storage post-FAT or commissioning testing.

7.3.18 Requalification

Qualified systems should be maintained in their qualified state. A program for doing so should include the following:

- preventive maintenance;
- calibration;
- evaluation of system performance;
- change control;
- investigation of process performance; and
- quality audit.

Requalification can range from complete data analysis to selected testing to full qualification. It can be triggered if an event occurs, putting in question the qualification of the system. Such events may include the following:

- changes in equipment or process;
- changes in components or materials run on equipment;
- changes in use of equipment;
- repeated failures;
- changes in procedures; and
- changes in regulatory expectations or requirements.

Requalification can also be performed on a periodic basis. The FDA process validation guidance addresses periodic reevaluation of qualification by stating that: "Maintenance of the facility, utilities, and equipment is another important aspect of ensuring that a process remains in control. Once established, qualification status must be maintained through routine monitoring, maintenance, and calibration procedures and schedules (21 CFR part 211, subparts C and D). The equipment and facility qualification data should be assessed periodically to determine whether requalification should be performed and the extent of that requalification. Maintenance and calibration frequency should be adjusted on the basis of feedback from these activities [17]." A reason for periodic requalification would be that the complexity of systems does not allow for complete knowledge of all changes, effects of wear, and unknown occurrences that may cause variations in performance.

Risk assessments and evaluations can be used to address the need for and extent of requalification. Factors that need to be considered are as follows:

- robustness or reliability of processes or systems;
- experience with systems;
- redundancy in process and system controls; and
- potential impact of system failure on product quality and patient safety.

Written procedures should be in place to address evaluation, decision criteria, justification, and requalification methodology and schedule.

7.3.19 Performance Qualification: Contamination Control

Product contamination represents a notable risk to product quality. Contamination from microorganisms and other foreign substances is often difficult to detect and may result in considerable harm to the patient. Contamination control process steps such as cleaning, disinfection, sanitization, sterilization, and holding of sterilized materials are almost always deemed as critical and direct impact. Risk assessment techniques can be used to help determine worst case configurations and conditions to be qualified and areas and items that are most difficult to clean or sterilize. It can also be used to help develop bracketing and family approaches to qualification studies.

A *bracketing approach* is one where equipment configuration at extreme ends of the range is qualified, with the assumption that the results will qualify all configurations in between. In order to do this, it is important to know the range and whether there are any differences in the process conditions in between the range limits, which could affect product quality or process performance. Risk assessment methods, designed to uncover process risk conditions and steps, such as fishbone and fault tree analysis, can be used to identify these risks and assist the study designer in determining where the bracketing approach can effectively be used. The most effective use of bracketing will occur when configurational conditions are. Some of these might include length of process transfer line, holding time durations, size of similar vessels, autoclave loads, etc.

A *family approach* is based on similar assumptions. The premise is that if all items are of the same design, installation, and operational capabilities, then they are essentially the same. If they are the same, then the PQ successfully performed on one item should qualify all others. In order for this to work, the commissioning and qualification of these items must show that all are equal and that there are no material differences that can affect quality-related performance. Risk assessment methods can be used to determine the critical aspects of the equipment, which must be similar.

Risk assessment can also be used to design test functions and acceptance criteria. Test functions should be set to the process-related objective of equipment function, rather than the overall functional capability of the equipment. An example is the washing of parenteral glassware or vials. Some vial washers claim endotoxin reduction capability. It is difficult to remove all endotoxin from the surfaces of smaller vials when the flow of rinse water is relatively low. Therefore, it is difficult to qualify this depyrogenation process. A risk assessment of the overall process would determine that the glassware washing process was not utilized to depyrogenate, but rather to remove foreign substances and particles. The depyrogenation step was the subsequent dry heat oven or tunnel exposure. Therefore, a failure of the glassware washer to remove endotoxin should not necessarily constitute a qualification failure.

Another area where risk assessment could be used might be the inclusion of overkill parameters in moist heat sterilization processes. Where controls and processes are in place to control bioburden and result in a more predictable bioburden level, time and temperature parameters beyond those needed to achieve 10^{-6} sterility assurance levels may not be as necessary as would be the case with a less well controlled and less predictable situation.

7.3.20 Change Control

Change management and change control have an important role in commissioning and qualification. Changes made in the design, construction, and installation of facility and equipment must be communicated to those preparing or executing the qualification protocols. Even noncritical changes to impact systems may result in deviations, errors, and delays during execution. In some instances, the changes may not be consistent with process requirements or may not be compliant with regulatory expectations. For this reason, a good engineering change control system should be in place as early as practical on the project. Changes should be evaluated as to their potential impact on product quality and process performance. Where they are found to have an impact, steps should be taken to mitigate or minimize any adverse affects on product quality or related process performance.

Chapter 13 presents an in-depth discussion of the use of risk management for change management and change control. Therefore, this chapter will not go into detail on methodology. However, care should be taken to address the effects or unintended consequences of change. Additional risks remaining after changes are made, sometimes referred to as *residual risk*, should be evaluated and reduced to acceptable levels. Once addressed, changes should be communicated to all parties who will make decisions based on the information related to the changed system, including those responsible for development of commissioning, qualification, and validation protocols. Some changes to equipment or process steps may alter the results of impact assessment, moving items from one category to another. For example, a process step change that adds a sterile addition to an upstream compounding procedure may move certain environmental conditions from indirect or relatively low impact to relatively critical direct impact. Likewise, a change from an elastic gasket to a more rigid gasket material, coupled with the need for sterilization may result in the need for additional testing to assure integrity of the sealed system after steaming.

7.3.21 Failure Investigation

Deviations or discrepancies may be the result of unacceptable study results, inaccurate information, or poorly executed protocols. Where deviations are the result of study failures, and investigation should be conducted to determine if and to what extent the failure has an impact on or represents a risk to product quality. Corrective actions should be determined and implemented before repeating the study. Actions should be commensurate with the level of risk the failure or systems function represents to product quality.

One should avoid repeating studies, without corrective actions, until results are satisfactory. If the deviation is a result of study execution or testing, then the study may be repeated. If the deviation is a result of inaccurate information transfer, then the study may not have to be repeated. In most cases, deviations should be investigated with the conclusion and subsequent actions documented. Avoid initiating an investigation with a predetermined conclusion as the objective. Any corrective actions that affect operating procedures or parameters should be communicated to the appropriate individuals. Any investigations that involve equipment, processes, or steps, which have an impact on product quality, should be reviewed by the quality unit.

7.3.22 Project Close Out

Once the qualification project is complete, the result should be reported to those individuals responsible for the review and approval of the study results. Commissioning and prequalification activities should be complete. The protocol and report should be complete. Investigations should be closed out. Deviations should be addressed. Corrective actions should be implemented. Changes should be identified and documented. All documentation should be available and suitable for review and approval. Information that is critical to the operation of the equipment and system, including preventive maintenance and operating parameters, should be transferred to operations and included in the appropriate procedures. PPQ protocols should reflect any changes in design or process uncovered and confirmed, during the qualification studies. These changes should be included in system descriptions and where applicable PPQ test functions, data sheets, and forms. The result of the qualification and conclusion should be stated. If the conclusion is that the equipment is suitable for release to the next step of process qualification or validation, then this should be clearly articulated.

7.4 CONCLUSION

QRM plays an important role in facility and equipment qualification. Risk assessment techniques can be used to help develop and execute effective facility and equipment qualification plans. Qualification efforts should be based in part and commensurate with the effect on product quality, which the facility system or equipment poses. Risk assessment techniques can be useful in providing information needed to make decisions related to which items to qualify, the prioritization of qualification efforts, setting of useful acceptance criteria, the effective role of the quality unit, leverage of prequalification and commissioning information, and the addressing of design changes and study failure.

A risk-based approach to qualification should provide confidence that the dependent process will perform in a consistent and predictable manner. Having a strong level of confidence of process performance before commercial manufacture commitment is logical, beneficial, as well as compliant. If the study is designed properly, then qualification failure is an indication that something is wrong with the equipment and must be corrected. If studies are not designed or executed properly, then successful completion may not be a complete indication of successful system installation and function.

It may be true that sometimes qualification appears to be more effort than necessary. This may be because our industry tends to be overly cautious. On the other hand, in order for processes to perform to expectations, systems and equipment must work properly. Inaccurate information, poor design/installation, and ineffective commissioning practices can also result in systems that cannot or do not function as expected. Qualification programs should be designed to assure the required function of any, all, and only facility and equipment features that could affect product quality. A good qualification program will do this in an effective and efficient manner, utilizing all reliable sources of information, including where applicable prequalification information and sources.

Risk assessment may be used to eliminate redundant and limited value efforts, such as those involved with the qualification of items that pose no risk to product quality or patient safety. It may also be used to provide guidance on the leveraging of information from prequalification efforts such as design, fabrication, construction, installation, acceptance testing, and commissioning. In doing so, risk assessments provide a valuable tool for developing and executing an effective and efficient qualification program.

It is an expectation of many regulatory agencies that companies consider risk to product quality and patient safety when making product manufacturing decisions. These expectations are reflected in recent FDA guidance and industry standard guides. The specific method for considering or assessing risk is largely left to the companies. In most cases, it is not essential that a particular method or tool be employed, rather that some logical method for considering risk be used. Many of the methods and tools discussed in other chapters may be found to be useful for the assessment of risk with respect to qualification, including FMEA and FMECA (failure mode effects and criticality analysis), Ishikawa or fishbone analysis, fault tree analysis, and impact assessments.

This chapter has attempted to present some suggestions and points to consider when developing a risk-based qualification program. It was not meant to give an exhaustive or prescriptive procedure for qualification or to cover all topics related to the qualification of pharmaceutical facilities, equipment, and systems. Rather, it is the hope of the authors that the reader will be able to use the information in this chapter to formulate better plans and more pragmatic programs for qualification efforts based on risk and science.

REFERENCES

1. Department of Health and Human Services, U.S. Food and Drug Administration, Code of Federal Regulation, Title 21, Chapters 210 and 211, Section 100 (CFR 211.100 of cGMPs), 2008.
2. Department of Health and Human Services, U.S. Food and Drug Administration, Guidance for Industry on the General Principles of Process Validation, January 2011.
3. Department of Health and Human Services, U.S. Food and Drug Administration, Guidance for Industry on the General Principles of Process Validation, January 2011, page 10.
4. Department of Health and Human Services, U.S. Food and Drug Administration, Warning Letter to Bell-More Laboratories, Inc., January 5, 2007.
5. ASTM (American Society for Testing and Methodology) E2500-07, Standard Guide for Standard Guide for Specification, Design, and Verification of Pharmaceutical and Biopharmaceutical Manufacturing Systems and Equipment.

REFERENCES

6. Department of Health and Human Services, U.S. Food and Drug Administration, Code of Federal Regulation, Title 21 Chapters 210 and 211.
7. International Society for Pharmaceutical Engineering, ISPE Active Pharmaceutical Ingredients Baseline® Guide Second Edition, June 2007.
8. PDA Survey of Quality Risk Management Practices in the Pharmaceutical, Device & Biotechnology Industries, PDA Journal of Pharmaceutical Science and Technology.
9. ASTM (American Society for Testing and Methodology) E2500-07, Standard Guide for Standard Guide for Specification, Design, and Verification of Pharmaceutical and Biopharmaceutical Manufacturing Systems and Equipment, page 2.
10. International Society for Pharmaceutical Engineering *ISPE Baseline*® Pharmaceutical Engineering Guide Series, Volume 5–Commissioning and Qualification, First Edition, March 2001.
11. Department of Health and Human Services, U.S. Food and Drug Administration, Guidance for Industry, Q9 Quality Risk Management, 2006, page 2.
12. ASTM (American Society for Testing and Methodology) E2500-07, Standard Guide for Standard Guide for Specification, Design, and Verification of Pharmaceutical and Biopharmaceutical Manufacturing Systems and Equipment, page 3.
13. Department of Health and Human Services, U.S. Food and Drug Administration, Guidance for Industry on the General Principles of Process Validation, January 2011, page 11.
14. Technical Report 44, Quality Risk Management of Aseptic Practices, PDA Journal of Pharmaceutical Science and Technology, 2008 Supplement Volume 62, No. S-1.
15. Department of Health and Human Services, U.S. Food and Drug Administration, Guidance for Industry on the General Principles of Process Validation, January 2011, page 11.
16. ASTM (American Society for Testing and Methodology) E2500-07, Standard Guide for Standard Guide for Specification, Design, and Verification of Pharmaceutical and Biopharmaceutical Manufacturing Systems and Equipment, page 3.
17. Department of Health and Human Services, U.S. Food and Drug Administration, Guidance for Industry on the General Principles of Process Validation, January 2011, page 15.

8

PROCESS LIFECYCLE VALIDATION

A. Hamid Mollah and Scott Bozzone

8.1 INTRODUCTION

Process validation (PV) is a requirement of the Current Good Manufacturing Practices Regulations for Finished Pharmaceuticals (21 CFR Parts 210/211 and EU GMPs) [1,2]. Since 1987, when the U.S. FDA issued guidance [3], the pharmaceutical industry approach to PV has typically been to evaluate three consecutive, prospective batches. This approach was used regardless of risks associated with aspects such as complexity of the process or dosage form, type of unit operation, or development history. There has historically been limited application of risk management in defining the amount of data or number of batches required for PV studies. In contrast, a science- and risk-based approach applied throughout the process lifecycle is a more holistic and robust approach. Table 8.1 shows a comparison between traditional, and science- and risk-based process approaches.

Risk management can be applied in several areas of PV, from early process design/development through maintenance of validated states during commercial manufacturing [4]. Some of the benefits of applying a science- and risk-based approach during PV are as follows:

- improves process understanding by proactive identification of failure modes (hazards), and managing the identified risks as early on in the product lifecycle as possible;

Risk Management Applications in Pharmaceutical and Biopharmaceutical Manufacturing, First Edition. Edited by A. Hamid Mollah, Mike Long, and Harold S. Baseman.
© 2013 John Wiley & Sons, Inc. Published 2013 by John Wiley & Sons, Inc.

TABLE 8.1 Traditional versus Science and Risk-Based Lifecycle Approach

Traditional Approach	Science and Risk-Based Lifecycle Approach
• One time activity focuses on final product testing • Demonstrates process based on current understanding • Process runs at established set points	• Facilitates increased process understanding • Proactive and iterative activities throughout the process lifecycle • Process monitoring is tied to process validation through the control strategy
• Arbitrarily selecting three batches for initial validation (e.g., qualification batches) • Selecting three batches typically for process changes	• Uses scientific rationale to determine number of batches/amount of data required • Determines scope of validation by analyzing data and significance of change • Uses statistical tools during development, initial validation, and process monitoring • Considers all potential risk factors in designing process validation studies • Prioritizes critical aspects and reduces effort on aspects that are not important
• Study design is based on template and/or existing practice • Deviation is seen as a setback	• Leverages historical knowledge and experience • Determines extent of validation testing/sampling • Analyzes validation data and determine controls • Deviation may provide an opportunity to increase process understanding and avoid failures during commercial manufacturing

- enables science-, data-, and risk-based discussions by a team of subject matter experts (SMEs) so that decisions and outcomes are sound and robust;
- ensures that high risk, critical aspects of the process are well understood by appropriately designed studies;
- reduces product and process failures;
- saves cost and time by focusing on the pertinent components of process and establishing priorities that separate essentials from "nice to haves."

The risk-based approach does not mean doing less, but doing the right amount at the right time and avoiding non-value-added activities. While equipment and systems are qualified before the start of qualification (also known as validation, demonstration, and consistency) batches, commercial-scale PV, and cleaning validation (CV) are performed during the qualification campaign. There are many similarities between PV and CV: both include a process development phase and have regulatory requirements to be qualified at the commercial scale, both have potential critical process parameters (CPPs), utilize a continued verification phase during commercial manufacturing, and are a significant aspect in establishing product quality. In addition, one major concern addressed with CV is the impact to product quality because of contamination or cross-contamination from other products and processes. Because of the similarities, this chapter discusses quality risk management (QRM) application in PV, and also describes relevant aspects of CV. The chapter is not meant to provide a detailed description on how to perform PV and CV. Instead, it discusses utilization of a risk-based approach to develop and implement a sound, efficient validation program. A risk-based approach can utilize various formal and informal risk assessment (RA) tools.

8.2 REGULATORY GUIDANCE FOR QRM IN PROCESS VALIDATION

QRM has been described in various recent regulatory guidances for several aspects of PV, such as the following:

Lifecycle approach
- QRM can be used at different stages during product and process development and manufacturing (e.g., risk analyses and functional relationships linking material attributes and process parameters to product critical quality attributes (CQAs)) [5].

Extent of validation
- An RA approach should be used to determine the scope and extent of validation and revalidation [2,6], including the sampling, testing, and amount of data required.

Critical quality attributes and CPPs
- QRM should be used to prioritize or rank the list of potential critical quality attributes (pCQAs and CPPs for subsequent evaluation and validation studies. An iterative process of QRM and experimentation can identify relevant CQAs and assess the extent their variation impacts product quality [5,7]. RAs and experimentation can be used to establish relationships between CQAs and CPPs in the manufacturing process. On the basis of these relationships, a control strategy can be designed to demonstrate that a product of uniform quality is produced consistently. Material attributes that may have an impact on product CQAs should also be evaluated [5].

Design of experiments
- RA tools should be used to screen potential variables for design of experiments (DoE) studies in process development and characterization to minimize the total number of experiments needed while maximizing knowledge gained [4].

Sampling plans and statistical confidence levels
- Risk analysis should be used to determine the confidence intervals (e.g., 80, 90, 95, 99, 99.5%) to be used in determining sampling and acceptance criteria [4], particularly for final dosage forms.

Training
- Risk assessors should be qualified and trained [6].

8.3 TYPICAL QRM TOOLS USED

QRM is a good way to proactively analyze a process to improve the process understanding needed for PV and CV [8–10]. Common QRM tools used in PV, process transfer, and CV can range from general qualitative-risk-based approaches (e.g., checklist, decision tree, risk ranking, and filtering, technical/scientific rationale based on historical data, prior experience, etc.) to specific risk scoring methodologies (e.g., preliminary hazard analysis (PHA), failure modes and effects analysis (FMEA), hazard analysis and critical control point (HACCP), etc.). Refer to Chapter 4 for details on various RA tools. Per ICH Q9, risk is defined as "the combination of the probability of occurrence of harm, if at all possible, and severity of that harm" [11]. Risk management should focus as much as possible on reducing the probability of occurrence, reducing the severity of harm, if at all possible, and fast detection capabilities. Some of the scoring RA tools involve the use of detection as a third component, in addition to severity and probability of occurrence. The detection component (although it is not a part of the ICH definition of "risk") factors in the ability to recognize a risk and manage it appropriately before it can cause harm, the detection component becomes especially important in risk management if the risk cannot be managed by further reduction of severity or probability of occurrence of harm.

NOTE: Probability of occurrence, frequency, and likelihood are used interchangeably in this chapter.

The detection component plays an important role in PV because of the nature of the drug inspection and quality review processes. For example, the ability to detect or monitor the drift in a CPP and CQA before it approaches a limit could be a factor in the evaluation. The degree of criticality aids in determining the amount of control and/or detectability of the CPP and CQA. If the risk can be found and the defects removed before they impact product quality and/or reach the user, then there is minimal risk to patient safety.

- Some techniques do not use detection [11]. In these two-component risk analysis, severity and probably are used. The degree of probability of a hazard or failure mode occurring and causing harm will be dependent on the degree of an existing detectability. However, the ability to detect does not change probability of occurrence, but it can help mitigate the overall risk to an acceptable level if the level of detection is sufficient. For example, with a known sensor or alarm in place, the probability of an OOS causing harm could be assessed as very unlikely.

GENERAL CONSIDERATIONS OF QUALITY RISK MANAGEMENT   183

- The use of detection in an overall risk score should increase as the project moves from development into the validation or process performance qualification (PPQ) stage. Detections are controls and the evaluation of the effectiveness of these controls is an essential part of validating a system. Thus, the value of using detection could be determined on a case-by-case basis.

Appropriate statistical methods should be applied throughout the PV program as a means of ensuring the robustness of the experimental design and resulting data, as well as supporting the application of QRM [12,13]. The statistical approach and methods, including justification and rationale, should be documented. Potential application of statistical methods includes the following:

- setting of acceptance criteria;
- design of testing and sampling plans;
- data analysis and trending; and
- process robustness and capability analysis.

8.4 GENERAL CONSIDERATIONS OF QRM IN PROCESS VALIDATION

A number of factors such as manufacturing equipment, raw materials, and processing conditions are likely to impact product quality. Assurance of product quality is derived from product and process design, adequate control of input process parameters, and testing of in-process and finished product samples. Each step of the manufacturing process must be controlled to maximize the probability that the finished product meets all quality and process requirements. The validation project plan (VPP) or master plan should document a complete list of PV studies required for product/process licensure. Note that any study that supports process parameter ranges in the license application and/or master batch record is identified as a PV study that includes development, characterization, and validation studies. The plan may incorporate a risk-based approach to rank activities based on risk and/or criticality and frequency of the activity. In addition, this will help identify an appropriate level of effort and timing of PV activities.

The following are some factors that may lead to unsuccessful PV; this is not an all inclusive list.

1. Inadequate risk management, inappropriate RA tool selection, lack of involvement of the SMEs, inadequate identification of hazards, inaccurate evaluation of risks, or insufficient or lack of timely control of unacceptable risks.
2. Poorly written protocol execution cannot be completed as written or repeated by other operators.

3. Lack of rationale for protocol acceptance criteria such that it is not defendable to a regulatory agency.
4. Acceptance criteria are too restrictive and likely results in unnecessary failures or deviations.
5. Lack of process understanding. No development data available before validation (e.g., no supporting data on equipment capability), hence no assurance of successful validation protocol design and execution.
6. Process is not robust and validation sampling is not representative of manufacturing conditions, hence increased likelihood of finding problems during commercial manufacturing.
7. Small-scale studies not representative of commercial-scale manufacturing: operating ranges were not challenged or verified.
8. Sampling plan is not appropriate and/or assays are not suitable.
9. No statistical consideration for sampling plan.
10. Inadequate training on validation and sampling procedures.

A science- and risk-based approach may alleviate some of these potential sources of failures, and potential factors that should be considered are described in the following sections.

8.4.1 Process Steps

Every processing step does not pose the same level of risk; hence, validation efforts need to be proportional to the level of risk to product quality and patient safety. For example, leachables from upstream operations would be a lower risk to product quality when there are subsequent purification steps compared to downstream operations. Hence, vendor-generated data for leachables may be sufficient for an upstream product contract surfaces, but robust leachables/extractables studies are required for downstream equipment. Risks during physical processing of drug substance and drug product (e.g., mixing components, and compression) are different from risks associated with chemical and biological processing, where the active molecule is formed via chemical or biochemical reactions and subsequently purified.

8.4.2 Closeness to Patient

As the process moves from upstream to downstream steps in manufacturing, product quality and patient safety risks increase [14]. For example, bioburden contamination in a bioreactor could be detected easily and likely result in discarding the batch, and the problem could be resolved before starting the next product batch. On the other hand, contamination in the final product vial is unlikely to be detected by limited batch release testing and poses a greater risk to the patient. Leachables from product vial stoppers may cause an adverse reaction to the patient [15].

8.4.3 Level of Testing and Number of Studies

The study type and testing required for PV will depend on a number of factors, such as the processing goal, input and output parameters, amount and type of data needed to demonstrate process control, process variability, routine testing of in-process and final product, risk control strategy, and process monitoring.

The number of studies required to fully validate a process will also depend on the complexity of the manufacturing steps. The biologic drug substance manufacturing process is very complex; hence, a large number of studies are usually required for its PV [16,9]. Application of an appropriate RA tool such as risk ranking of the activities based on complexity, robustness, previous laboratory and pilot studies, and knowledge from similar products (e.g., platform technology for biologics) can be helpful in prioritization of risks. The decision to omit a study creates its own risk and should be assessed for risk of limited (or lack of) process knowledge and risks to product approval.

A sampling plan (i.e., number of samples and locations) for a validation study should be based on statistical consideration, process/equipment design, and/or potential worst-case location considerations. For example, samples in a filling process should include first and last vials in addition to other samples to account for variation at the start and end of processing. A mixing study needs to be performed in product pool tanks to ensure homogeneous conditions. For lyophilization processes, worst case location(s) should be established from temperature mapping and that location(s) must be part of the lyophilizer sampling plan. One key consideration during sampling is patient (or product quality) risk versus business risk. For example, a nonrepresentative sample is a potential patient/product quality risk in that it is not able to capture worst-case conditions. Bioburden sampling for CV in a noncontrolled area may pose a business risk for possible false failure (i.e., false positive).

8.4.4 Validation Approach

Grouping strategies (e.g., family, bracketing, worst-case, modular, and generic approach) may be justified through a scientific and risk-based approach to determine an appropriate level of testing. The modular approach used in biologics (also known as platform strategy) is the use of data from a study performed on a specific unit operation for one product to support the manufacturing process for a different product. A science- and risk-based approach related to the process parameters should be performed to apply the modular approach. To accomplish this, a set of scientific criteria should be developed to compare the process parameters for the process under development with the process parameters of previously validated product(s). A unit operation in the manufacturing process should meet the following conditions/requirements in order to apply modular validation.

- The unit operation is robust and comparable raw materials are used.

- The unit operation has been validated for other product(s) of the same type (e.g., monoclonal antibodies, water-soluble mixing), and comparable results are obtained.
- Process parameters of the new process fall within the limits of the previously validated processes.

8.5 CHALLENGES IN USING RISK-BASED VALIDATION

Owing to the increasing regulatory expectations, companies appear reluctant to implement new approaches that they are not fully familiar with and that may delay product approval. Before the FDA's 2002 initiative on risk-based approach "Pharmaceutical GMPs for the 21st Century," most firms used the traditional validation approach of selecting three batches, which has been generally acceptable and incurred little inspectional scrutiny. Risk-based approaches are currently well accepted by the regulatory agencies. The challenge is determining the appropriate amount of validation data to ensure product quality and patient safety as well as to deliver confidence to the company and regulators. If sound science and reasoning is used along with documented rationale, the confidence goal is attainable. However, the risk-based approach/strategy should not be an excuse to avoid or minimize validation; even if a risk is deemed acceptable with current controls and no new controls are required, the validation requirements still need to be assessed and focused.

One approach to achieve that goal is by using a team of SMEs and fully documenting the RAs and the scientific rationale used. In addition, it is essential for risks to be communicated to appropriate stakeholders and decision makers at various stages of the risk management process.

The following issues should be addressed when conducting formal RAs for PV in order to make them successful and acceptable to the regulatory authorities.

1. Minimizing subjectivity during risk ranking that can lead to uncertainty and bias in the results obtained. An unbiased facilitator who is trained in the application of the formal RA tool plays an important role in ensuring the success of the RA. Note that regulators may challenge the risk scoring and the outcomes of the RA if the risk scores are inadequately justified. Hence, the main focus should be on documenting the risks, controls, and any supporting rationale that are truly scientific, knowledge based, and/or data based.
2. Clarity of business versus GMP risks and patient versus compliance risks is essential.
3. Document GMP controls and the risk acceptance and risk control decisions made from the risk management exercise.
4. Robustness, flexibility, and documentation of the individual risk evaluations (assigning of numerical scales and values). Use of appropriate RA tools and training is essential.

5. Consider and document all the risk factors, not prematurely ruling out while identifying risks because they may appear negligible.
6. Ensure management support for risk management program, with adequate resources and investment at the beginning.
7. Have a process for ensuring communication of risks to appropriate stakeholders including decision makers at various stages of the risk management process.
8. Ensure that the RA establishes connections between process monitoring and the initial validation as appropriate.
9. Ensure RA documents are usable at multiple sites.
10. Maintain RA documents current.

8.6 LIFECYCLE APPROACH

With the issuance of the FDA guide on PV, the activities of PV should be viewed from a lifecycle approach [4]. The three applicable stages are process design (e.g., process development, characterization, and validation), performance qualification (e.g., equipment/utilities qualification and process performance qualification), and maintenance of validated state (e.g., process monitoring, change evaluation, and verification) (Table 8.2). Note that the FDA guide includes both system (equipment and utility) qualification and PV. (Refer to Chapter 7 for QRM application for system commissioning and qualification.) This chapter discusses process development, characterization and validation, process performance qualification (PPQ), and process monitoring as well as CV and cross-contamination risks. Examples of QRM application in this chapter encompass process design through continued process verification stages, including cleaning. PV studies add to the knowledge throughout the lifecycle.

8.7 PROCESS DESIGN

Process design, such as process development, scale-up and characterization, starts with the identification of the properties of a target molecule or new chemical entity. As the product moves through preclinical and clinical studies (Phase I to III), scale-up and manufacturing process optimization takes place. The process is finalized before qualification batches at the commercial scale (also known as PPQ batches). Before the start of PPQ, studies supporting commercial-scale manufacturing can be performed at any scale as long as it is representative of commercial-scale manufacturing. Scaled-down models are used in process development and characterization to increase process knowledge through increasing the number of conditions tested. Any planned, documented study that adds process knowledge and supports product licensure is considered part of PV in this chapter (e.g., development, characterization, validation, comparability, and compatibility).

TABLE 8.2 Three Stages of Process Validation Lifecycle

Lifecycle Stage	Implementation Elements	ICH Q10 [17]
Stage 1: process design	• Formulation, process development, and process understanding (primarily R&D activity) • DoE, process modeling • Process characterization • Initial process ranges (averages and variations) • Determination of process variables (e.g., CQAs, CPPs, KPIs) • Initial control strategy established	Knowledge Management/ Quality Risk Management
Stage 2: process qualification a) Design of facility; qualification of equipment and utilities b) Process performance qualification	• Equipment and utilities qualification • Systems verification (suitability) • Conformance phase or initial manufacturing • Scalability and process performance • Primarily manufacturing-scale activity • Confirm/verify data at commercial scale	
Stage 3: continued process verification	• Product and process monitoring systems (e.g., ongoing monitoring, trending, continuous quality verification, annual product review) • Continuous improvement	

PROCESS DESIGN 189

Process development and characterization is performed to establish a reliable manufacturing process and to evaluate the effects of selected process parameters and critical material attributes (CMAs) on process performance and product quality. Quality by design (QbD) principles should be used when designing a process. QbD is a risk- and science-based concept for the development of products and processes which recognize that quality cannot be tested into products but should be built in by proper design. QbD principles increase product and process understanding through the application of RAs and innovative tools and methodologies [18]. QbD results in robust manufacturing processes with boundaries that can consistently deliver the desired product quality. The basis for the selection of parameters or evaluation in process studies should include scientific rationale and the use of risk-based approaches where relevant. The highest risk characteristics (i.e., parameters with the greatest likelihood of product compromise) should be evaluated first. All CPPs and CMAs for that unit operation or process step need to be evaluated in PV studies. Screening experiments, fractional factorial design, and response surface design studies can be used to determine CPPs [19,17].

Process development studies may be performed at laboratory, pilot, or manufacturing scale. The cost of running experiments is exponentially higher at the commercial scale compared to small scale. On the other hand, scaled-down systems can be completed independent of manufacturing operations. The studies performed at the small scale must be representative of the manufacturing-scale process in appropriate conditions and rationale, including equipment type/scale, instrument calibration, and assay must be documented. Risk-based approaches could be used to determine study type (e.g., generic or product-specific), number of experiments, and scale. In the example given for roller compaction process of a solid oral dosage product, the relative importance of inputs into the DoE is based on an FMEA. Polymer concentration received the highest score and excipient particle size the lowest, as shown in Figure 8.1. Hence, polymer concentration, roller gap width, and roller gap force were subsequently evaluated in a DoE.

8.7.1 Risk-Based Study Design

Process input and output parameters must be identified for developing study plans/protocols. Knowledge from other similar products and processes may be leveraged to help design the study for a new product. At a minimum, the process unit operations established during development (studies) should be defined before the start of characterization or validation studies. pCQAs are identified using scientific and clinical data before the start of formal PV [20]. When assessing risks, all relevant data/information should be considered, such as product-specific process development data, process knowledge from similar products (i.e., modular data), manufacturing history (clinical and commercial), and scientific knowledge. Process mapping is performed to document all input and output parameters. Figure 8.2 shows the steps and considerations for PV in a risk-based approach. A new study is not required if the process parameter range does not pose any risk to product quality and patient safety. RAs can be used for study design, e.g., DOEs

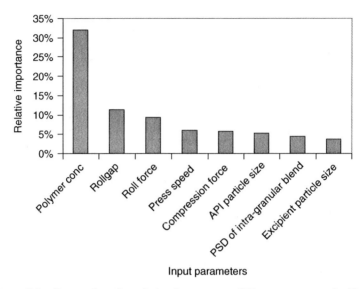

Figure 8.1 Pareto chart for relative importance of input parameters for DoE.

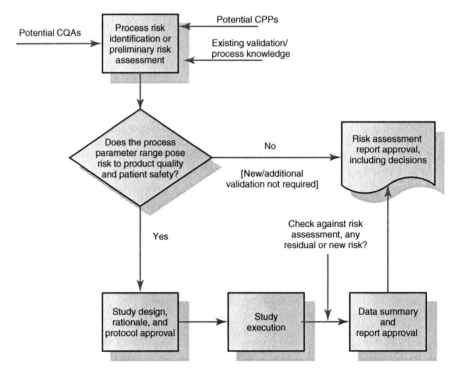

Figure 8.2 Incorporating risk assessment into study design.

and multivariate or univariate approaches [8]. Residual risk should be assessed and documented in the RA report.

The risk-based approach uses risk ranking to classify process variables based on their potential impact to CQAs, process performance, and possible interaction with other parameters. Process parameters can be divided into three groups: (i) parameters warranting multivariate evaluation; (ii) parameters whose ranges could be supported by univariate studies; and (iii) parameters that do not require new studies, but instead would employ ranges based on knowledge space or modular claims established from prior knowledge. Each parameter was assigned two rankings: one based on the potential impact to CQAs or other attributes (main effect) and the other based on the potential of interactions with other parameters (interaction effect). Main and interaction effects were multiplied to calculate the overall severity score. The severity score serves as the basis for identifying study types: multivariate studies (high scores), univariate studies (medium scores), or no further study needed (low scores).

The following factors should be considered when performing RA:

- level of existing knowledge on effect magnitudes and potential interactions;
- historical data and scientific rationale that may indicate minimal risk and justify that no additional studies are required;
- higher risk associated with impact on CQAs;
- potential of an unexpected interaction effect for a parameter with a high impact main effect.

Table 8.3 shows a template to score main and interaction effects. On the basis of the severity score (equal to main effect × interaction effect), studies can be divided into various categories as shown in Table 8.4.

8.7.2 Critical Process Parameters

CPP is defined as a process parameter whose variability has an impact on a CQA and therefore should be monitored or controlled to ensure the process produces the desired quality [5]. CQA is a physical, chemical, biological, or microbiological property or characteristic that should be within an appropriate limit, range, or distribution to ensure the desired product quality. A process parameter that

TABLE 8.3 Template for Effects and Risk Scoring

Process Parameter	Proposed Range		Main Effect Rank (M)	Interaction Effect Rank (I)	Severity Score (M×I)	Process Outputs Affected	Potential Interaction Partners	Rationale

TABLE 8.4 Severity Score and Study Strategy

Severity Score	Experimental Strategy
High	Multivariate study (DoE)
Medium	Univariate study
Low	No additional study required

influences important process performance outputs (e.g., titer, step yields or other key performance indicators), but does not influence product quality is called a key process parameter (KPP) by many biological firms [16]. Note that KPP could influence product quality if an extreme deviation in that parameter were to occur. Given here is an example of a risk-based approach for the identification of CPPs. It is recommended that if an RA is used, it should follow a defined work process (guidelines, expectations, milestones during development), which provides support and training for conducting appropriate RAs for selecting potential CPPs.

8.7.2.1 New Products For new products, an RA can be incorporated in the study design to determine whether there is a direct relationship between the process parameters and a CQA. Also, the use of pilot, scale-up and manufacturing-scale data will help determine if there are more CPPs.

8.7.2.2 Existing Products RAs and experimentation may need to be performed to understand the nature of the change and the impact of process parameters and material attributes on CQAs in the case of product/process improvement projects that are in commercial manufacturing (Stage 3), such as:

- improving an existing quality characteristic;
- reducing or eliminating complaint(s);
- resolving an inspection observation;
- instituting a more efficient process (e.g., continuous process);
- instituting a real-time verification or monitoring program; and
- instituting new technologies such as PAT or real-time release testing.

This would include refining the criticality continuum (i.e., complete range of scoring) in the initial relationships between CQAs and CPPs. On the basis of an enhanced understanding of these relationships, an initial control strategy can be designed or redesigned to demonstrate that a product of uniform quality is produced consistently throughout the manufacturing process.

8.7.2.3 Effect of Processing Steps Some of the steps of drug product manufacturing are notoriously more difficult than others. For example, blending a powder mixture containing a very low percentage of an active ingredient such

PROCESS DESIGN   193

as 1%–2% would be a critical step unless there is a downstream blending step that assures content uniformity, especially if the particle sizes of the ingredients are not well controlled. Mixing a sparingly soluble ingredient into a solution at nearly saturated concentrations is typically a critical step and parameters such as mixing speed, rate of addition, and solution temperature may be CPPs. In an RA for tablet manufacturing [21], several processing steps were viewed as high risk (scores of 4–5 on a 5-point scale): compression (tablet), granulation (wet and dry), mixing-blending, and pelleting. Measuring or weighing and primary packaging were viewed as lower risk (scores of 3 or less on a 5-point scale).

8.7.2.4 *Normal Operating Range and Proven Acceptable Range* For an existing product, comparing the normal operating range (NOR) to the proven acceptable range (PAR) should be considered when performing an RA of potential CPPs. The assessment should also include consideration of the interaction of potential parameters. The NOR is typically the range specified in the master batch record, whereas PAR is demonstrated during development or characterization/validation studies. An acceptable product is produced within the PAR range [5]. It is paramount to know the variability of both the control parameters and measurements at the target (set-point) and the limits. The variability of control parameters and measurements will determine the level of risk. The probability of exceeding the NOR/PAR limit and the consequence (i.e., severity) will determine the level of criticality. A comparison of the NOR and PAR will typically reveal one of the following general situations:

(I) *When PAR is unknown*: In this case, the PAR has not been identified or historical information does not provide substantiation of acceptable ranges broader than the NOR. It may be possible to establish the PAR from historical experience with the process (e.g., from investigations). Another possibility is to assume that the NOR and PAR are the same.

Figure 8.3 illustrates the case when NOR equals PAR and for two different distribution patterns around the set points, A and B. This example shows how control variability around the set-point is important. The likelihood or risk of exceeding the operating limit for scenario A is much less than scenario B based on the variability of the control from the set-point.

A higher risk of reaching the limit is apparent when the parameter's variability is greater and/or the set-point approaches the limit. For an existing process, if PAR is unknown and variability of the parameter is

a) Lower (i.e., scenario A in Figure 8.3) compared to the limit then it is safe to assume that PAR is equal to NOR. There is little value in determining the PAR value as the risk or probability is extremely low of ever reaching the approved limit.

b) Higher and/or close to the limit (i.e., scenario B in Figure 8.3), the likelihood of reaching the limit becomes high and the parameter is considered a CPP. In this case, it would be value added in determining the PAR, which may be beyond the NOR.

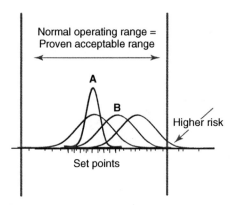

Figure 8.3 When PAR is unknown and considered equal to NOR. (*See insert for color representation of the figure.*)

(II) *When PAR is known*: Compare NOR to PAR and size of the Δ (difference between NOR and PAR limits) (Figure 8.4). It is also important to know the variability of the parameter within the NOR near the limits (e.g., normal distributions). This will impact the likelihood or probability of maintaining control within the NOR. When the risk of exceeding the PAR is negligible or none, knowledge of the edge of failure (EOF) is insignificant. The EOF could be near or far from PAR, but would be inconsequential as it would never be reached. Also, it is possible to tighten the NOR based on historical results and then ascertain the level of risk.

- NOR is a significantly smaller range than PAR (as depicted in Figure 8.4, scenario A). The value of Δ is relatively large and/or the variation of the

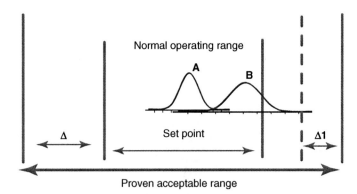

Figure 8.4 Determining level of risk-based on the spread between PAR and NOR, variability of the operating parameter, and the location of the set-point. (*See insert for color representation of the figure.*)

parameter at the limit is relatively small. One can conclude that such parameters are not critical to product quality as the magnitude of Δ minimizes the risk of exceeding PAR.
- NOR is close to one or both limits established by PAR and the value of Δ is relatively small or the variability is large. In these cases, the parameter is a potential CPP as the likelihood and risk of exceeding the PAR is higher. In Figure 8.4, the higher variability of the parameter in scenario B compared to A along with the smaller difference Δ (i.e., the distance between the dashed line and solid line limits of PAR, $\Delta 1$) would increase the risk and thus lead to categorizing it as a CPP.

Using RA of process parameters and selection of CPPs is consistent with strategies used in industry publications [16,22–24]. The risk analysis used to select the CPPs may be influenced by the ability of the equipment and supporting systems to control process variables (e.g., temperature, pressure, agitation, compression force, etc.). The equipment's capability to control process parameters within defined limits is typically demonstrated by commissioning and qualification of the process equipment. Figure 8.5 shows a decision tree for evaluation of CPPs and the role of RAs. A risk-based approach for CPPs is applicable in both QbD and traditional approaches, but is slightly more extensive in the QbD approach.

Other aspects of process control that are not operational parameters (e.g., output of an intermediate step) should be evaluated as part of the RA. These may influence equipment qualification, method validation, and/or additional studies that may be needed because of their importance to product quality, for example, a performance parameter such as an in-process control (IPC) that may impact a product CQA:

- endpoint of reaction of an API process;
- blend homogeneity of a drug product;
- level of insoluble particulate matter after filtration;
- environmental condition (e.g., temperature or humidity);
- equipment set points and configurations (nonoperational parameters but they impact CQA);
- processing time limits, if the probable adverse consequence of exceeding a time limit results in a risk of unacceptable final product quality, such as the following:
  - permitting an excessive reaction time in a synthetic API process when this allows formation of an unacceptable amount of a process impurity not adequately controlled by other means;
  - delay in the processing of a mixture;
  - other hold time limits that should be identified to understand process capabilities.

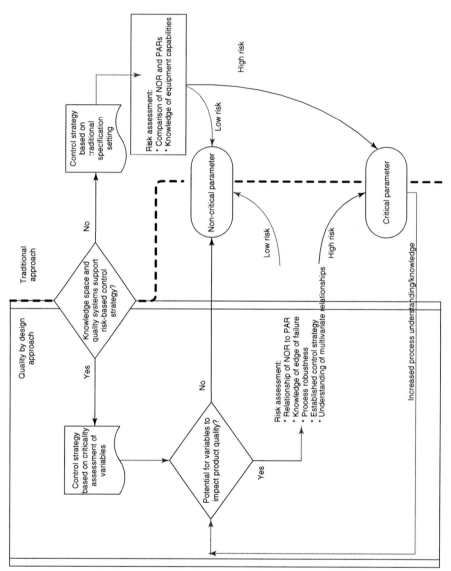

Figure 8.5 Decision tree for evaluation of CPPs using risk assessments.

PROCESS DESIGN 197

8.7.3 Example 1: Defining Controls for Tablet Compression/Coating Process

In this example, risk scoring is used to select controls for tablet compression/coating process. In the beginning, scales for severity, frequency of parameter being out of range, and action thresholds are defined (Tables 8.5 and 8.6).

8.7.3.1 Determination of Overall Risk The overall risk is determined by a risk index (RI). The RI is calculated as follows: RI = Severity (S) × Frequency (F). Using the risk scoring described, an RI threshold was established to classify a parameter as a CPP (Table 8.6).

Thresholds for action (or for determining criticality) based on risk scoring should be agreed on by reviewers before performing RA. An example of action thresholds based on this scoring strategy is shown here. Justification of values assigned to severity and frequency for each evaluated risk should be provided in the RA documentation. Table 8.7 shows an example of the output of the RA.

TABLE 8.5 Four Level Scales of Severity and Frequency for a Process Parameter

Severity (S)	Definition	Interpretation
8	High	Predicted to cause severe impact to quality
4	Moderate	Predicted to cause significant impact to quality
2	Low	Predicted to cause minor impact on quality
1	None	Predicted to have no impact on quality of product

Frequency (F) or Probability of Occurrence	Definition	Interpretation
10	High	Problem likely to occur frequently (expected or has occurred multiple times in the past)
7	Moderate	Problem has occurred in the past and can be expected to reoccur if action is not taken to correct or prevent
3	Low	Problem unlikely to occur but is possible
1	Remote	Highly unlikely to occur (probability of failure occurring is so low that it can be assumed that the failure will not reoccur)

TABLE 8.6 Three Action Thresholds

Risk category	Risk Index, Maximum 80	Interpretation
Intolerable region	RI ≥40: intolerable risk	The risk is so severe that it must be mitigated and/or controlled
Broadly unacceptable level of risk	RI <40 and >24: risk is tolerable only if reduction is impractical (as low as reasonably practical-ALARP), or costs of risk reduction are disproportionate to the benefit	Parameters with risk in this region are potential CPPs and should be evaluated bearing in mind the benefits of accepting the risk and the costs of further risk reduction. Risk acceptance is based on a case-by-case basis
Broadly acceptable level of risk	RI ≤ 24: negligible risk	The risk is negligible (non-CPP). Further risk reduction is not necessary, however for business reasons, or decision makers may decide to reduce the risk further

8.7.4 Example 2: Mixing Study for Solutions and Product Pool

Biopharmaceutical manufacturing processes typically involve pooling of different liquids or dissolution of powders into solution. The mixing process must assure consistency and homogeneity without product degradation or other detriment to product quality characteristics. The appropriate input parameters (e.g., temperature, mixing speed, time) and output (e.g., density, conductivity, clarity of solution, pH, and product degradation) parameters for the solution should be identified and measured during the mixing study. The mixing procedures that may impact CQAs should be evaluated extensively to ensure the consistency of manufacturing performance. QRM can be used to design and prioritize mixing studies. Using science- and risk-based principles, a family approach (using the hardest to mix solutions) or a modular approach (applying mixing times obtained from other processes) can be utilized when justified. Relevant characteristics and variables should be included in determining worst-case scenarios such as solubility [25].

Factors that may impact solution mixing performance include tank design and size, impeller type and position, liquid volume, temperature, mixing speed, and time. On the basis of the expected outcome, mixing studies can be divided into two types: 1) mixing that is likely to cause no product impact (e.g., salt

TABLE 8.7 Example of Risk Assessment for Cause/Effect Process Parameter and Identification of CPPs

Parameter/ Control	Acceptable Range	Failure Mode	Effect	Cause	Controls	Severity (S)	Frequency (F)	RI (S*F)	Recommended Risk Control Actions	Decision
Press speed	30–70 rpm	Out-of-range speeds (high or low)	Can give nonuniform tablet weights, thicknesses, friability, and hardness, impacting product potency, and dissolution	Machine speed controlled by operator (manual)	1) Defining press set-point and monitoring of press and tablets, 2) Press verification or qualification, 3) Tablet tests at various press speeds	8	3	24	Perform continuous monitoring, alarms	Not CPP
Feeder speed	60–100 rpm	Out of range (high or low)	Impacts tablet weight, but no impact to product quality	Machine speed controlled by operator (manual)	1) Feeder verification or qualification, 2) Tablet tests at various feeder speeds, 3) Defining feeder set-point and monitoring of feeder and tablets	2	7	14	Perform continuous monitoring of tablet characteristics with adjustments made to ensure required product characteristics are met	Not CPP

(*continued*)

TABLE 8.7 (*Continued*)

Parameter/ Control	Acceptable Range	Failure Mode	Effect	Cause	Controls	Severity (S)	Frequency (F)	RI (S*F)	Recommended Risk Control Actions	Decision
Overload setting	Maximum 40 K Newtons (NOR)	Higher	Potential impact to tablet weight, hardness, thickness, friability	Machine set-up error	Compare tooling specs versus press overload capability and controls; Manuf. Batch Record check	8	3	24	Determine specific maximum force allowed to avoid tooling damage; verify proper machine set-up	Not CPP
Spray rate	Total 380–420 g/min for all guns	Out of range (high or low)	High spray rate may impact tablet appearance and dissolution. Low spray rate extends coating process times but not critical to product quality.	Spray rate governed by automated controls with limit alarms	Spray gun nozzle size, and number of guns placement defined and controlled	8	7	56	Rate periodically verified within range by operator; confirm control strategy and regulatory commitment	CPP

| Pan load weight | 260–340 kg | Out of range (high or low) | High load weight may exceed film-coating equipment working capacity and process performance. Low load weight may cause nonuniform coating to tablets. | Impact of Pan load charge is unknown | monitoring pan load versus film-coating acceptability | 8 | 3 | 24 | Adjust parameters for each pan load to ensure appropriate process performance | Not CPP |

solution); 2) mixing with potential product impact (e.g., shear-sensitive product). Undermixing (i.e., inadequate) would be a concern in both cases because of a nonhomogeneous condition that may cause a nonrepresentative assay sample, quality, or yield impact; however, overmixing (i.e., excessive) should be prevented for shear-sensitive products because of potential product degradation or physical breakdown. The factors that impact mixing and could contribute to product impact because of overmixing are power per volume, mixing time, agitation style, temperature, product type, and concentration.

The first step in a risk-based approach is to identify mixing equipment, size and configuration, and mixer type by the process steps. Risk analysis of the mixing process parameters (i.e., input variables) and their impact on product quality would then help design the mixing study to determine appropriate mixing parameters for manufacturing operations. Figure 8.6 shows mixing behavior (i.e., homogeneity) in a 15,000 liter tank for solutions, where the sample was collected from a sample port. As expected, concentration varied initially with time, but a homogeneous condition is achieved within 6 min of time for working volumes of 2001–14,997 liters (Fig. 8.6). Sampling at various locations would have revealed additional data on mixing dynamics at the prehomogeneous state; however, sampling from the top, middle, and bottom of the tank is not required to determine a homogenous condition. Solution samples from any location should provide the same results (except measurement variability) once a homogeneous condition is reached. As shown in Figure 8.6, a homogeneous condition is independent of sample locations; hence, there is no need to create a special sampling device to

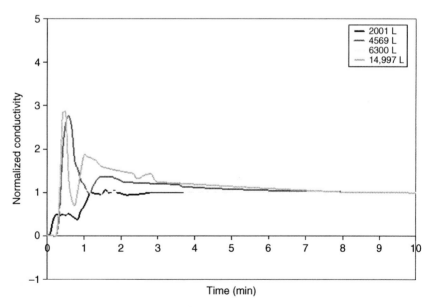

Figure 8.6 Mixing in a 15,000 L tank. (*See insert for color representation of the figure.*)

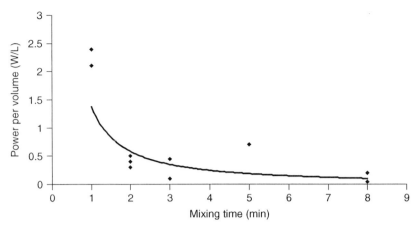

Figure 8.7 Mixing time as a function of power/volume.

sample from the top, middle, and bottom for liquid mixing. However, samples at various locations may be required for mixing with potential product impact or nonsteady-state condition.

Figure 8.7 shows a relationship between power/volume and mixing time to establish mixing time and speed for a mixing process. An appropriate mixing speed can be selected using the data on equipment capability and manufacturing needs, including product characteristics (e.g., physical processing, shear-sensitive solution). Higher speed and shorter time is appropriate for physical mixing, whereas low speed and longer time would be desirable for shear-sensitive mixing. Table 8.8 shows the RA of mixing parameters and their impacts for product pool mixing in biologics manufacturing to develop a small-scale study. Historical data showed that there was a decrease in filterability because of protein aggregates formation or precipitation.

8.7.5 Example 3: Hold Times Study for Solutions and Product Pool

The chemical/biochemical stability of process solutions and intermediates should be studied for storage time and condition. The validation approach should include physico–chemical/biochemical stability and contamination control (e.g., bioburden, and endotoxin). In-process hold stability studies should be performed to demonstrate product stability throughout specified in-process hold times and manufacturing conditions (e.g., temperature, pH, etc.).

Contamination control and stability studies can be combined into one study or performed separately. Stability of the solutions is typically performed using scaled-down models, and assays include the measurement of chemical attributes, such as pH, conductivity, or component concentration, over time. The scaled-down container must be made of identical material as used in the

TABLE 8.8 Risk Analysis of Mixing Parameters to Design Small-Scale Studies

Factor that may Impact Mixing	Impact or Effect on Product Quality	Approach for Study Design
Type of mixing equipment	1) Filterability decreases with long exposure to bottom mounted magnetically driven mixers, 2) little or no impact with top mounted mixers or bottom mounted shaft driver mixer	Develop scale-down model for both top mounted shaft driver and bottom mounted magnetically driven mixers
Power/volume	As agitation rate and power/volume increases, filterability decreases	Use worst-case power/volume target for each type of mixer
Time	As mixing time increases, filterability decreases	Evaluate the impact of holding up to and past validated hold time at temperature range
Temperature	As temperature increases, filterability decreases	Evaluate the impact of temperature range at holding up to and past validated hold time
Product pool	As concentration increases, downstream filterability decreases	Use pool with highest product concentration (e.g., diluted ultra filtration/diafiltration (UF/DF) pool)
	Known vulnerability to precipitation caused by pH and temperature changes	Test pools that have a tendency toward precipitation

commercial process and must represent a worst-case scenario with respect to the solution–container and the air–liquid interfaces.

In a traditional hold time study, the solution is held at a desired condition for a specific duration in triplicate and samples are collected at the beginning and end. The validated hold time is then established on the basis of the end points meeting acceptance criteria, such as 5 hours in Figure 8.8. This study does not provide any information about the behavior of the solution after the maximum time of 5 h; hence, an additional study is required for instances where any time-validated hold time is exceeded during manufacturing.

In a risk-based hold time study, the impact of hold on process parameters (e.g., time, temperature, concentration, etc.) is analyzed to determine the potential effect on product or solution characteristics. Then the solution is held at a worst-case condition (e.g., temperature and mixing speed) for an extended period of time (i.e., beyond 5 h in this example). Samples are taken at various time points and a maximum hold time without significant change in output characteristics is considered the maximum validated hold time (Fig. 8.9). Note that the decision

PROCESS DESIGN   205

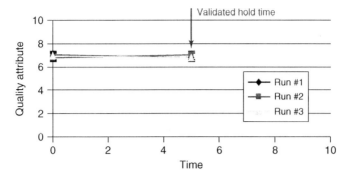

Figure 8.8 Traditional hold time validation (performing three experiments). (*See insert for color representation of the figure.*)

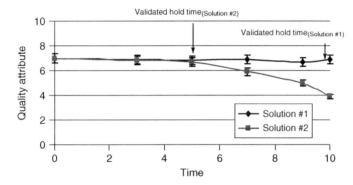

Figure 8.9 Risk-based hold time study (samples at various time points).

on testing of total number of samples may be taken on the basis of the results of end point samples. More data points in this case would help with making batch release decisions when the hold time exceeds the NOR and for determining the rate and trend with time. For example, the product should be rejected when a time of 5 h is exceeded for solution #2 as a significant measurable decline occurred after 5 h. However, the hold time may be extended for solution #1 for up to 10 h based on solution stability (i.e., lack of decline in the trend line). This risk-based approach provides more information about the solution behavior with time compared to the traditional method.

8.7.6 Risk Prioritization in Large-Scale Experiments

Running commercial-scale experiments is expensive and requires considerable plant time and resources. RAs can be used to identify the processing parameters that pose the highest risk to product quality [26]. Given here are the steps for risk prioritization for commercial-scale experimental design at sterile fill/finish and packaging operations.

Step 1: Gather Information (e.g., Process Mapping, Flow Charts, Historical)
Perform process mapping and list all processing steps, operating parameters, and IPCs.

Step 2: Identify Potential Sources of Harm, Hazards, or Parameters to be Ranked
Identify all potential risk factors, failure modes, hazards, or parameters in the system. At this step, it is important not to judge or evaluate the risk but to brainstorm all potential hazards regardless of severity or frequency of occurrence. Potential risks for fill/finish operations include patient safety; product degradation and misbranding; microbial, endotoxin, viral, and chemical contaminations; product quality and stability; vial integrity; and appearance.

Step 3: Define Scales (Severity, Occurrence, Detection) and Thresholds of the Ranking Evaluate each risk factor, hazard, or parameter against the predetermined scales and perform the ranking. FMEA or any other RA tool can be used to determine risk ranking. Severity scales would be the consequences to product quality or patient safety if the hazard were to result in harm. What are the failure modes/hazards and their effects, pathways/sources/controls, and detection mechanisms? Then, prioritize the risk based on the risk scores.

Step 4: Identify Area that Poses Highest Risk From the risk ranking data identify steps or areas with highest risk scores using a Pareto analysis (Fig. 8.10). Scores reflect cumulative risk profile per hazard/risk type to ensure controls are aligned accordingly. Similarly, ranking can be performed to identify manufacturing steps with highest risk, or alternatively an RA tool such as HACCP can be used to identify, rank, monitor, and control the critical points in the manufacturing process. Note: Critical control points (CCPs) may not be the same as CPPs.

Step 5: Experiment Design and Execution A DoE should be performed to determine design space and control range for CPPs. For parameters that pose minimum

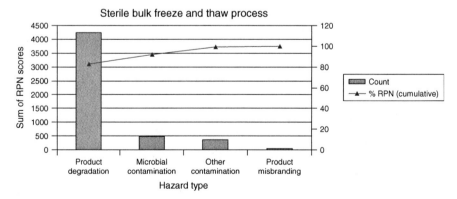

Figure 8.10 Pareto chart of hazard analysis for bulk thaw and freeze process.

PROCESS DESIGN 207

risk, commercial-scale experiments are not required. Here are the steps for experimental design and execution.

1. Collect input from SME(s) to identify input parameters that may impact product quality attributes.
2. Identify potential interactions between two or more parameters.
3. Define preliminary operating ranges for each mechanical parameter.
4. Design and execute specific experiments (e.g., DoEs) to study parameter interactions and their impact on product quality.
5. Use the DoE results to define key mechanical parameter operating windows for optimal equipment set points.

Benefits of commercial-scale mechanical DoE study include the following:

- improved understanding of production processes and equipment through hands-on experimentation;
- identification of high risk activities;
- definition of optimal process windows (parameter operating ranges);
- discovery of equipment issues/mismatches;
- more reliable production output; and
- more consistent product quality.

Step 6: Risk Control Strategy and its Effectiveness Develop and document a risk control strategy. It should include a summary of the results obtained from the RA, the actions that were or need to be taken to control the identified risks, any risk acceptance decisions, and the estimation of the residual risk remaining in the product/process, together with the rationale as to why this level of risk is acceptable.

The effectiveness of the risk reduction strategies developed and implemented as a result of the QRM process needs to be monitored on an ongoing basis following its implementation. This monitoring can take the form of changes to existing production documentation to ensure the appropriate testing implemented becomes part of the standard manufacturing process for the product. The following tools, data and information can be used to determine the effectiveness of a control strategy:

- inputs process variables—statistical process control (SPC) and process capability indices for critical input parameters;
- outputs (quality control assay results)—SPC and process capability indices for critical and key quality attributes; and
- discrepancies, CAPA (corrective and preventive action), deviations, and product complaints.

8.8 PROCESS PERFORMANCE QUALIFICATION

Data must exist to demonstrate that the commercial-scale manufacturing process is reproducible, can be maintained within established parameters, and consistently produces products that meet specifications. This could be done by actual commercial-scale batches or by process knowledge and data from PV studies, which must be statistically and scientifically sound. Successful PPQ is reflected by the level of robustness of the manufacturing formula, process recipe and SOPs. Equipment and supporting systems to be used for the PPQ need to be qualified in accordance with GMPs [4,6,27]. In most cases, PPQ will have a higher level of sampling, additional testing, and greater scrutiny of process performance. Consistency beyond the initial qualification batches is demonstrated through continued monitoring of a subset of the original parameters evaluated during PPQ.

In the past, it has been generally considered acceptable for three consecutive batches under routine operating conditions to constitute initial validation of the process or changes to an existing process. This became the norm in the industry with limited or no inspectional scrutiny from the agencies. Under the new concept of using a science- and risk-based approach, when selecting the number of batches in PPQ, appropriate use of scientific data, risk management, and statistical tools should be considered. Three batches should not be selected without taking these considerations into account. A FDA colleague has stated that between 2 and 30 would be a range probably used to satisfy the FDA's new paradigm [28]. Note that international guidances currently maintain an approach that is less risk based and typically follow a three-batch approach [6,7].

Robustness of contamination controls that prevent bioburden, endotoxin, or foreign contaminants into a process must be demonstrated. This is accomplished through a combination of process controls such as raw material specifications and testing, equipment cleaning and sanitization, operational control, and in-process monitoring during the production of drug substances and drug products. Cleaning, sanitization, sampling, sterilization, and depyrogenation processes should be qualified or validated to limit the introduction of bioburden into the process. The manufacture of sterile drug products is subject to specific validation requirements, depending on whether the product is terminally sterilized or aseptically processed, in order to minimize the risk of microbiological and pyrogen contamination. Aseptic processing operations, such as sterile sampling, filtration, filling, and lyophilization are considered by regulators as higher risk processes. For mammalian cell culture processes, viral clearance studies must be performed to assess effectiveness of the process steps in inactivating/removing viruses and to measure quantitatively the overall level of virus clearance by the process.

The scope and extent of validation-related testing of qualification batches may exceed that of routine commercial manufacturing. In addition to standard specification tests, comparability studies may include those additional tests needed to document that the processes operate comparably and produce comparable products. Evaluation of historical data from multiple commercial manufacturing sites may be included in this assessment.

8.8.1 Technology Transfer

The goal of technology transfer activities is to transfer product and process knowledge between development and manufacturing, and within or between manufacturing sites to achieve product realization [17]. The transfer must demonstrate comparability of the product and process between the donor and recipient sites. Changes to the recipient site facility, equipment, or automation systems, and process should utilize quality approved change control processes during all phases of the transfer. Verification of GMP readiness of the equipment and systems is required before manufacturing of commercial-scale batches. The process transfer should follow a structured process with defined milestones as described in a master transfer plan (MTP). The level of detail in the MTP documentation should be commensurate with the level of complexity of the transfer and should include the following elements: objective, scope, governance structure (i.e., management and decision-making responsibilities), transfer responsibilities, transfer acceptance criteria, transfer deliverables, and target dates.

Transfer deliverables should include (but not be limited to) the following:

- QC methods transfer report;
- raw materials transfer report;
- equipment/systems assessment, including materials of construction and capabilities;
- risk assessments;
- recipient site manufacturing batch records and SOPs;
- validation summary reports (e.g., equipment, automation, cleaning, process); and
- MTP summary report confirming that acceptance criteria for the transfer have been met and the process has been successfully transferred.

There may be multiple RAs associated with a technology transfer (e.g., RAs for process changes, facility and equipment changes, method transfer, multiproduct operations, etc.). Therefore, it is important that QRM activities are adequately planned; this QRM planning can be incorporated into the MTP. As noted earlier, the risk-based approach may be qualitative (e.g., technical rationale, checklist, decision tree, etc.) or may involve the use of specific scoring tools (e.g., PHA, FMEA, etc.). When risk scoring methodologies are used, it is important to ensure consistency in the scoring and risk evaluation (e.g., risk priority number (RPN) matrix) criteria used for all transfers. Given here are some risk management activities that may be useful in a typical technology transfer. QRM application for multiproduct operations in order to avoid cross-contaminations is shown in Example 4.

Multiproduct Facility Risk Assessment This RA should be performed during the introduction of a new product to assess and review any potential cross-contamination or nonroutine risks of mix-up between existing products at the facility and the new product. An example of a nonroutine risk is loading of an incorrect automation recipe as the operator may have recipes for multiple products to select from. On review of these risks, appropriate multiproduct controls should be identified and implemented, as determined through the RA. If no new multiproduct risks are identified with the introduction of the new product, the existing multiproduct RA does not require further update.

Process Change Risk Assessment If any process changes are being made as part of the transfer, an assessment should be performed to ensure that the process changes do not introduce any unacceptable quality risks and that proper controls are in place for the manufacturing process. A PHA, FMEA or FTA may be used to evaluate the hazards or failure modes associated with differences in the manufacturing process between the donor and recipient site.

Facility/System/Equipment Modifications for Process Fit The risks associated with modifications to existing facility, systems, or introduction of new equipment to ensure process fit at the recipient site should be assessed and managed adequately. Where RAs exist for the impacted system/equipment, they should be reviewed and updated for any new risks associated with the change being made. This RA should identify if additional controls or commissioning/qualification activities are needed to support the implementation of that change.

Analytical Methods Transfer Analytical methods are transferred to a receiving site to support in-process and release testing. The methods transfer should be performed in accordance with a methods transfer protocol. The risks associated with the methods transfer can be assessed through a technical rationale or by the use of an RA tool similar to a PHA. The results of the RA should help determine if additional controls are required for transferring and executing the method at the recipient site.

Risk Control After completion of the RA, decisions need to be taken on further risk reduction or risk acceptance. It is essential to identify the decision makers (person with the competence and authority to make appropriate and timely QRM decisions and document the risk control decisions including the associated rationale where needed [11].

When a risk scoring methodology is used, the following approach is an example for determining the order of risk control actions (NOTE: Risk control includes risk reduction and risk acceptance).

1. High risk scores (RIs or RPNs): address those with highest severity first for risks with identical RIs or RPNs.
2. High severity: consider controlling items of extreme severity even if total RI or RPN is not very high.
3. High occurrence: recurring problems indicate improvement opportunities.
4. Low detection: improving detection can allow the risk to be noticed when it occurs and managed before it impacts product quality or patient safety.
5. Risk acceptance: if it is not practicable to reduce the risk further the rationale for why the risk is being accepted (e.g., a risk benefit analysis) and the persons accountable for the risk acceptance decision, as well as monitoring plan, should be documented.

Results of the RA and risk control implementation should be documented and residual risks should be evaluated to determine if further risk reduction is required. Formal acceptability of the residual risk should be approved by appropriate decision makers.

Risk Management Report The risk management report summarizes the results of the RA, risk control strategies, residual risk, and rationale for why the level of risk is acceptable and the frequency for periodic review. The report should be approved by the appropriate decision makers including process owner and quality unit.

Risk Review On the successful technology transfer and operational startup of the process, completion of the initial RAs and implementation of risk control strategies, residual risks need to be monitored to ensure that risk controls are appropriate and that no new risks have been introduced without appropriate consideration. The RA and risk control strategies should be reviewed and updated to maintain risk management as a "living" process. One or more of the following instances will require updates to the RA document:

1. identification of new risks or changes to existing risk profile during routine operations;
2. introduction of new controls or changes made to existing controls; and
3. need for change in previous risk control decisions.

Some instances where changes to risks or controls can occur are changes that could potentially impact system/process configuration (e.g., changes to open vs closed processing), changes to QC testing requirements, changes that could potentially impact filter design, use, or operation, or in case of significant events such as investigations (e.g., contamination, etc.) or adverse monitoring trends.

8.8.2 Example 4: QRM Application for Multiproduct Operations

The scope includes assessing the risks of cross-contamination and nonroutine operations (e.g., mix-ups between product operations such as loading of an incorrect automation recipe) pertinent to multiproduct manufacturing, including risks associated with difficult-to-clean equipment, breaches of closed systems, and incremental controls in addition to single product controls for the manufacture of multiple commercial products.

In this example, a cross-functional team of SMEs performs a touch point assessment using the FMEA tool to quantify the relative risk and document current controls used to minimize the risk for multiproduct operations. The touch point assessment performed for this RA is a comprehensive evaluation of shared personnel, space, equipment, tools, and utilities. The FMEA includes risks typical to a multiuse facility and baseline controls. Severity is scored based on cross-contamination between commercial products. Occurrence and detection scoring are based on operations and current detection methods.

Step 1: Risk Identification A process map is created to identify all steps in the manufacturing process. For each process step, failure modes related to cross-contamination and nonroutine events are identified for the following categories:

- cleaning;
- maintenance;
- waste handling;
- storage;
- sampling;
- small parts;
- shared corridors and common areas and rooms;
- utilities;
- shared equipment and instruments; and
- personnel/gowning.

For each cross-contamination category and identified risk, the multiproduct RA team conducts a touch point assessment. The touch point assessment is based on the assumption that contamination is a three-step process:

- release of a contaminant out of a process or product stream;
- transport of a contaminant between process or product streams;
- entry of a contaminant into a process or product stream that is different from the process or product stream that released the contaminant.

For contamination to occur there must be a point of contact, or touch point between a contaminant, a mechanism of transport, and an uncontaminated production process. This touch point allows the possibility of transporting components from one product stream into another product stream. Without a touch point, the cross-contamination process flow is interrupted and transport cannot occur.

Certain general facility risks, which are inherent to the facility and are not specific to cross-contamination risks, may be excluded from this evaluation. General facility risks include the following:

- environmental factors inherent to biochemical manufacturing (e.g., bioburden);
- activities that occur outside of the facility (e.g., transport of personnel and materials);
- receipt of raw materials; and
- QC testing of multiple products.

Step 2: Risk Analysis Following the identification of cross-contamination and nonroutine event risks, a detailed RA is performed. The FMEA tool is used for this analysis of all multiproduct touch points. Each identified risk is analyzed using the severity, occurrence, and detection scoring criteria. The severity of failure, probability of occurrence, and the ability to detect a failure that could result in cross-contamination or nonroutine event is determined for each risk identified and scored. The severity of the cross-contamination event is determined based on the type of cross-contamination or nonroutine event between two commercial products.

Probability of occurrence is based on the general multiproduct controls as well as facility-specific controls that help reduce the occurrence of the failure mode. The detection score is based on the ability to detect the failure mode or potential cause of the failure mode. The cross-contamination risk is estimated for each identified risk and a RPN is calculated.

Step 3: Risk Evaluation and Risk Control Strategy The cross-contamination risk prioritization matrix is used to determine risk acceptability and assess mitigation controls needed on the basis of the calculated RPN value. For example, critical risk, high risk, medium risk, and low risk. The critical risks must be mitigated immediately and the high risks need active mitigation projects. Low and medium risks are those considered acceptable and no further action is required.

The RA also identifies additional risks classified as acceptable with controls, such as the following:

- transport of waste in shared corridors and common areas;
- transport of soiled equipment and parts in shared corridors and common areas;
- cleaning supplies and equipment that are not product dedicated; and
- cross-contamination via small parts, shared equipment, and shared areas.

For each of these items, further evaluation of the controls is required. Additional corrective and preventative actions may be implemented on the basis of these evaluations. When a new product is being introduced to the facility, it will be evaluated against the approved RA to determine if additional incremental controls need to be considered.

8.9 CONTINUED PROCESS VERIFICATION

Assurance that validated processes continue to perform as originally validated is achieved through IPCs, testing against in-process limits and product specifications, change control, ongoing process monitoring, annual product reviews, and ongoing risk management. Ongoing process monitoring and review should be performed to assure that processes are operating in the validated state and confirms initial decisions on PPQ acceptance criteria. The monitoring strategy should be documented (e.g., parameter monitoring, SPC).

Various changes that occur during a manufacturing process lifetime include normal process variability over time, changes in equipment and control system, and process improvement. Trending of process monitoring data at established intervals is useful in the ongoing evaluation of the process. Statistical analysis and trending of the data should be applied to alert any undesirable process behavior. Where no significant changes have been made and process monitoring or continuous verification confirms that cumulative changes have not impacted the validated state of the system or process, and that they are consistently producing results or the product meeting its specification, then there is no need for process revalidation, unless explicitly required by regulatory authorities. Changes to processes, raw materials, specifications, methods, procedures, labeling, and packaging systems must be evaluated per approved change control procedure. RAs and other actions that support the implementation of a change should also be documented and approved.

The evaluation of process changes can determine the significance of the change and define the necessary supporting validation study requirements and any potential requirements for regulatory reporting. Validation of changes may require a repetition of studies originally used to validate the process or new studies to test specific aspects of a change. An evaluation of the potential impact of a change on that unit operation, as well as potentially impacted steps downstream of that unit operation, is made to determine the scope of required validation. When evaluating the potential impact of a change, the associated risks for the parameters that impact product quality should be reexamined. Revisiting the RA should include all parameters that may impact product quality, not just those identified earlier as CPPs. While overall it may appear that no quality impact is likely, some aspects of a change can alter the potential risk of deviation, perhaps leading to a different conclusion of whether a parameter should be classified or reclassified as a CPP. See Chapter 13 for details on QRM application in change control.

8.9.1 Determining the Significance of a Process Change

When defining a need to conduct additional validation for a process change, the risks associated with that change should be evaluated to determine the impact of the change (e.g., the degree of significance) and the likelihood of that change leading to any potential failure. The following are examples of questions that could be asked relating to the significance of the change and answers would lead to the ranking of the change based on associated risks:

- degree of sameness or similarity to the existing conditions, formulation, process, etc.;
- impact on a CPP, CQA, AQL, or release criteria;
- nature of the change or correction: permanent or temporary, severe or mild;
- location of the change in the overall process (e.g., final product—early or late in the process);
- type of dosage form;
- impact on validation: is the change within validated range?
- impact on compliance with a regulation, GMP guideline, regulatory-filed condition or parameter;
- impact on the patient or customer; and
- impact on the firm's reputation, supply, or business.

The likelihood of occurrence of a failure or problem could be based on the following:

- amount of process understanding and knowledge;
- past history of the area, part, or item that is being changed;
- deviations in the past with respect to this change or similar type of change;
- ability to detect failure as quickly as possible and no later than before the product leaves the site or manufacturing area, if there is a problem with the change (i.e., detection component of RA);
- comparison of before and after the change, whether it includes any studies that may demonstrate reduction in further likelihood of occurrence; and
- history of inspectional or regulatory observations, citations, etc.

Where a risk scoring tool is used, risk scores obtained by combining the two components, degree of severity and frequency, would lead to the final RI score, which could then be used to determine the extent and timing of any additional validation studies. Note that the detectability component may be added on the basis of the RA tool used (e.g., FMEA). Similar to Example 1, scales and predetermined thresholds need to be established before scoring the risks for severity and frequency (and detectability).

8.9.2 Defining a Periodic Review Schedule of Validated Process/Systems

When defining a review schedule of validated systems/process, a risk-based approach could be used to rank the systems or classify them according to common features. For example, potential sources or factors influencing the schedule could be as follows:

- criticality of the quality attribute being measured;
- deviations or past history of the system; and
- production volume or use of the system.

Four levels of severity and probability are defined in this example. Severity would be a measure of the criticality or impact on the validated system. Table 8.9 shows that severity depends on four levels of control of CQAs and CPPs. Frequency is based on percentage of batches with deviations and production volume, also with four levels. Those systems that directly control or impact a CQA or CPP would be the most severe, whereas monitoring a noncritical parameter would be the least. If a system has frequent deviations (>2%) or use (>50 batches/year), the system would have the highest frequency.

In this example, risk scores (i.e., Severity × Frequency) would range from 1 to 16 (Table 8.9). Predetermined thresholds of review periods could be ranked or grouped into periods of time. A risk score was used to rank systems into three

TABLE 8.9 Definition of Severity and Probability for Review Scheduling of Validated Systems

Ranking	Severity	Probability or Frequency
1	Monitors a quality attribute(s) that is classified as noncritical	No deviations for 5 years and/or low production volume (≤ 3 batches/yr)
2	Monitors critical quality attributes	No deviations since last review (2–5 years) and/or low production volume (4–8 batches/yr)
3	Indirectly controls critical process parameter/critical quality attribute	Deviations $\leq 2\%$ of the batches produced since the last review (2 years) and/or moderate production volume (9–50 batches/yr)
4	Directly controls critical process parameter/critical quality attribute	Deviations observed in $>2\%$ of the batches produced since the last review (2 years) and/or moderate production volume (>50 batches/yr)

predetermined thresholds, leading to review periods of 1–5 years. For example, a risk score above 12 (out of 16) would be reviewed annually, intermediate scores 6–9 would be reviewed every two years and no periodic review is required for scores <6.

8.10 CLEANING VALIDATION AND CROSS CONTAMINATION RISKS

CV is establishing documented evidence through a collection and evaluation of data that will provide a high degree of assurance that a specific cleaning process will produce cleaning results that are consistent and reproducible, meeting its predetermined level [29]. The purpose of CV is to:

- ensure that cleaning procedures are adequate for cleaning new products and/or new equipment;
- ensure that residues after cleaning of equipment are reduced to an acceptable level before the manufacture of the next batch and/or the next product in the same equipment;
- assess a new product and/or new equipment for cleanability before GMP production;
- provide ongoing assurance of a state of control of validated cleaning procedures through monitoring and periodic revalidation; and
- evaluate changes to cleaning processes, other manufacturing processes, and equipment to maintain these validated cleaning processes in a state of control.

The design and cycle development for an automated cleaning process is considered a prerequisite for CV. Maximum allowable time intervals for dirty or noncleaned hold times (time between equipment use and cleaning or sterilization) and clean hold times (time elapsed between cleaning and equipment use) should be established [7,14]. The dirty and clean hold times should be reasonable and not excessive and lengthy; risk-based approaches may be used to determine the extent of validation of these parameters.

According to the ICH Q9 Guide, risk-based approaches are acceptable in differentiating efforts in cleaning of equipment based on intended use [11]. A risk-based approach could be used to determine validation strategy, sample sites and testing, acceptance criteria, small-scale study, number of commercial-scale experiments, and appropriate control to prevent cleaning failures [30]. There are several elements of cleaning process development, control, and validation including but not limited to cleaning solution, cycle parameters, equipment type, hold times, sampling techniques, analytical assays, and acceptance criteria [31,32]. Small-scale cleaning studies may provide a useful model to evaluate cleanabiltiy of new products relative to the worst case and determine the need to perform validation [33]. Unlike cleaning development studies on a commercial scale, which require large amounts of material, small-scale cleaning studies can be performed with very low material requirements and well in advance of technology transfer to the manufacturing facility. They also provide the benefit of performing cleaning evaluations under controlled simulated conditions, thereby offering a useful tool to characterize the cleaning process. However, small-scale data should be verified at the commercial scale. Authors observed that cleaning without any detergent could be as effective in some instances (such as buffer tank), thereby eliminating the risk of product contamination by residual detergent.

Absence of a risk-based approach makes the validation unnecessarily complex and/or time consuming. Finally, the extent of revalidation after a change should be justified on the basis of the risks associated with the change.

The following are some of guidances/references related to QRM application in CV:

- Product grouping can be accomplished through RAs [6]. Worst-case products, equipment, or limits can be determined on the basis of solubility, potency, toxicity, difficulty of detection, cleaning behavior, etc. Determining the most-difficult-to-clean product (i.e., worst-case product) from a group of products should involve a risk ranking based on cleaning behavior, history, and solubility.
- Cleaning limits and sampling locations should be determined on the basis of RA [10]. For example, limits could be based on risk for potential carryover, toxicity, detectability, past history, safety factors, etc. For sampling locations, areas or equipment could be ranked or classified on the basis of severity/impact to product quality (e.g., nature of operation, toxicity, extent and impact of contamination), likelihood of occurrence (e.g., possibility of contamination), detectability (capability to detect, inspect), or any combination of these three elements.
- Level of containment based on severity (i.e., toxicology, acceptable daily intake, hazardous nature of the compound(s)) and the frequency (degree of exposure) of contamination of drug product and/or manufacturing personnel [34]. Contamination and cross-contamination issues may be directly related to CV and would be good areas for use of a risk-based approach [14].

8.10.1 Equipment Dirty Hold Time in Cleaning Validation

There is little guidance on how to establish or extend equipment hold times in CV, other than three commercial-scale validation experiments. Using a risk-based approach is one possibility. For example, a site producing multiple products used an FMEA to establish the number of experiments required for establishing the dirty equipment hold time (DEHT). Severity, occurrence, and detection were defined on a scale of 1–10 (Table 8.10). RPN (severity × occurrence × detection)

TABLE 8.10 Severity, Occurrence, and Detection Rating for Cleaning Studies

Extreme Risk ↓ No Risk	Rating	Severity	Occurrence	Detection
	10	Extremely high, cause harm to the patient	Very high, failure almost inevitable	No detection, defect caused by failure is not detectable
	7	High	High	Significant risk of no detection
	5	Moderate, customer experiences some dis-satisfaction	Moderate, e.g., once a month	Probable detection, process is monitored with manually inspection
	3	Minor	Low	Very high chance of detection
	1	None, no effect	Remote, almost never	Almost certain to detect, defect is obvious and can be kept from affecting the customer, all units are automatically inspected

TABLE 8.11 Determining Number of Hold Time Experiments or Extent of Sampling Using RPN

RPN	Number of Experiments (Extent or Sampling Required)
<125	0–1 (minimal or none)
125–150	1–2 (confirmation or moderate)
>150	>3 (multiples or extensive)

thresholds were established to determine the experimentation needed to justify dirty hold times (Table 8.11).

The following risk factors should be considered for establishing the DEHT:

1. Drying on product surface: Certain organic compounds, APIs, waxes, or polymeric formulations may harden or dry on standing, making them more

difficult to remove. Examples are polymethylacrylates as coating polymers and residues in bioreactor after production campaign. In some cases, it is possible that after drying of the residue during normal manufacturing, further increase in hold time will have no effect on the difficulty of cleaning to remove product residue. For example, this may be the case when processing conditions are significantly more severe than idle hold time conditions (e.g., drying a product in a Rosenmund filter for three days at $70°C$ versus idle hold time of the empty noncleaned filter at room temperature). In some cases, it may become easier to clean when dried as residues flaking onto plastic-type surfaces.
2. Adhesion of material on surface: If exposed to humid conditions, hygroscopic materials may become sticky and more difficult to remove (e.g., starch).
3. Solubility of residue in cleaning agents: In some cases, where the solubility of the residue in the cleaning agent is very high, dirty equipment hold time does not affect cleanability.
4. Potential for degradation: Degradants may have different solubility, toxicity, and cleanability characteristics than the original compound. These may be easier or more difficult to clean and should be evaluated on a case-by-case basis.
5. Equipment surfaces: The nature of the equipment surface such as aluminum, steel, hastelloy, plastic, or rubber may be affected by the duration of the contact or porosity of materials. For example, staining may result from certain APIs or dyes that can cause this quickly on contact. Also, residues dried onto plastics may completely flake off, making it easier to vacuum or blow down before cleaning.
6. Microbiological accumulation and proliferation: Organisms may grow exponentially if wet residues or stagnant water is left in equipment. Microbiological considerations should be evaluated when water rinsing is used. A microbiological assessment of residual product over time on the noncleaned equipment should be considered. Conditions of temperature, exposure, time, and product history should be evaluated with respect to the ability of microorganisms to proliferate. The ability of the cleaning process to reduce microorganisms would also be a factor, along with the subsequent dosage form of the next product to be produced in the equipment.
7. Data on cleaning and failure history: A product or equipment piece with a history of cleaning problems may be an indication that dirty equipment hold times is a factor that should be considered. On the other hand, data may point to a noncritical condition by having the following:
    - Routine verification after each changeover cleaning (e.g., routine rinse sampling and testing for major equipment, routine visual inspection for

minor equipment where the visual limit is at or below the residue acceptability limit) may provide adequate data to support the conclusion that dirty equipment hold times for product residues are not critical. For example, rinse checks after cleaning are required for changeover cleaning of major equipment that cannot be visually inspected in API manufacturing.
- Laboratory recovery study data for the product residues may have been generated when residues are dried on representative sample surfaces (coupons). These data may also support noncleaned equipment hold time rationales or data demonstrating that DEHT is not critical. This may be more applicable for chemical APIs, where only a single component is typically removed during cleaning (versus active and excipient mixtures in drug product).
8. Firm's Standards: Company requirements were reviewed to ensure compliance.
9. Regulations: Local GMP regulations and inspectors' expectations.

Product groups and equipment groups can also be used in the analysis. In the risk evaluation, products should be evaluated for each of these risk factors for their potential failure mode. If necessary, additional mitigation can be achieved by initiating cleaning immediately after equipment use to reduce or eliminate the significance of dirty hold times with respect to product residue. This can include post-campaign/batch flushing of the equipment. The type, complexity, and amount of disassembly of equipment (valves, lines, and hoses) may influence the amount of sampling needed to extend the hold time. Table 8.12 shows a simplified example of FMEA.

8.11 SUMMARY

QRM is currently recognized by the industry and regulatory authorities as a valuable tool to define the type and extent of PV or performance qualification required for product licensure. Performing new RAs involves upfront costs with potential pay-off later both in terms of product quality and cost saving. To minimize any subjectivity of the outcome, QRM application including tool selection should be appropriate and robust, facilitated by trained QRM personnel, and the outcomes and decisions documented and communicated.

QRM is not only a good business practice but it also meets regulatory expectations. With early investment made in the validation lifecycle, the risk-based approach focuses on performing appropriate studies, scientific data, and rationale, and avoids non-value-added activities. It can be used to make decisions, prioritize activities, investigate deviation, and achieve a level of confidence in the process.

TABLE 8.12 Simplified of FMEA for Noncleaned Equipment Hold Time

Process Step	Risk Factor	Potential Failure Mode	Potential Effect on Patient	S E V	Potential Cause for Failure	O C C	Current Controls	D E T	RPN
Dry product mfg.	Drying of product and adhesion to surface	Product will dry on surface	Chemical contamination (major adverse reaction)	10	Powder hidden	3	Vacuum, visual inspection	5	150
	Solubility in cleaning agent	Product partially soluble	Chemical contamination (moderate adverse reaction)	5	Cleaning agent only partially effective	3	Cleaning validation performed successfully	3	45

As it is used throughout the process lifecycle, QRM documents are living and should be maintained in the current state through timely review and update.

This chapter presents QRM application in PV, typical tools, risk factors, and challenges in applying risk-based validation. PV activities at all three stages of PV lifecycle are discussed, along with examples of QRM application in process design, PPQ and continued process verification. It shows risk-based validation in study designs, such as univariable and multivariable studies, and mixing and hold time studies. QRM tools can be utilized in determining CPPs, identifying high risk areas, defining controls and risk control strategies, determining the significance of process changes and the periodic review schedule. Considering the higher cost of running experiments at the commercial scale, the risk prioritization tool can be used to identify areas that pose the highest risk, design space, and control range for CPPs. Integration of QRM in technology transfer would ensure successful process transfer.

CV is an integral part of successful licensure and commercial manufacturing. This chapter also discusses risk factors for CV and multiproduct operations. Finally, a list of applicable regulatory guidances for CV and PV was provided.

DISCLAIMER

The information provided in this paper reflects the authors' view and is not intended to represent the official position of our companies. Actual processes previously or currently implemented by the company may differ from those disclosed in this paper. Author makes no representations or warranties regarding these processes as may be implemented by readers of this chapter.

REFERENCES

1. 21 CFR 210 and 211: Current Good Manufacturing Practice In Manufacturing, Processing, Packing, or Holding of Drugs: General. April 1, 2009 (revised).
2. EudraLex by the European Commission, Volume 4—Medicinal Products for Human and Veterinary Use: Good Manufacturing Practice; Annex 15 to the EU Guide, Qualification and Validation, July 2001.
3. FDA Guideline on General Principles of Process Validation, May 1987.
4. FDA Guideline on General Principles of Process Validation, January 2011.
5. Q8 (R2) Pharmaceutical. Development, ICH Harmonised Tripartite Guideline, August 2009.
6. Supplementary guidelines on good manufacturing practices: Validation, Annex 4, WHO Technical Report Series, No. 937, 2006.
7. PIC/S Guide to Good Manufacturing Practice for Medicinal Products Annexes, September 2009.
8. CMC Biotech Working Group, Product Development and Realisation Case Study: A-Mab, Version 2.1, 30th October 2009.

9. Mollah AH. Risk Analysis and Process Validation. Bioprocess International, oct 2004;2(9):28–35.
10. Mollah AH. Application of FMEA for Process Risk Assessment. Bioprocess International; Nov 2005;3(10):12–20.
11. Q9, Quality Risk Management, ICH Harmonised Tripartite Guideline, Nov 2005.
12. Validation by Design: The Statistical Handbook for Pharmaceutical Process Validation, Lynn D. Torbeck, Co-Published PDA Books, 2010.
13. Wang X, Germansderfer A, Harms J, Rathore AS. Using statistical analysis for setting process validation acceptance criteria for biotech products. Biotech Prog 2007;23:55–60.
14. Q7, Good Manufacturing Practice Guide for Active Pharmaceutical Ingredients, ICH Harmonised Tripartite Guideline, November 2000.
15. Boven K, Stryker S, et al. The increased incidence of pure red cell aplasia with an Eprex formulation in uncoated rubber stopper syringes. Kidney Int 2005;67:2346–2353.
16. Process Validation of Protein Manufacturing, PDA Technical Report No. 42, 2005.
17. Q10, Pharmaceutical Quality System, ICH Harmonised Tripartite Guideline, June 2008.
18. Amit Banerjee, Designing in quality: approaches to defining the design space for a monoclonal antibody process: how to use risk assessment strategies to integrate operations, BioPharm Int, May 2010;23(5):26–40.
19. Steven Walfish and Thomas Harrington, Practical Application of Design of Experiments in Biopharmaceutical Process Characterization, IBC's 2009 Process Validation, March 2009.
20. EMEA/CVMP/598/99, Note for Guidance on Process Validation (March 2001) CPMP/QWP/848/96.
21. Tran, N.L., Hasselbalch, B., et al., Elicitation of expert knowledge about risks associated with pharmaceutical manufacturing processes, Pharm Eng, 25(4):24–38, July/Aug 2005.
22. Ganzer WP, Materna JA, et al. Current Thoughts on Critical Process Parameters and API Synthesis. Pharm Technol; July 2005. http://www.pharmtech.com/pharmtech/article/articleDetail.jsp?id=170114.
23. Potter, C., PQLI application of science- and risk-based approaches (ICH Q8, Q9, and Q10) to existing products, J. Pharm Innovation; 4:4–23, March 2009.
24. PQRI Workgroup Members, Process robustness—a PQRI white paper, PQRI working group, Pharm Eng, Nov/Dec 2006, vol 26, No. 6. Exclusive On-line article, p.1–11.
25. Hospira warning letter by FDA, 12 April 2010 (http://www.fda.gov/ICECI/EnforcementActions/WarningLetters/ucm208691.htm).
26. Louis Johnson and and Sarah Burrows, For Starbucks, It's in the Bag: How the Java Giant Fine-Tuned its Sealing Process and Improved Product Quality, Quality Progress, p. 19–23, March 2011.
27. PIC/S Guide to Validation Master Plan, Installation and Operational Qualification, Non-Sterile Process Validation and Cleaning Validation, 25 September 2007.
28. Brian Hasselbalch, FDA at the PDA/FDA Workshop Oct 26, 2009, Bethesda, Maryland, Presentation "Process Validation: A Lifecycle Approach" and Q&A on Stage 2– Performance qualification.

29. Mollah, A. H., Cleaning Validation for biopharmaceutical manufacturing at Genentech, Part I/II, February/March 2008;21(2);21(3).
30. Mollah AH. Risk-Based Cleaning Validation in Biopharmaceutical API Manufacturing. BioPharm International; November 2005;18(11):54–67.
31. Destin Leblanc, Validated Cleaning Technologies for pharmaceutical manufacturing, Interpharm Press, 2000.
32. Gil Bismuth and Shosh Neumann, Cleaning Validation: A Practical Approach, PDA, 2000.
33. N. NitinRathore, Wei Qi, Cylia Chen, Wen changJi, Bench-scale characterization of cleaning process design space for biopharmaceuticals, a method to evaluate the relative cleanability of new products, BioPharm Int, Volume 22, Issue 3, Mar 1, 2009.
34. ISPE Baseline® Guide: Risk-Based Manufacture of Pharmaceutical Products (Risk-MaPP), Sept 2010.

9

ASEPTIC PROCESSING: ONE

James P. Agalloco and James E. Akers

9.1 INTRODUCTION

Sterile products are given to millions of patients daily around the world, and in every instance there is a belief that the drug being administered is actually "sterile." This might seem like a valid assumption considering that every container is labeled "sterile"; the product has passed a sterility test (or been parametrically released); and appropriate control measures have been taken throughout to assure that the product is "sterile." Attaining "sterility" for parenteral products has been the goal of both industry and regulators for many years.

It is useful to understand the origins of "sterility" to fully understand pharmaceutical industry approaches and how existing expectations evolved. In the 1800s as canned foods became more prevalent, deaths were attributed to foods contaminated with the anaerobic pathogen *Clostridium botulinum*. It was determined that treatment at temperatures near 250°F could make the canned goods safe for human consumption. The knowledge gained from that experience with retort processing of canned foods led to sterilization concepts still in use in pharmaceuticals and other industry. [1,2] A major component of that knowledge transfer is the establishment of the sterility assurance level (SAL), which is an estimation of a process' effectiveness against the target microorganism.* The minimum expectation for a sterilization process across the global healthcare industry is 1×10^{-6} [3]. This value is essentially a

---

*Many practitioners prefer to use an alternative term—probability of a nonsterile unit (PNSU), which is considered substantially more intuitive.

Risk Management Applications in Pharmaceutical and Biopharmaceutical Manufacturing, First Edition. Edited by A. Hamid Mollah, Mike Long, and Harold S. Baseman.
© 2013 John Wiley & Sons, Inc. Published 2013 by John Wiley & Sons, Inc.

maximum potential contamination of the sterilized materials and establishes a risk tolerance level for them. As such, it defines a minimum acceptable level of safety for sterilized materials. The validation of the sterilization process in conjunction with production monitoring and routine controls on the materials being processed allow for confirmation of the SAL on an every cycle basis.

Aseptic processing is a substantially different process in which individually sterilized items—product, container, closure, and product contact parts—are assembled under appropriate environmental conditions. Aseptic processing is subject to adventitious contamination during its execution. The absence of a lethal step subsequent to the aseptic assembly means that a SAL determination for aseptic processing is presently impossible. Aseptic processing capability is seemingly established using process simulations where the absence of positive units in a large population suggests a maximum contamination rate during aseptic processing. The oft cited acceptance criterion of 0.1% positive units in a study is actually a maximum contamination rate and not a SAL [4].

At present, the certainty with which aseptic processing can be accomplished cannot be precisely measured. The existing monitoring mechanisms cannot be directly correlated with any estimation of either the SAL or maximum contamination rate in a specific lot. We believe that microbial sampling results, whether from air, surface, or personnel, in manned cleanrooms lack adequate sensitivity to detect the low levels of microorganisms expected to be found [5,6]. Nonviable monitoring has no direct correlation to microbial contamination even in potentially heavily contaminated environments [7]. Process simulations (media fills) provide point-in-time assessments of process capability, and cannot be utilized to define the "sterility" of materials produced under different circumstances [8]. The sterility test has such severe limitations that its use is perhaps more ceremonial than anything else [9]. Even when contamination is detected during sterility testing, the levels of contamination within the lot must be egregiously high given the inherent sampling limitations. The increasing utilization of advanced aseptic processing systems will further reduce the utility of all of these monitoring/evaluation methods [10].

In light of the fundamental uncertainty associated with the aseptic processing monitoring methods, it would be appropriate to consider what does establish its acceptability for use. The answer is neither straightforward nor immediate. Success in aseptic processing operation is derived through the use of appropriate facilities, equipment, components, personnel practices, and procedures and is somewhat supported by the monitoring methods, inadequate though they might be. When replicated over a period of time, a degree of confidence is developed. Moreover, while the direct impact of refinements cannot be measured, we believe a general sense of improvement can be implied from the results. Our industry over the past 30 years has witnessed a steady progression toward superior performance as evidenced by the ever tightening expectations associated with process simulation. In the mid-1970s, the World Health Organization suggested a maximum contamination rate of 0.3% [11]. The Food and Drug Administration adopted a criterion of 0.1% in their 1987 Aseptic Processing Guidance [12]. ISO and PDA developed expanded guidance in this area during the 1990s [13,14]. ISO's effort relied heavily on statistical treatment for the definition of contamination rate acceptance criteria, while PDA's approach was more holistic and recommended a goal of zero contamination in the filled containers.

In parallel with the regulatory/organizational efforts to define ever more restrictive expectations, industry dramatically altered the technologies it employed for aseptic processing.[†] In the 1970s, flexible and incomplete barriers were the norm in many facilities; yet by the end of the century, the first production isolators had already completed a decade of operation. The changes evidenced in aseptic processing were myriad and impacted all of the supportive elements. The absence of major recalls or compliance initiatives during the 1980s suggest that while "sterility" may not have been definitively established, sufficient controls were in place to assure patient safety with aseptically produced drug products.[‡]

9.2 ASEPTIC PROCESS DESIGN RESPONSE

Those charged with the preparation of sterile drug products by aseptic processing have long understood the risks associated with aseptic processing and the central role that personnel play with respect to contamination potential. Acknowledgement of the adverse consequences of personnel involvement resulted in a steady progression of improvements, all of which served to reduce the impact of personnel. The brief history of aseptic processing that follows demonstrates an awareness of the contamination risk associated with the aseptic operator, and the means to minimize their impact on the aseptic process evolved without formal risk analysis.

The earliest aseptic processes utilized rather crude environments and controls, and these were replaced in some instances by gloveboxes that separated the workers from the aseptic field where the assembly of the product containers was performed.[§] The US government's declassification of the HEPA filter in the early 1950s led to the development of the pharmaceutical cleanroom, in which equipment could perform a majority of the aseptic process, eliminating the manual steps that had been previously necessary. The period from 1960 through 1990 witnessed refinements to cleanroom practices including barriers of increasing reliability, improved HVAC systems, equipment refinements, automation and robotics, improved components, and others that brought demonstrable improvements in performance [15,16]. The next big advance in aseptic processing came in the late 1980s with isolation technology, in which the operator was no longer present in the same environment as the sterile materials. The restricted access

---

[†]This may be, in fact, a question of which came first—the chicken or the egg: regulatory expectations would not have been raised unless there was a realistic belief that they could be met.
[‡]'Sterility' is defined as the absence of viability; something as unprovable in 1970 as it remains today.
[§]These were true gloveboxes as they lacked the defining features of isolators (rapid transfer ports and automated decontamination).

barrier system (RABS) came somewhat later in an effort to eliminate some of the more difficult technical challenges with isolators, while endeavoring to maintain the operational excellence attained by keeping the operator access to the critical environment to a minimum.

PDA and ISO independently developed guidance documents that endeavored to define the spectrum of aseptic processing systems [17,18]. Each of these documents included a continuum that strove to visually clarify the various technologies and methodologies. A more contemporary continuum that essentially converts the ISO 14644-7 continuum into recognizable pharmaceutical technologies is presented in Figure 9.1.

The primary source of microbial contamination in aseptic processing is universally acknowledged to be personnel [19]. The improvements in aseptic performance have been driven by designs that endeavor to move personnel away from sterile articles during the processing. Ascending the continuum, the various systems provide for increasingly robust separation of personnel from the critical processing area. The systems near the center incorporate physical barriers that further increase the separation, partially in RABS systems and more fully in isolator designs. There are no currently available means to measure the sterility assurance in any of these systems; thus, there is likely considerable overlap in their performance capabilities. The uncertainties of environmental monitoring, infrequent and equally uncertain media fill, and the almost useless sterility test cannot adequately differentiate the performance capabilities of the various aseptic processing technologies. Microbiological monitoring methods have not kept pace with the advances in aseptic process technology; their sensitivity is too limited and sample sizes are too small to provide meaningful assessments, especially in the most advanced technologies. Perceived improvements such as

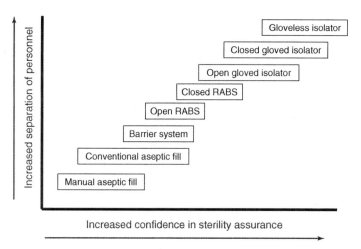

Figure 9.1 Aseptic processing continuum. (*See insert for color representation of the figure.*)

rapid microbiological methods only provide nondefinitive information sooner rather than later. Sampling can never prove the absence of something. The placement of one system higher in the continuum relates solely to the system features that have moved personnel away from critical activities. With isolators, that separation is near absolute, a situation that can only be attained in aseptic processing systems operating without personnel.

9.3 RISK ASSESSMENT

The advances in aseptic processing technology as well as the various continuums were not derived using any formal risk assessment methodology. As humans, we evaluate risk daily in our personal lives, and almost never invoke any sophisticated means for doing so. Choosing whether to carry an umbrella on a cloudy day, eat sushi for lunch, or stop when the light turns yellow are choices relative to risk we make more or less intuitively, and sometimes almost instantaneously. The technological advances described previously were made without formal consideration of risk in aseptic processing, most likely because defined methods for doing so did not yet exist.

While risks and human assessment of them have existed for millennia, the use of formalized methods is a much more recent development. The first of these is failure mode effects analysis (FMEA), which was originally developed for the US military in the late 1940s. The evaluation keys to FMEA are estimates of severity, occurrence, and detection; these are multiplied to determine a risk priority number [20]. Inherent in its application is the ability to identify the rating given in each category. Numerous other formal risk assessment tools have been developed including HAACP, HAZOP, FTA, and others. Each of these has been applied with success within the healthcare industry for various applications—process reliability, bioreactor contamination, process safety, etc. The use of risk assessment for aseptic processing is newer still.

The Food and Drug Administration essentially reinvented its role as a regulator and radically altered its vision of expectations in "Pharmaceutical CGMPs for the 21st Century—A Risk-Based Approach [21]." This document essentially challenged the pharmaceutical industry to rethink its approaches and attitudes with respect to pharmaceutical quality. Among the fundamental precepts were the following ones.

- "Encourage the early adoption of new technological advances by the pharmaceutical industry.
- Facilitate industry application of modern quality management techniques, including implementation of quality systems approaches, to all aspects of pharmaceutical production and quality assurance.
- Encourage implementation of risk-based approaches that focus both industry and Agency attention on critical areas.

- Ensure that regulatory review, compliance, and inspection policies are based on state-of the-art pharmaceutical science.
- Enhance the consistency and coordination of FDA's drug quality regulatory programs, in part, by further integrating enhanced quality systems approaches into the Agency's business processes and regulatory policies concerning review and inspection activities."

This document has had a profound impact on the pharmaceutical industry, and not the least of it was the acknowledgement that evaluation and minimization of patient risk would play a much larger role in future compliance considerations. Concurrent with the release of this major CGMP document, FDA released new aseptic processing guidance [22]. The aseptic processing guidance, while claimed by FDA to be risk-based, lacks clarity as to the principles of risk assessment and mitigation to be followed.¶ Nevertheless, risk-based thinking is now the order of the day within FDA, and increasingly across the global pharmaceutical community [23].

Aseptic processing is a logical candidate for formalized risk assessment, and due credit must be given to Dr. Whyte of the University of Glasgow. His landmark papers in this area set the stage for virtually all of the subsequent efforts [24–26]. Central to much of Dr. Whyte's work is the following equation:

Number of microbes deposited on a product = $C \times S \times Pd \times Pa \times A \times T$, where:

- C = concentration of microbial contamination on, or in, a source (number/cm^2, or number/cm^3)
- S = surface material, or air, that is dispersed or transferred, from the source in a given time (cm^2/s for surfaces, and cm^3/s for air dispersion); could also be concentration per frequency of occurrence
- Pd = proportion of microorganisms dispersed that are transferred to the area adjacent to the product
- Pa = proportion of microorganisms that arrive at the adjacent area carrying microorganisms in the concentration C that are deposited per unit of the product area (/cm^2)
- A = area of surface onto which microbes are deposited (cm^2)
- T = time, during which transfers occur(s); could also be frequency of occurrence

The logic inherent in this calculation is irrefutable; and provided the various parameters can be accurately measured, the contamination rate can be precisely determined. Unfortunately, the first four variables in the equation defy estimation, let alone determination with any precision. The inability to place metrics on these elements dramatically reduces the utility of the equation

---

¶The 2004 Aseptic Guidance had been initially developed well before the risk-based CGMP initiative, and lacks meaningful risk-based thinking. Its publication on the same day appears to be more a forced coincidence than a meaningful effort on the part of FDA to provide a truly risk-based aseptic guidance.

for real-world application. In effect, while the guiding principles are correct, utilizing this method would rely on estimates of these values. As noted early in this chapter, the reliability and precision of microbial monitoring methods in the most critical environments are severely limited, and thus these risk analysis methods must be considered more theoretical than anything else. The greatest value in Dr. Whyte's work is as an indication to others of where the risks lie, and that quantification of those risks may be possible.

The concepts in Dr. Whyte's work were adapted for application in conjunction with a Monte Carlo simulation of contamination derived from microbial levels in aseptic processing operations by Tidswell and McGarvey from Eli Lilly & Co [27]. The potential for microbial ingress onto sterilized items was considered in a case study process sequence consisting of the following:

1. filling line setup;
2. vial transfer from depyrogenation tunnel to accumulator;
3. conveying empty vials to filling;
4. emptying stoppers into stopper bowl;
5. stopper handling in stopper bowl and conveying to stoppering;
6. filling;
7. stoppering;
8. conveying stoppered vials to accumulator;
9. loading accumulator with stoppered vials;
10. conveying stoppered vials to capping; and
11. capping of stoppered vials.

In this model, the potential contributions in the early steps are weighted by a number of factors, and, as might be expected, only minimal increased risk is associated with the steps subsequent to stoppering. The data utilized to develop this model was drawn from a specific operational facility, and thus is not transferable to a different facility or process design. A Monte Carlo estimation was applied to the environmental monitoring results from the evaluated facility to estimate the contamination potential. This model assesses relative risk within a specific process, but does not appear to be well suited for comparison of different aseptic operations or consideration of design alternatives.

The work done by Tidswell and McGarvey served as the basis for a sterility assurance risk management model developed in 2009 by G. Berrios, also from Lilly. Berrios redefined the approach taken in the earlier model and developed a risk assessment method that is suitable for use in any manufacturing facility or process design. In the revised method, the risk of microbial, endotoxin, or particulate contamination is evaluated by ranking the following:

- amount of challenge;
- likelihood of the challenge's ingress into the product stream;

- likelihood of the contamination's survival, proliferation, and retention (if introduced);
- level of personnel activity and/or equipment traffic;[‖] and
- probability of detecting the hazard due to testing or other method of detection.

These factors are ranked to provide a risk score for each operation of a given manufacturing process. This risk score is then compared to a pre-established threshold, beyond which a process step is considered unacceptably risky and subject to improvement via the implementation of additional protective measures [28].

The authors of this chapter initially developed an aseptic processing risk method in 2005, and modified it substantially a year later [29,30]. The Agalloco–Akers (A–A) method endeavored to evaluate aseptic risk using only quantifiable metrics, a significant departure from Whyte, but one that eliminated the uncertainties associated with estimate of proportions of microorganisms transferred. The central premise focus of the A–A method is that contamination in aseptic processing is almost exclusively associated with human activity. The metrics in the A–A method focuses on the operator's activities, their complexity, duration, and proximity to sterile surfaces being primary drivers in the estimation of risk. Other involved factors are weighted by the extent to which they entail human intervention. The original method had some inherent redundancies and lacked flexibility for varying environments across the aseptic processes, which are corrected in the second version of the method.**

Unlike the other risk assessment methods, which rely on subjective determinations, the A–A method uses only easily identifiable and/or quantifiable metrics; thus, it affords a truly objective assessment of risk. In addition, the A–A method is actually three separate methods in one. The intervention risk, which estimates the number of operator touches per container, is an integral part of the method that can be utilized independent of the rest.†† The remaining elements address processing technology and system design and can be used without determination of the intervention risk, allowing for technology and procedural assessment independent of the operator's impact. This is useful in a design context when

---

[‖]Multiple post-conference communications with Berrios have revealed that the Lilly method has continued to evolve. The risk factor ranking "Personnel Activity/Equipment Traffic" was replaced in 2010 by a risk factor that rates the impact of "Personnel Interventions/Manipulations" at each process step. This change was prompted by the perceived need to evaluate the role of personnel in the manufacturing process more objectively.

**The publications are not identical and in order to understand and apply the A–A method both should be considered.

††Intervention risk (IR) is a measure of the need for operator intervention. Manual procedures and extensive intervention during machine-based filling result in higher numbers of operator contact per container and are thus considered higher risk.

considering what a planned aseptic processing might look like and what are the consequences of design choices from a risk perspective. Using both the intervention risk and processing technology components together affords the most comprehensive use of the A–A method.

The Parenteral Drug Association developed a technical report on risk assessment for aseptic processing that utilized an FMEA-type approach [31]. The use of an FMEA type approach in this effort is inherently limiting, in that many of the nuances and choices associated with aseptic processing cannot be considered with precision. In some ways, this effort merely codifies the thinking that industry went through during the years of greatest improvement in aseptic processing. Its overall utility is rather limited compared to the more evolved methods described previously.

Warren Charlton developed a quantitative risk evaluation method of extraordinary simplicity [32]. This method follows a classical FMEA type approach with severity, frequency, and detectability rankings, but incorporates a defined risk value above which change to the process design is expected. This method makes a strong case for consideration of design elements across the process to provide greater confidence in the acceptability of the outcome.

The inclusion of a definitive limit in risk assessment that would drive process improvement is inherently desirable; however, where that line should be drawn is of course open to discussion. Most of the above cited risk models include subjective assessments of the contributing elements to aseptic risk. Given that subjectivity, the utility of a defined limit can be questioned. The key questions relative to a defined expectation are as follows:

- Would that defined limit be suitable in all process situations?
- What would constitute "proof" of the acceptability of a specific limit?
- Is the method of "proof" broadly applicable?

These questions and others like them will likely remain unanswered for some time.

An evaluation of several different risk models for aseptic processing was conducted by Katayama et al. based on operations at several operating facilities in Japan [33]. This article highlighted the difficulties in using microbial monitoring and media fill results as means for discerning differences in perceived performance of the facilities. As a result of this belief, they expressed preference for the A–A method; however, that may be an artifact of the facilities involved in the comparison and not an indication of the true superiority of that method.

The importance of aseptic processing is such that there are likely to be continued efforts to develop improved risk assessment methods. Anything that contributes positively to understanding and potentially alleviating the inherent contamination hazards with aseptical processing should be considered as a means to reduce the associated risk.

9.4 RISK MITIGATION

The goal of any risk assessment is not merely to understand the risks associated with a particular situation. One is expected to use the assessment to reduce those risks to the extent possible through suitable means. In truth, much greater emphasis should be placed on that aspect of risk management as it provides the desired benefits of the entire exercise. The use of the aseptic process entails careful consideration of the contributing factors that influence the outcome of the process. In 1987, the FDA offered this definition/description of aseptic processing:

"In aseptic processing, the drug product, container and closure are subjected to sterilization processes separately and then brought together. Because there is no further processing to sterilize the product after it is in its final container, it is critical to the maintenance of product sterility that containers be filled and closed in an environment of extremely high quality [34]."

Within this definition are found the essential elements necessary for aseptic processing: facility, environment, equipment, containers/closures, product procedures, and personnel. This can be visualized in Figure 9.2.

Each aspect is a necessary part of the overall process, and inattention to them can adversely impact outcome. The contamination potential associated with each of these elements is certainly variable, but nevertheless attention must be paid to each to achieve success. The foundation for these controls can be found in the CGMP regulations, with some of the key elements associated with each outlined in the following summaries. The following list is not intended to be all inclusive;

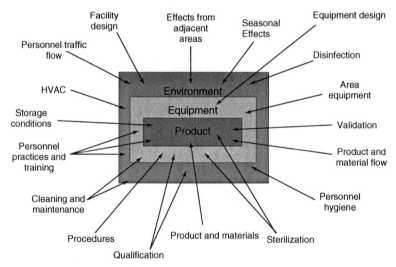

Adapted from Leonard Mestrandrea.

Figure 9.2 Factors influencing aseptic processing. (*See insert for color representation of the figure.*)

RISK MITIGATION

the considerations necessary for success with aseptic processing are far more extensive than this summary provides.

9.4.1 Facilities

Contamination is avoided in any operation by proper design of the facility in which it will be executed. This includes aspects such as the following:

- material, personnel, and other flows that avoid mix-ups, cross-contamination, etc.;
- adequate protection from the building surroundings including temperature, humidity, and dust control;
- arrangement of operations such that soil and other potential contaminants are minimized in preparation and processing areas;
- separate and dedicated areas for cleaning, waste treatment, and similar activities; and
- sufficient space for the operations within, including access for operation, cleaning, and maintenance.

9.4.2 Environment

Within the overall facility, environmental controls must be provided over those materials and activities vulnerable to contamination,

- Progressively cleaner environments should be utilized as products, components, and equipment are cleaned, prepared for sterilization, and eventually assembled into the finished product.
- The environments should be decontaminated on a periodic basis to minimize microorganisms present during operations.
- The environments should be periodically monitored to establish their suitability.
- Appropriate qualification activities should be performed to establish suitable change rates, differential pressures and air flow that best maintain control over conditions.
- The environment should provide a comfortable working environment for the operators at all times with respect to temperature, humidity, noise level, and lighting.

9.4.3 Equipment/Utensils

The equipment utilized for the process should be designed to minimize its contamination potential and properly maintained in that state.

- Product contact surfaces and utensils must be cleaned and sterilized using validated procedures.
- Equipment should be selected for reliability of operation, and ease of adjustment.
- Equipment should be tolerant of container and closure variability to the maximum extent possible.
- Equipment should be designed to minimize interventions in the critical environment.
- Equipment should be designed so that operators are ergonomically situated to minimize stress, discomfort, extended reach, etc.

9.4.4 Containers/Closures

Containers and their closures are intended to maintain the product's quality attributes.

- Container/closure integrity must be demonstrated to be integral throughout the product's intended shelf life.
- Containers, closure, and other final container items should be prepared and sterilized using validated methods to assure their acceptability in the primary container assembly process.
- Containers, closure, and other items should be selected for ease of handling in the assembly process.
- These items should be maintained under conditions that best preserve their sterility and cleanliness until just before use.

9.4.5 Product

The product represents the reason for the entire aseptic process and while its chemical, physical, and microbiological characteristics are unaltered in the context of the aseptic operation; some important concerns should be addressed.

- The product should be sterilized using a validated process, and delivered to the aseptic processing environment in a manner that protects its sterility.
- Any connections necessary to deliver sterilized products should be made in the critical zone of the aseptic processing area.

9.4.6 Procedures

The sequence of activities that comprise the overall aseptic process must be considered with care to assure their acceptability.

- Processes should be defined to eliminate/optimize/minimize interventions throughout the process.

- Environmental monitoring generally requires interventions as well, and their impact must also be considered.
- All interventional activities should be carefully defined, practiced, and executed using proper aseptic technique.
- Particular consideration should be given to corrective interventions with the goal of reducing their impact and frequency.‡‡
- There is no "safe" intervention; contamination potential is a part of each [35].

9.4.7 Personnel

The aseptic operator is *the primary source of contamination* in aseptic processing [36]. In large part, the concerns identified previously for the other contributing factors in aseptic processing are intended to reduce the impact of the operator on the materials being produced. Concerns relating solely to personnel include the following:

- Operators must be trained in a variety of relevant subjects including microbiology, aseptic processing, CGMP, gowning, and any job-related tasks.
- Operators must demonstrate proficiency in aseptic gowning (where necessary) and aseptic activities such as equipment assembly, inherent and corrective interventions (including environmental monitoring).

9.5 STERILITY (SAFETY) BY DESIGN

Proper consideration of the preceding elements constitutes the design phase for an aseptic processing operation. As described, it is clearly quite different from the quality by design (QbD) expectations that are becoming prevalent in drug substance or drug product process development [37]. In those areas, correlation between the independent and dependent variables can often be established with some degree of certainty. The goal is to establish operational controls for the independent parameters, which will ensure that the dependent quality attributes associated with the process can be appropriately controlled. The lack of sufficiently sensitive metrics for what is in effect "sterility" precludes the application of conventional QbD thinking to aseptic processing. The linkage between the independent variables and successful outcomes resulting from the aseptic process is much less distinct and subject to variations for which there are no means of detection. Risk assessment, as described earlier in this effort, is perhaps the only effective means for objective evaluation of the suitability of many of the various design decisions.

---

‡‡Corrective interventions as defined in PDA TR#22 are those utilized to correct faults in the equipment, components, or procedures requiring operation correction. In theory, a process could operate without the need for corrective interventions. It is inappropriate to consider any corrective intervention as a 'routine' activity necessary in every batch.

The goal of the aseptic processing design activity is assurance of the abstract goal of "sterility" for the materials being produced. The extent to which a firm attains success with a specific aseptic process is a consequence of the attention to detail given to the contributing elements. In effect, the goal is "sterility by design," although it is more correct to consider it as "safety by design" or SbD. Each aspect within the overall process potentially contributes to contamination and, therefore, risk; inadequate consideration of any individual component risks failure overall. It must be recognized that the systems necessary for success are essentially a chain, and the overall process is no more robust than the weakest link in that chain. It is sometimes (and wrongly) believed that it is possible to compensate for deficiencies in an individual area of aseptic processing through extraordinary care in other areas; however, that is not a desirable, compliant, or sustainable way to operate.[§§]

9.6 CONCLUSION

This chapter has endeavored to review contemporary thinking relative to aseptic processing risk assessment and mitigation. Pursuing the elusive goal of "sterility" for pharmaceutical products has been a continuing process in this industry since the first parenteral drug was formulated. Early measures incorporated refinements and approaches that were perhaps largely instinctive; nevertheless, the intervening years brought forth performance advances that have made these products increasingly safe. The current methods are so evolved, that intuition may no longer realize further improvement. Contemporary performance has reached such a high level that the best systems can perhaps no longer be evaluated by classical means of sterility testing, environmental monitoring, or process simulation. If further refinement is to be made, the risk assessment tools described herein may offer the most objective means for considering the impact of potential technological changes. Aseptic processing is likely to remain a means for the production of sterile products for the foreseeable future. Formalized risk assessment, as described in this chapter, and its essential counterpart, risk mitigation, will play an increasing role in the design, operation, and maintenance of aseptic operations. Nevertheless, we would be remiss if we did not indicate that we are unlikely to attain a situation of "zero" risk with respect to aseptic processing. We may never attain the "sterility" sought since the origins of parenteral drug administration; however, through application of these methods we can certainly further improve their "safety."

[§§] It is believed by some that highly proficient operators might be able to conduct critical aseptic operations under less than adequate environmental conditions; however, the contamination potential will always be higher than it would were the same operators to perform the same activities under more appropriate conditions. There are no valid arguments for doing something less capable than the available technologies.

REFERENCES

1. Ball CO. Thermal Process Time for Canned Food. Washington (DC): National Research Council, Bulletin No 37; 1923.
2. PDA, Technical Report #1, Validation of Moist Heat Sterilization, 2007.
3. Evans K, Sterilization Options. Presentation to OPS Advisory Committee, 10/22/02. Available at http://www.fda.gov/ohrms/dockets/ac/02/slides/3900S2_10_Evans.ppt. Accessed 2010 May 27.
4. Agalloco J, Akers J. Aseptic Processing for Dosage Form Manufacture: Organization & Validation. In: Agalloco J, Carleton FJ, editors. Validation of Pharmaceutical Processes. 3rd ed. New York, USA: Informa; 2007. pp. 317–326.
5. USP, 1116, Microbiological Control And Monitoring Of Aseptic Processing Environments.
6. EMEA, Annex 1, Sterile Medicinal Products, 2008.
7. Reinmuller B. Dispersion and Risk Assessment of Airborne Contaminants in Pharmaceutical Clean Rooms. Stockholm, Sweden: Royal Institute of Technology, Building Services Engineering, Bulletin No. 56; 2001. p 38.
8. Agalloco J, Akers J. Validation of aseptic processing. In: Agalloco J, Carleton FJ, editors. Validation of Pharmaceutical Processes. 3rd ed. New York, USA: Informa; 2007.
9. Akers J, Agalloco J. A critical look at sterility assurance. Eur J Pharm Sci 2002;7(4): 97–103.
10. Akers J, Agalloco J, Madsen R. What is advanced aseptic processing? Pharm Manuf 2006;4(2):25–27.
11. World Health Organization, Technical Report Series, No. 530, 1977.
12. FDA, Guideline on Sterile Drug Products Produced by Aseptic Processing, 1987.
13. ISO 13408, Aseptic Processing Of Health Care Products–Part 1: General Requirements, 1992 draft.
14. PDA, Process Simulation Testing for Aseptically Filled Products, PDA Technical Report #22, PDA J Pharm Sci Technol 1996;50(6), supplement.
15. Agalloco J, Gordon B. Current practices in the use of media fills in the validation of aseptic processing. J Parenter Sci Technol 1987;41(4):128–141.
16. PDA, TR# 36, Current practices in the validation of aseptic processing–2001. PDA J Pharm Sci Technol 2002; 56(3).
17. PDA, TR#34, Design and validation of isolator systems for the manufacturing and testing of health care products, PDA Technical Report #34. PDA J Pharm Sci Technol 2001;55(5), supplement.
18. ISO 14644-7, Separative Devices.
19. Agalloco J, Akers J, Madsen R. Current practices in the validation of aseptic processing–2001. PDA J Pharm Sci Technol 2002;56(3) PDA Technical Report #36.
20. http://en.wikipedia.org
21. FDA, Pharmaceutical CGMPs For The 21st Century–A Risk-Based Approach, September 2004.
22. FDA, Guideline on Sterile Drug Products Produced by Aseptic Processing, 2004.
23. International Conference on Harmonization, Quality Risk Management, 2006 June.

24. Whyte W. A cleanroom contamination control system. Eur J Parenter Sci 2002;7(2): 55–61.
25. Whyte W, Eaton T. Microbial risk assessment in pharmaceutical cleanrooms. Eur J Parent Pharma Sci 2004;9(1):16–23.
26. Whyte W, Eaton T. Microbiological contamination models for use in risk assessment during pharmaceutical production. Eur J Parenter Pharma Sci 2004;9(1):11–15.
27. Tidswell E, McGarvey B. Quantitative risk modeling in aseptic manufacture. PDA J Pharm Sci Tech 2006;60(5):267–283.
28. Berrios G. Sterility Assurance Risk Management. Arlington (VA): Presentation at IVT Aseptic Processing Conference; 2009.
29. Agalloco J, Akers J. Risk analysis for aseptic processing: the akers-agalloco method. Pharma Technol 2005;29(11):74–88.
30. Agalloco J, Akers J. Simplified risk analysis for aseptic processing: the akers-agalloco method. Pharma Technol 2006;30(7):60–76.
31. PDA Technical Report No. 44, Quality Risk Management for Aseptic Processes 2008;62(S-1).
32. Charlton W. Applying quality by design to sterile manufacturing processes. Pharma Technol 2009;33(5):S28–S32, S42.
33. Katayama H, Toda A, Tokunaga Y, Katoh S. Proposal for a new categorization of aseptic processing facilities based on risk assessment scores. PDA J Pharm Sci Tech 2008;62(4):235–243.
34. FDA, Guideline on Sterile Drug Products Produced by Aseptic Processing, 1987.
35. Agalloco J, Akers J. The truth about interventions in aseptic processing. Pharma Technol 2007;31(5):S8–S11.
36. Agalloco J, Akers J. Validation of aseptic processing. In: Agalloco J, Carleton FJ, editors. Validation of Pharmaceutical Processes. 3rd ed. New York, USA: Informa; 2007.
37. FDA, Draft Guideline: Guidance for Industry–Process Validation: General Principles and Practices, 2008 Nov.

10

ASEPTIC PROCESSING: TWO

EDWARD C. TIDSWELL

10.1 INTRODUCTION

One area of the healthcare industry especially vulnerable to hazard and therefore a potential risk to patient health is the provision and administration of parenteral therapies and medical devices. The omnipresent microbial challenge typically inhabiting sterile manufacturing, aseptic manufacturing facilities, compounding, hospital, healthcare, and community healthcare environments represents a significant hazard. Recoverable and culturable microorganisms, their products, the vestiges of microorganisms (for example, endotoxin, exotoxin, and peptidoglycan) and the more controversial dormant/nonculturable microorganisms all fall into this category of microbial hazard. In each context, the microbial hazards pose a risk to product quality and the potential to realize risk of infection to the recipient patient population. The absence of microorganisms from products processed via adequately validated and controlled terminal sterilization processes is scientifically and statistically recognized [1]. In contrast, the absence of microorganisms in products generated by the aseptic pharmaceutical manufacture of parenterals, aseptic intervention during admixing, compounding, and administration in the healthcare setting are perhaps not so equally (scientifically and statistically) assured. By the fundamental nature of aseptic processing, the successful generation of sterile products is primarily governed by effectively excluding potentially contaminating microorganisms. The ingress of contaminating microorganisms must follow the mechanics of vectors and transfer from originating source to vulnerable product in accordance with the laws of physics. We can therefore recognize that there is some recognizable predictability; however, this is

---

Risk Management Applications in Pharmaceutical and Biopharmaceutical Manufacturing,
First Edition. Edited by A. Hamid Mollah, Mike Long, and Harold S. Baseman.
© 2013 John Wiley & Sons, Inc. Published 2013 by John Wiley & Sons, Inc.

accompanied by uncertainty and variability far exceeding that for the established mechanics and statistics of spore/cell death inherent in terminal sterilization processes. Although sterility has been described as an abstract concept [2,3], and there is merit in this thesis, there remains significant opportunity in the application of sophisticated, mechanistic, and statistically valid risk assessment to define the likelihood that one unit of a product is free from microorganisms. Such a strategy does demand the combination of sophisticated process analytical technology (PAT) to measure hazard (microorganisms), risk assessment, quality management systems, and quality by design (QbD). Assessments for the purpose of evaluating and managing risk in the broad field of aseptic processing (here I include pharmaceutical manufacturing, healthcare admixing, compounding, and patient administration) is therefore deserving of special consideration. Risk assessment in the context of aseptic pharmaceutical manufacture is the principal focus of this chapter, in particular, with a description and explanation of a quantitative, statistical tool of risk analysis that permits a more exacting evaluation of risk.

10.2 PATIENT RISK

In the broadest sense, microorganisms represent an extrinsic form of hazard [4] and may conceptually achieve ingress (contamination) at any number of stages within the aseptic processing life cycle of a product. At each stage or step, there may be a variable number of different sources, associated routes of contamination, and an uncertain magnitude of microbial challenge. At any point post aseptic pharmaceutical manufacture, the access of microorganisms across the physical sterile barrier (container closure) of a parenteral product or device also renders the item nonsterile and has the potential to introduce microorganisms into the patient during administration. Numerous locations in the aseptic pharmaceutical manufacturing process potentially permit the contamination of a product from "resident" bioburden. Table 10.1 inventories many of the typical locations and originating sources of bioburden in aseptic manufacturing environments. Although, there are usually numerous locations from which bioburden might eventually contaminate a product, there are usually only a few sources. A technique to generate comprehensive lists of the locations and origins of bioburden has previously been described by Whyte [5]. Within the aseptic manufacturing environment opportunities and locations for microbial ingress, post final active (critical) control points (Fig. 10.1) represent the greatest risk to product "sterility." Here, the term active control points describes those purposefully instituted mechanisms that are specifically designed to address the adverse affects of the hazard. An example would be a submicron sterilizing grade filter to remove microorganisms from a formulated product before filling the final sterile container. Within Figure 10.1 the active control of hazards are the washing process of containers, closures, and equipment to remove endotoxin; sterilization of container, closure, and equipment to remove bioburden; and sterile filtration of a product to remove microorganisms.

TABLE 10.1 A Nonexhaustive Inventory of the Locations of Bioburden in the Aseptic Manufacturing Environment and the Respective Origin of Bioburden

Location of Bioburden	Original Source of Bioburden
Personnel gloves	Human body flora
Personnel garments	Human body flora
Cleanroom ceilings	Cleanroom air, human body flora
Cleanroom walls, floors, doors	Cleanroom air, human body flora
Tables, chairs, mobile carts	Cleanroom air, human body flora, external (uncontrolled) surfaces
Cleanroom air	Supply air, human body flora
Spillages (sterile items)	Cleanroom air, human body flora
Spillages (nonsterile items)	Spilled material, air, human body flora
Equipment (e.g., vessels, filling lines)	Cleanroom air, human body flora
Ancillary equipment and items (e.g., forceps)	Cleanroom air, human body flora
Paperwork and paper records	Cleanroom air, human body flora
Control panels (e.g., human–machine interfaces)	Cleanroom air, human body flora
Container closures and components (before sterilization)	Specific material, cleanroom air, human body flora
Packaging	Specific item, cleanroom air, human body flora
Raw materials and formulated product (before sterilization)	

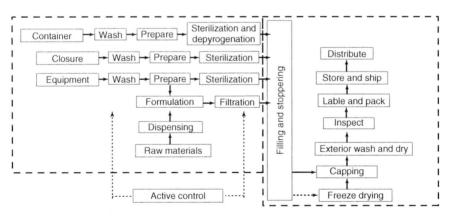

Figure 10.1 Generalized schematic for the aseptic manufacture of parenteral product presentations (vial). Reproduced with permission from Ref. [4].

The process steps and operations that actively reduce or remove bioburden and endotoxin and are recognized as active control points are specifically identified in Figure 10.1 [4]. In almost all manufacturing scenarios, the final mechanisms are those steps that are the final means of actively removing or destroying any resident microbial hazard. Generally, these are sterilization and depyrogenation processes executed before aseptically combining the container, drug/solution, and closure and regarded as active critical control points. In this case, critical control points are authentic critical control points that genuinely adhere to established hazard analysis and critical control point (HACCP) criteria, that is, those points with no subsequent risk-mitigating control points downstream, and which if they were to fail would result in loss of product quality (loss of sterility). Evidence from sporadic field alerts, product recalls, and adverse events would suggest that the control of microbial hazards, analysis, evaluation, and management of the risk of microbial contamination remains an area for improvement. One analysis notes that nonsterility of pharmaceutical products accounted for almost 8% of 215 product recalls in 2006, [6] and between 1998 and 2006 there were a total of 115 recalls of sterile products because of the lack of sterility assurance [7]. The potential for wide-scale impact to patient health from nonsterile product risk is exemplified by the 40,000 adverse reactions (infection) noted in 2002–2003 from the administration of contaminated vaccine. Over 200,000 contaminated units of vaccine entered the Spanish market [8] as a consequence of microbial contamination during manufacture; in this event, the vaccine was an animal health product administered to ruminants.

Aside from microbial contamination of sterile products during aseptic manufacture, the ingress risk of microorganisms during subsequent admixing, compounding, preparation, and administration of parenterals is significant and represents a substantial risk to patient health. Fundamentally, a single microorganism is the minimum numerical amount of a microbial agent which, when accessing a sterile product, has the potential to elicit a potentially deadly infection. In 2002, bloodstream infections accounted for approximately 14% of 1.7 million hospital-acquired infections; such healthcare-associated infections continue to have dramatic impact with significant morbidity and a mortality rate as high as 27% [9]. The attendant cost equates to an additional 3.5 million patient days, costing an additional $3.5 billion as it works out to approximately $29,000 per episode [10]. There has been some conjecture that up to 5% of blood stream infections are due to intrinsic sources of microbial contamination from the parenteral infusate, that is to say contaminated products [11,12]. The preliminary conclusion here is circumspect; contamination from the infusate is likely not intrinsic per se but rather originates from identifiable opportunities for microbial ingress, most often aseptic interventions and during any manual admixing preparation [13]. Regardless, there is and remains an acute need and attendant benefit of applying risk assessment and risk management in aseptic manufacture.

10.3 RISK ASSESSMENT AND RISK MANAGEMENT

This chapter seeks to address risk assessment in aseptic pharmaceutical manufacturing. In this context, risk assessment encompasses risk analysis and evaluation of those determined risks by applying objective and quantitative methods. This, of course, is only a fraction of the complete risk management process, which must be applied in its entirety to truly benefit the patient population. There are many industry standards and guidelines available detailing risk management tools, processes, and techniques. Despite the broad number of publications on risk management, the same fundamental terms, processes, and concepts apply. A cursory glance through the risk management standard [14] generated by the Institute of Risk Management (IRM), the Association of Insurance and Risk Managers (AIRMIC), and Alarm the National Forum for Risk Management in the Public Sector will clearly illustrate to the reader the level of generic commonality. Furthermore, the International Standards Organization (ISO) has a family of international risk management standards that will act as a universally accepted paradigm for the application of risk management practices [15].

10.4 MEASURING AND CONTROLLING RISK

The benefits of a formal assessment of risk to help assure product sterility or product quality during aseptic pharmaceutical manufacture, afforded to the manufacturer, are universally recognized by the regulatory agencies [16–18]. Risk assessment and risk management are arguably facilitating the genesis of the next phase in the continued evolution of GMPs. Risk assessment (risk analysis and risk evaluation) as part of an integrated risk management program permits the practitioner to adroitly measure and control risk as a common discipline within the diverse spectrum of aseptic environments and scenarios. Tidswell and McGarvey [19] originally listed a number of means by which measuring and controlling risk could be applied to achieve specific objectives benefiting product quality; these include, but are not limited to, the following:

- *Prospective analysis of designs.* Analysis and evaluation of risk performed in the early stages of any process or product design can be used to drive specifications, design criteria, or refine attributes. Applying risk management at the early phase or when "on the drawing board" likely means the absence of performance data but does necessitate inclusion of a greater number of assumptions and increased uncertainty. Understanding the magnitude and scope of these constraints may direct choice of the most pertinent risk analysis and decision tools. Repeated and consecutive cycles of risk assessment performed on evolving product prototypes or conceptual process models prospectively refine the design "on the drawing board" to an acceptable level of risk. Recently, the application of a quantitative evaluation of microbial ingress risk into medical devices during aseptic admixing exemplifies how this technique may be used to drive device design with contextual consideration of the clinical or in-use environment [13]. In an analogous manner, any proposed aseptic manufacturing process or facility design can be subjected to rigorous and repeated cycles of evaluation. Here, the risk of product or process contamination can be evaluated with cycles of refinements made to the likes of personnel flow and process and equipment flow, with contextual consideration of the product and process containment. Clearly, both quality and economic factors support the application of this technique to optimize a facility design and prevent the need for post build retrofit or modification early on in the facility's life span. These examples illustrate that the exploitation of risk assessment in the application of QbD has a profound opportunity for improving sterility assurance [20] and demands far wider adoption. Truly quantitative risk analysis, methodically and simultaneously considering the multiple contributors germane to contamination risk, realized the full potential of QbD; however, it necessitates a sophisticated computation. QbD is further discussed in Chapter 14.

- *Comparative analysis of processes and products*. Understanding product and patient risk is paramount to making the most appropriate decision regarding the provision of aseptically manufactured products. Comparative examination, in a consistent manner, of the risks associated with multiple established processes or products permits distinction and comprehension of inherent strengths and vulnerabilities. Here, the application of available historical data can be employed to execute more accurate assessments of risk and adoption of quantitative tools and techniques, than the sole use of qualitative methods. In the aseptic manufacturing environment, such an analysis may facilitate decisions regarding the utilization or preference of available manufacturing strategies. This underlines a much under-valued aspect of risk assessment. Analysis and evaluation not only focuses our attention toward vulnerabilities and jeopardy but concurrently to those beneficial elements of processes, systems, and products permitting their further advantageous use [21]. Comparative risk assessment may also be adopted to direct or prioritize process improvements to focus resources to truly benefit product quality.

Optimization of monitoring, bioburden, and particulate management. In the management and monitoring of aseptic environments, it has been a historic misconception that more equates to better. Many aseptic manufacturing environments have been driven toward increased frequency and quantity of environmental monitoring and sampling, which have seemingly added little to the control of risk. Although the application of tools such as statistical process control (SPC) has merit in contributing to measuring and controlling risk from bioburden, [22] it could drive to a strategy untenable in terms of sampling and control limits if applied in the purest sense. For example, the population frequency and quantity of microorganisms annually recovered in a clean room environmental monitoring program will drive the program control limits when SPC is stringently adopted in the purest Deming sense. Inevitably, year-on-year application will drive these control limits to unobtainable levels. In terms of measuring and controlling risk, it is far more value adding to apply risk analysis and risk evaluation tools and techniques to assist the rational choice of monitoring methods, sampling locations, and frequency. Aseptic environmental monitoring strategies benefit particularly well from this approach. Akers and Agalloco, [23] Whyte, [5] and Whyte and Eaton [24] have innovated effective risk analysis tools and techniques to evaluate risk to product from bioburden and which have the additional utility to be utilized for science- and engineering-based rationalization of environmental monitoring programs. Akers and Agalloco, [23] and Whyte and Eaton [24] choose the FMEAC tool through which to apply the fundamental risk equation (risk = severity of occurrence × frequency of occurrence). Risk factors representing the quantity of microorganisms (criticality of occurrence), the probability of contamination (frequency of occurrence), together with any applicable controls mitigating the risk are individually identified by appropriately experienced and qualified subject matter experts (SMEs). Each risk factor is assessed and a descriptor (risk score) of its magnitude agreed on by the consensus of SMEs. Risk factors are assigned individual risk scores by cross-reference to a predetermined table of surrogate numerical values; Table 10.2 illustrates a predetermined table correlating surrogate value risk scores to perceived magnitudes for risk factors. In Table 10.2, each risk factor is assessed in terms of arbitrary but relative terms; each risk factor is assigned a risk score by comparison of term (e.g., nil, low, medium, high, etc.) to the numerical surrogate descriptor, 0, 0.5, 1, 1.5, or 2 (adapted from [24]).

- The risk factor scores are then multiplied, consistent with FMEA to generate an overall risk rating; comparison of risk ratings permits the evaluation of risk and appropriate decision. Table 10.3 illustrates the calculation of a risk rating for an identified source of microbial hazard (air from a corridor) using this methodology. The strength of this technique is that it is relatively simple, easily trained, and a sufficient understanding of risk can be achieved with a relatively small amount of information.

Within this technique, Whyte and Eaton [24] introduced the concept and incorporation of "transfer coefficients" to aseptic manufacturing risk assessments. This was an innovative development and provides a thorough means of

TABLE 10.2 Assignment of Risk Scores to Risk Factors

Risk Factor	A	B	C	D
Risk Factor Description	Amount of Microbial Contamination on or in a Source	Ease of Dispersion or Transfer of Microorganisms	Proximity of Source to Critical Area	Effectiveness of Control Measure
Risk score	0 (nil)	0 (nil)	0 (remote)	0 (full barrier control)
	0.5 (very low)	0.5 (very low)	0.5 (in outside corridor)	0.5 (very good control)
	1 (low)	1 (low)	1 (periphery of cleanroom)	1 (good control)
	1.5 (medium)	1.5 (medium)	1.5 (general area of cleanroom)	1.5 (some control)
	2 (high)	2 (high)	2 (critical area)	2 (no control)

TABLE 10.3 Calculation of the Risk Rating for the Air Originating Outside a Cleanroom (from a Corridor), Potentially Contaminating the Cleanroom (Adapted from [24])

Risk Factor	A	B	C	D	Risk Rating
Risk Factor Description	Amount of Microbial Contamination on or in a Source (Air)	Ease of Dispersion or Transfer of Microorganisms Area	Proximity of Source to critical	Effectiveness of Control Measure	
Risk score	1.5 (lower classification area)	1	0.5	1 (positive air outflow from the cleanroom)	0.75

including an evaluation of more accurate product risk. Transfer coefficients are numerical values defining the relative transfer efficiency of a microbial hazard into a vulnerable product, in other words, over time, considering the area of an open product, the quantity of bioburden in the context of the manufacturing environment, and the rate of ingress of microorganisms. This is a fundamentally important development that permitted a link between the microbial hazard in the manufacturing area to product risk; this incorporation of transfer coefficients has permitted the innovation of truly quantitative risk assessment and risk modeling (see Section 10.8.3). Systematic application of these forms of risk assessment effectively distinguishes in arbitrary but relative terms the risk, and therefore permits a rational decision regarding monitoring or bioburden management.

- *Determination of worst-case conditions.* Throughout the aseptically manufactured product or aseptic process, life cycle validation exercises have traditionally incorporated worst-case conditions. Risk assessment permits a data-driven, science- and engineering-based rationale for the expedient identification and choice of those conditions that truly represent the greatest challenge and jeopardy to a process. One area where this has been effectively accomplished is in the validation of microbial control during equipment cleaning and hold [25]. The rational choice of worst-case conditions in the equipment cleaning and hold process in the aseptic manufacturing arena can be complex. This is especially the case within multiproduct, multiprocess facilities utilizing nondedicated equipment of varying design, material of construction, fabrication, cleaned, and dispositioned by multiple means, cycles, and processes. Identification of worst-case equipment, product residues, cleaning cycles, and hold processes (pre- and postcleaning) can be accomplished by assessment of the microbial hazard. In this application, risk of microbial retention and proliferation, in addition to contamination (of equipment), are essential considerations. For example, some equipment surfaces lend themselves as more effective substrates for microbial adhesion and also augment the capability for microorganisms to maintain their physical juxtaposition enduring cleaning (retention). The relative risk of microbial retention on equipment surfaces can be determined and described in arbitrary but representative terms by ascribing risk scores to those equipment features that are contributing risk factors. Such risk factors may include surface hydrophobicity and surface roughness; those risk factors lending themselves more to microbial adhesion and retention are assigned higher risk scores. Table 10.4 summarizes the risk factors associated with products and equipment that lend themselves to elevated risks of microbial adhesion to surfaces. Typical characteristics and attributes of associated bioburden and inherent features permit the truly rationale assignment of risk factor values. These values can be used to determine the worst-case combination of product and equipment for expediting cleaning and hold validations. This exemplifies how the subjectivity of risk factor value (or risk factor score) assignment can be diminished by the tabulation of criteria and risk factor features that genuinely lend to risk. Products and raw materials will have varying microbial-growth-supportive capabilities. The selection of aseptic interventions for incorporation into process simulations might be equally benefited from the refinement and adaptation of techniques reported by Akers and Agalloco [23] and Whyte and Eaton [24]. Aseptic interventions represent events that clearly bring a large and diverse source of microbial hazard, harboring upon the human body close to the product. The degree of movement, the duration of the aseptic intervention, and the proximity to vulnerable product vary between interventions. No one intervention is identical and therefore elicits its own individual level of risk to product contamination. Describing risk factors (for example, duration, effort, proximity to product) and assigning risk scores in a manner consistent with the previously described methods permits the ranking and grouping of interventions based on their potential impact to product sterility. These data may then be used to decide the selection and incorporation of worst-case aseptic interventions into a process simulation program.

- *Assistance in batch disposition*. Formulaic risk assessments must be regular activities regularly revisited, maintained as a living document, and perpetuated in a state of currency. Furthermore, risk assessments must not only be applied under exceptional circumstance to assist with decisions of product quality. Exceptional circumstances include those occasions involving unplanned events, leading to uncertainty over the quality or sterility of a product. Often, the analysis of risk is applied as a one-off means of evaluating the nature of a critical quality attribute and justifying the disposition of the product. Such preferential and exceptional analysis of risk associated with a product can send an unfortunate message to regulatory agencies and undoubtedly color their opinion. There is, however, significant, and as yet widely unrealized, merit in the routine application of formulaic, quantitative

TABLE 10.4 Assignment of Risk Factor Values by Correlation to Associated Risk Factor Characteristics (Adapted from [24])

Risk Factor Value	Risk Factors				Microbial Adhesion Risk
	Amount of Bioburden (Typically Recovered Quantity)	Solute Composition (Concentration and Solute Valency)	Presence of Proteins, Peptides or Amino Acids	Typical Microflora Features	
1	<0.1 cfu/ml	<1 mM monovalent or divalent cations	<0.1 mg/ml	No appendages, no extracellular polysaccharide, low hydrophobicity	Product
2	0.1–1.0 cfu/ml	1–150 mM monovalent cations 1–50 mM divalent cations	0.1–1.0 mg/ml	Appendages or extracellular polysaccharide, hydrophobic	
3	>1.0 cfu/ml	>150 mM monovalent cations >150 mM divalent cations	>1 mg/ml	Spore former	

Risk Factor Value	Risk Factors				Microbial Adhesion Risk
	Amount of Bioburden (Typically Recovered Quantity)	Surface Smoothness	Surface Charge	Hydrophobicity	
1	<0.1 cfu/cm^2	Electropolished, <0.30 μm Ra	Net negative	Low	Equipment /item
2	0.1–1.0 cfu/cm^2	0.30–0.70 μm Ra	Neutral	Medium	
3	>1.0 cfu/cm^2	Visible imperfections, >0.70 μm Ra	Net positive	High	

risk analysis to aid assessment of batch quality; assisting batch disposition in a manner more methodical and data driven than is currently the norm. Consistent and routine application of risk assessment of every batch can be used to assist in a deterministic estimation of batch quality, instilling a higher degree of product quality assurance. Furthermore, it warrants the question as to whether such an application should be implemented to continuously maintain the validation status of an aseptic manufacturing process and associated controls in a state of currency. Agalloco and Akers rightly point out that process simulations cannot directly validate an aseptic process [26]; however, a systematic quantification of risk, furnished with batch-associated data performed in a structured manner on every single batch, represents a means of establishing the highest level of sterility assurance [27].

In this context, it is essential to recognize the attendant, significant complexity inherent in virtually all aseptic and clinical processes, circumstances, and scenarios vulnerable to the omnipresent microbial challenge. The interdependencies of factors contributing to risk of microbial contamination during manufacture are not always straightforward. Any application of routine risk formulaic and systematic risk assessment to facilitate batch disposition must recognize and account for such interdependencies. This cannot be adequately described or computed (to an exacting quantitative level) using single-dimensional (e.g., traditional FMEA) risk assessment. This can be illustrated in the simplest terms by simultaneously considering the combination of participating aseptic operators and interventions for two parenteral batches manufactured in an identical process within identical aseptic cleanroom filling processes (Fig. 10.2). In the manufacture of each batch, the same aseptic operators participate, and each executes an identical number of aseptic interventions. Each batch incorporates the same number and type of aseptic intervention. Although, there is a high degree of commonality, the risk of bioburden ingress into each product batch is not identical. Here, the unique combination of intervention duration, proximity to product, and the aseptic operator-associated bioburden levels combine to quite

Batch #1				Batch #2			
Interventions				Interventions			
Operator	A	B	C	Operator	A	B	C
1			2	1	1	1	
2		3		2	2	1	
3	5			3	2	1	2

Figure 10.2 Combination of aseptic operators and aseptic interventions associated with two aseptically manufactured parenteral batches [27].

different quantifiable risks. Using truly quantitative, formulaic, and systematic risk assessment to assist product disposition is described in greater detail later.

In contrast to the numerous reasons for executing an analysis of risk, there are only a finite number of events or triggers (planned, scheduled, or unforeseen) which elicit this activity and which should be systematically recognized within any risk management program. The circumstance, purpose, available data, and required output associated with a risk assessment may all vary, warranting judicious choice of the appropriate risk analysis and risk evaluation tools. Risk analysis and risk management are currently embedded in industry and regulatory agency expectations, the rational choice of tools, techniques, and their consistent data-driven application will remain key decisions for exponents of risk assessments. Equally, the manner in which a risk assessment has been performed in terms of objective and unbiased risk analysis (e.g., the assignment of risk scores), risk evaluation, and risk management must be beyond contestation. Table 10.5 lists several circumstances and stages within a product or process life cycle (including those described earlier) with suggested aspects to be considered for choice of the best tool. A number of tools and techniques are listed on the basis of their reported and potential application at each life cycle stage. This certainly is not an exhaustive inventory of tools and techniques, but provides the reader with benchmark applications.

10.5 ASEPTIC PROCESSING HAZARDS

Any perceived or substantiated existence of risk to an aseptic process or aseptic manipulation is necessarily contingent upon the presence of a tangible hazard. The World Health Organization (WHO) has clearly defined a hazard as "any circumstance in the production, control, and distribution of a pharmaceutical which can cause an adverse health effect [28]". Although ISO14971:2007(E) specifically refers to the risk management of a medical device, it does provide a list of example hazards that serve not only for illustrative purposes but also provides an excellent starting point to understand potential hazards associated with aseptic processes, admixing, or administration [29]. Within this definition, hazards can be separated into hazards that are either integral or inherent elements of processes or systems (termed intrinsic) or those entities that originate externally and are therefore not predesigned constituents of the process, that is, extrinsic [30].

10.5.1 Intrinsic Hazards

Intrinsic hazards are frequently associated with discrete events such as the risk of equipment, process, systems, or control failures. In addition, chemical, physical, physicochemical, biochemical, or biological hazards inherent in a therapeutic and which are not necessarily coupled with such "failure modes" might also be interpreted as intrinsic hazards. In contrast to inherent intrinsic hazards, extrinsic

TABLE 10.5 Typical Applications of Risk Assessment, Applied Throughout the Product or Process Lifecycle

Stage in Lifecycle	Event or Trigger	Objective	Purpose	Data Available	Risk Assessment Tool
Design	Initiation of design, part of design management	Prospective analysis of designs	Evolve the product or process to design out opportunities for harm	Minimal. qualitative, semi-quantitative	Fault tree analysis FMEA FMECA
Development	Approved design	Establish control strategy	Define critical quality attributes	Qualitative	Fault tree analysis FMEA FMECA
		Optimization of monitoring	Choose monitoring locations, frequency	Qualitative and quantitative	FMEA HACCP FMECA Akers & Agalloco Method [23] Whyte & Eaton Method [24] LR method [32]
Validation	Approved validation plan	Optimized validation	Choice of worst-case conditions	Qualitative and quantitative	FMEA FMECA Akers & Agalloco Method [23] Whyte & Eaton Method [24] LR method [32]
Routine Operation	Change control	Prospectively evaluate product quality risks	Ensure patient safety	Qualitative and quantitative	FMEA FMECA HACCP Akers & Agalloco Method [23] Whyte & Eaton Method [24] Quantitative risk assessment [19]
	Excursion or exception	Product disposition	Evaluate patient safety risk	Qualitative and quantitative	FMEA HACCP Akers & Agalloco Method [23] Whyte & Eaton Method [24] Quantitative risk assessment [19]

hazards are not an integral characteristic or predesigned constituent of the process or therapeutic.

10.5.2 Extrinsic Hazards

Extrinsic hazards necessarily originate externally to the product, item, or operation and must gain access across or through a physical barrier or zone designed to maintain sterility. Risks to product quality from extrinsic hazards during aseptic manufacture may manifest in two related forms: the hazard of ingress, accessing, and contaminating and secondly the risk of hazard retention. The hazard retention risk is associated with the sustained physical presence or juxtaposition within, or residing on, the product or aseptic process. On once accessing a product or process there may be no mitigating process or conditions that physically remove or destroy the contaminant. If the risk is constituted by microbial contamination, conditions may permit proliferation and even greater potential impact. An example here might be the cleaning of container closures before sterilization; a risk of extrinsic microorganisms (not associated with the closures) is foreseeable during the cleaning cycle, accompanied by a risk that these microorganisms may remain on or with the closures. The retention of these microorganisms or their vestiges (potentially endotoxin) may continue to contaminate and jeopardize product quality.

Invariably all aseptic manufacturing is vulnerable to varying degrees of risk exacted by extrinsic hazards. For any hazard to exact harm, a hazardous situation must first be encountered; initiating events, and a sequence of circumstances of varying complexity are necessary [29]. Complexity of technology, processes, procedures, and the almost ubiquitous involvement of human-mediated tasks makes aseptic processes particularly prone to risks from extrinsic hazard ingress [31]. Despite the diversity and complexity of aseptic processing, there exists only a finite number of extrinsic hazards with the potential to exact a risk to product quality. Invariably, the extrinsic hazards that contribute the greatest risks to product quality are particulates [32] and microorganisms [5,33]. Pharmacopoeia standards [34–37] for both large and small volume aseptically manufactured parenterals permit the presence of quantitative levels of particulates; in comparison, the requirement for microorganisms is their complete absence. The levels of tolerance for these two critical quality attributes in finished products is mirrored by the degree of control and requisite maximum quantities permissible within the aseptic and clinical environments. The contrast between the required degree of control for particulate and microbial hazard can be no more clearly exemplified than the levels defined in the EU GMP's Annex 1 [38] for Grade A (ISO 4.8) cleanroom environments. Within these Grade A conditions, the CGMPs permit no more than 20 particles (≥ 5 μm) and <1 cfu/m^3 of air. An amount of particulate is permissible in a product (and therefore the environment does not necessarily need to be devoid of them), and no microorganisms must exist within the product (and therefore the environment should be more stringently controlled with respect to this extrinsic hazard). In any risk analysis of aseptic environments assessing extrinsic risks from particulate and microbial hazards, the fundamental risk Equation 10.1 is best applied in two ways.

$$\text{Risk} = \text{Severity of occurrence} \times \text{Probability of occurrence} \tag{10.1}$$

First, the risk of product contamination from extrinsic particles in the manufacturing environment can be adequately described with the fundamental risk equation in its previously given form (Eq. 10.1). Here, severity of occurrence, that is to say the amount of particulates within a product, can acceptably vary from 0 to pharmacopoeia-defined quantities. It is thus appropriate to use a quantitative value of particulates that might exist within the product as the severity of occurrence. In comparison, the loss of sterility by ingress of a single microorganism is not permissible. Ingress of a single microorganism always results in loss of sterility, a two-parameter logic system is applied; an item is either sterile or it is not sterile, that is, it either contains a microorganism or it does not. Hence, the severity of occurrence can only be nonsterile, it will always be invariable and therefore the numerical contribution to the overall risk of extrinsic bioburden contamination will always be unity. Accordingly, the most appropriate means of applying the fundamental risk equation defining the risk of bioburden ingress is defined by Equation 10.2:

$$\text{Risk} = \text{Probability of occurrence} \tag{10.2}$$

A wide variety of physical, physiological, and biochemical characteristics make microbial hazards the most significant of hazards to aseptically manufactured products. Virtually all known environments possess a resident microflora, with microorganisms existing as individuals, collectives, or complex social consortia. The magnitude of the global microflora can be appreciated by the consideration of a single gram of soil in which there exists 10^8-10^9 microorganisms; assuming uniform distribution, a microgram of material accessing any clean room or controlled environment could conceivably carry 1000 microorganisms. Furthermore, 10^{14} microorganisms exist in association with the human body compared to only 10^{13} human eukaryotic cells; the biogeography of the bacterial communities associated with healthy individuals is astoundingly rich [39]. Irrespective of location of aseptic manufacture, a variable and uncertain quantity of microorganisms will unquestionably exist and represent a reservoir of hazard with potential to access and contaminate the product. Microorganisms are exceptionally adaptive in both morphological and physiological dimensions, employing phenotypic and genotypic mechanisms to endure environmental insult, proliferate, replicate, and perpetuate their existence. Many inherent mechanisms (including their capability to swiftly reproduce, some doubling every 20 min or so) permit species of microorganisms to expediently evolve. New species possessing different characteristics and survival traits evolve quickly to permit survival and successful population of new and changing environments. The magnitude of microbial species diversity is testament to the success of such adaptive survival strategy; more than 1.5×10^6 species of microorganisms are believed to exist [40].

Only a small portion (<1%) of this total amount of microorganisms have been successfully grown in axenic culture [41,42]. Microorganisms from all environments have frequently been demonstrated to be unculturable. Gao et al. [43] identified that approximately 15% of all species recovered from the human forearm are not currently culturable. Wade [44] similarly reports that only 50% of the microorganisms in the buccal cavity are culturable. A recent study of classified and controlled cleanrooms has identified that $>10^6$ cells/m^2 nonculturable cells exist, easily outnumbering culturable cells [45]. It is therefore likely that within environments associated with aseptic manufacturing, admixing, and administration, the actual magnitude of a microbial hazard significantly exceeds that determinable by traditional microbiological culturing techniques. Quantitative risk assessment incorporating such considerations, therefore, may well represent a means of assuring product quality (in microbiological terms), which exceeds that obtainable with currently available measurement technology.

10.5.3 Risk of Endotoxin

The adverse health effects elicited by endotoxins (the vestiges of bacterial cell walls) within parenteral products are very well described; endotoxins are clearly a hazard and need to be assessed within risk assessment. Endotoxins may realize a risk to product quality either as an intrinsic or an extrinsic hazard. For example, endotoxins existing as part of the "natural" load inherent in or originating from a raw material or active pharmaceutical ingredient derived from a recombinant gram-negative bacterial fermentation might be regarded as intrinsic. In contrast, endotoxins contaminating a parenteral product presentation via, say, a container closure (e.g., vial or stopper, sealing ring or plunger) originate externally and therefore constitute an extrinsic hazard. Endotoxins within a parenteral product presentation or derived from gram-negative bacteria entering the patient can significantly and acutely jeopardize patient health. Lipopolysaccharide is regarded as the main initiator of sepsis via the triggered release of inflammatory cytokines, tumor necrosis factor alpha, and interlukin-6 (IL-6). Within the past 10–15 years, peptidoglycan has also been recognized as a major contributor eliciting an IL-6 response and patient sepsis [46], warranting careful evaluation in any assessment of risk to aseptic processes. Moreover, peptidoglycan appears to act synergistically with lipoteichoic acid and lipopolysaccharide to cause organ injury [47]. The complexity of synergism and likelihood of a wider variety of microbially derived entities with the capability to agonistically elicit IL-6 response are some of the reasons for introduction of the monocyte activation test [35]. This is a clear acknowledgement that far more microbially derived molecules than those we have detailed knowledge of or the ability to specifically test for may adversely affect product quality. Any risk analysis and risk management strategies in aseptic manufacturing must consider this and evaluate these risks derived from intrinsic and extrinsic hazard ingress.

The detrimental effects on patient health caused by peptidoglycan, lipoteichoic acid, and lipopolysaccharide may be exacted as molecules, packets of molecules (frequently as micelles within a fluid), coating as a layer upon substrate (equipment or product) surfaces or when associated with the cellular form either as viable microorganisms, dormant or nonreplicating or nonculturable microorganisms. Therefore, to thoroughly evaluate the risk to product quality posed by the various vestiges or components of microorganisms, careful consideration of the nature, form, and origin of the hazard is imperative.

10.6 MODEL FOR MICROBIAL INGRESS

A conceptual model for microbial ingress during aseptic pharmaceutical manufacture (Fig. 10.3) appropriately describes the main origins of bioburden, from which comprehensive determinations of risk can be achieved [19]. Within this model, primary sources of bioburden include personnel, healthcare professionals, and the surrounding facility air supplied from HVAC systems. Primary sources of bioburden also represent potential reservoirs of bioburden capable of continually propagating bioburden to elicit a sustained risk to product quality. The surrounding facility routinely possesses surface-borne bioburden originating either from primary sources (air or personnel) or from extraneous sources and are therefore most appropriately regarded as secondary sources of bioburden. These sources of bioburden and the surrounding environment lead to a complex and dynamic exchange or physical bidirectional transfer of microorganisms between environment (facility and equipment), air, and personnel. This dynamic interchange, movement, and transfer is likely to be highly variable from circumstance to circumstance; the quantity, frequency, rate, and direction of hazard movement in the aseptic manufacturing clean room will be quite distinct from that of the clinical environment. Even so, such a single model describing hazard ingress is both adequate and necessary.

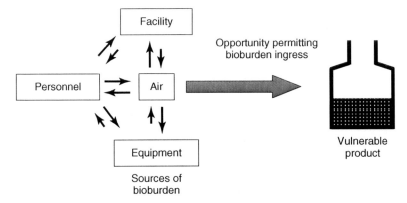

Figure 10.3 Conceptual model for air-borne bioburden ingress into aseptically processed (manufactured, admixed, administered) items.

10.7 RISK HIERARCHY

Although the conceptual bioburden ingress model is intrinsically simplistic, the precise mechanisms by which microorganisms access a product during aseptic pharmaceutical manufacture is a complex and multifactorial process. Any risk assessment must therefore adequately account for and incorporate consideration of all participating factors within a structured architecture or hierarchy. Risk hierarchies (of varying design) have been successfully incorporated into contemporary formulaic and quantitative assessments of risk in aseptic manufacture [23,48]. Irrespective of the preferred process or system of risk assessment adopted, a common hierarchical description of overall risk can be envisioned, which satisfies all risk scenarios and the model for bioburden ingress (Fig. 10.4).

This hierarchical relationship is fundamental to any assessment of risk and is suitable for application of any type of the recognized risk assessment tools and techniques including FMEA, HACCP, and quantitative risk modeling. This hierarchy can be applied at each step, circumstance, or more simply encompass the entire assessment of risk. The overall quantitative estimate for risk is derived from the numerical sum of individual risk components that are generated from individual risk factors. Risk factors have some level of interdependency in that their magnitude may directly affect other risk factors and which combine to a value contributing to the overall risk. Risk factors that affect each other are therefore multiplied together. The assignment of numerical values for each risk factor can be achieved by the means previously described (via arbitrary surrogate but representative relative values) or by actual empirically derived measurements (see quantitative risk modeling, later). Risk components (and therefore their numerical value) have no interdependency; they have no influence or propensity to influence each other's contributors to the overall risk. Typical examples of risk components include the following [31,48]:

- facility contribution to risk;
- transfer of equipment to filling area;
- equipment setup; and
- personnel interventions.

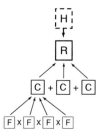

Figure 10.4 A risk hierarchy describes the relationship between hazards [H], risk [R], risk components [C], and risk factors [F] [30].

RISK HIERARCHY

One or several risk components may be perceived for each stage, process, step, circumstance, or scenario under evaluation; risk components are likely to vary in number and type. Each risk component is quantified from risk factors that interdependently influence and impact each other; therefore, it is the numerical product of risk factor values that define each risk component. Typical risk factors commonly adopted have included the following [31,48]:

- amount of equipment transfer into a critical area;
- equipment setup complexity;
- complexity of a critical area;
- number of personnel interventions;
- type of interventions;
- number of personnel in the area;
- environmental conditions; and
- surface contamination levels.

Irrespective of the risk hierarchy, risk components must always include risk factors that represent the likely occurrence of an event (causing harm) and severity of the harm imparted; choice of salient risk factors and components is ultimately discretionary based on expert opinion to genuinely represent the model and mechanisms of ingress risk. This is most easily illustrated through a previously reported example application [19]. In this application, all likely contributors (risk components and risk factors) associated with the overall risk of microbial contamination of an 11-step parenteral filling process were determined; a fishbone diagram succinctly illustrates these (Fig. 10.5). Risk components and risk factors participating and contributing in the risk for each of the individual 11 steps of the parenteral filling process were listed; note that the combination of risk components vary from process step to step (Table 10.6). The parenteral filling process manufactured a liquid, biological, protein-based product presentation filled, stoppered, and capped by machine-based processes. Sterile, depyrogenated empty vials are transferred from a depyrogenation tunnel onto an incoming accumulator table. Vials are then conveyed to a filling head and filled with the liquid, biological, parenteral formulation before being transferred to stoppering and subsequently to a second accumulator before capping. The sterile vial stoppers are initially emptied into a stopper bowl from a sterile container before being transferred to the stoppering equipment. These operations are performed in a qualified, controlled, and classified environment maintaining bioburden and particulate levels to Grade A [38] and ISO class 5 requirements. A clean room environment controlled to Grade B [38] and ISO class 5 requirements surround this critical zone. For a convenient and meaningful analysis of risk of bioburden ingress throughout the filling process, the filling line is divided into individual sections or steps. These divisions are representative of the main functional tasks and activities of the filling process, designated 1–11 and listed in Table 10.6. Overall

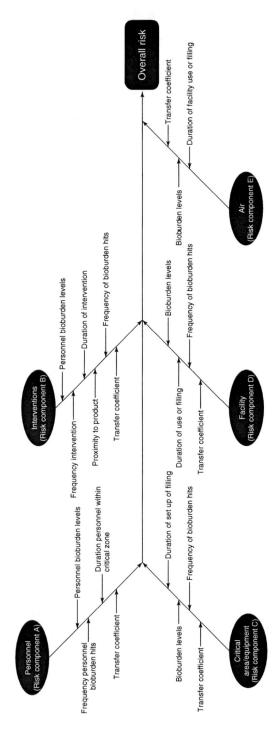

Figure 10.5 Risk components and their constituent risk factors describing the risk of bioburden ingress during the eleven steps of an aseptic parenteral filling process [19].

TABLE 10.6 The Eleven Steps of the Aseptic Machine Filling Process of a Parenteral Product[a]

No.	Filling Process Step	Risk Components Contributing To Step Risk
1	Filling line setup	A, C, D, E
2	Vial transfer from depyrogentation tunnel to accumulator	C, D, E
3	Conveying empty vials to filling	A, B, C, D, E
4	Emptying stoppers into stopper bowl	A, B, C, D, E
5	Stopper handling in stopper bowl and conveying to stoppering	A, B, C, D, E,
6	Filling	A, B, C, D, E
7	Stoppering	A, B, C, D, E
8	Conveying stoppered vials to accumulator	A, B, C, D, E
9	Loading accumulator with stoppered vials	A, B, C, D, E
10	Conveying stoppered vials to capping	A, B, C, D, E
11	Capping of stoppered vials	A, B, C, D, E

[a]Source: The risk components (see Figure 10.5) contributing to the risk of bioburden ingress are listed for each step.

risk at each filling step is calculated from the combination of risk components and the overall risk of microbial contamination to the process determined from the sum of all process step risks.

10.8 QUALITY BY DESIGN, QUANTITATIVE RISK ASSESSMENT, AND PRODUCT DISPOSITION

Inevitably, all aseptic processes are vulnerable to varying degrees of risk posed by extrinsic hazards accessing the product or device across a sterile barrier (or zone), rendering the product contaminated. Typically, sterility assurance programs adopt an integrated range of technologies, practices, tests, and monitoring systems incorporating critical evaluation whose aggregate contributions establish a level of sterility assurance. Generally, numerous environmental controls are integrated into a strategy that includes (but is not limited to) facility (design, finish, materials of construction, cleaning, and sanitization), personnel (number, traffic flow, movement, dress code), and equipment (design, finish, materials of construction, operation, maintenance, cleaning, and sanitization/sterilization).

10.8.1 Improving Assessments of Microbial Risks

A successful strategy combines these controls to minimize and measure the number of microorganisms within the immediate (or extended) vicinity of a product and must be considered in detail as a fundamental part of any risk assessment. Whyte [5] and subsequently Whyte and Eaton [24] clearly illustrate the necessity of thoroughly documenting the primary sources, vectors, and potential routes of microbial transfer within aseptic environments as the foremost and pivotal step of any risk assessment. Assessment of risk to a product is dependent on the comprehensive identification and inclusion of all those salient contributors to risk and concomitantly on a thorough knowledge of the process, product, and environment. Brainstorming and application of mind maps clearly assist and are useful tools [5,24]. Fishbone tools (see previously) and the methodical determination of all means by which microbial hazards can generate a risk in aseptic manufacture (essentially a failure mode analysis) permit the comprehensive inventory of all risks. An organization's design history files, process flows, and change controls associated with the aseptic manufacturing process must be used to detail the process flow and assist in the identification of all vulnerabilities permitting microbial hazard ingress, proliferation, or retention. As with any assessment of risk, the accuracy of the magnitude of the hazard contributes to the level of certainty of the predicted risk. Historically, comprehensive monitoring of the environment has adopted growth-based technologies to recover a variable and uncertain portion of the bioburden hazard present [49]. Sutton [50] exemplifies this and points out that the limit of detection of an agar plate is 1 but the limit of quantification is 25! Uncertainty and variability implicit to traditional growth-based measurement of the magnitude of microbial hazards likely diminish the accuracy of predicted risk and confound the effectiveness of any assessment, irrespective of its specific objective.

Any evaluation of the environmental levels of microbial hazards with the objective of determining adequacy of control and product impact is most usually arbitrary in nature. Measurement uncertainty and variability restrict evaluation of the control of microorganisms in the environment, which at best can only be judged in terms of trend analysis or with respect to levels perceived to represent adequate environment of control [51]. Furthermore, there is routinely no data linking the magnitude of a microbial hazard in the immediate or extended environmental vicinity to product quality and the risk of product contamination. Moreover, to the author's knowledge there has been no detailed published data that strictly links environmental levels of microbial hazard to nonsterility of a product. However, constrained by measurement uncertainty and the absence of a correlation between the magnitude of microbial hazard and product impact, environmental (facility, personnel, equipment) levels of the microbial hazard are universally considered as key parameters to facilitate the disposition of an aseptically processed product. The rigor and objectivity of such analysis for product disposition is profoundly opportune for improvement via a truly quantitative risk assessment, which necessarily captures and considers associated measurement uncertainty and variability. Two strategies can be used to reduce measurement uncertainty and variability. First, more accurate and rapid microbial measuring technologies permit a realistic determination of the magnitude of the microbial hazard in the aseptic manufacturing environment. Quantification of the levels of microorganisms in the same time frame of filling and associated with each step of aseptic manufacture generates an accurate assessment of the magnitude of the bioburden, which can be linked to true product risk. Flow-cytometric-based measurement platforms are commercially available and are being validated within aseptic manufacturing facilities. This technology does allow us to apply actual bioburden data to describe hazard magnitude and when combined with transfer coefficients permit calculation of the quantitative risk (transfer of bioburden into vulnerable product) to product. Application of true quantitative values of risk is far superior to the employment of surrogate descriptors. The second strategy is the use of population distributions to describe the risk factors. This does require some empirical data to enable the definition of the risk factor in terms of the likely minimum and maximum mode and shape of distribution. Environmental monitoring data lend itself especially well to this form of describing the magnitude of bioburden as a risk factor. With a high level of assurance, we can say that in controlled aseptic manufacturing environments the level of resident bioburden is a population between less than one and a maximum value (identifiable from environmental monitoring program data), is non-normally distributed at a value close to one (again identifiable from the environmental monitoring program data). The difficulty with using population distribution data in quantitative risk analysis is the necessity for a sophisticated software program to interpolate the data.

10.8.2 Culture-Based Microbial Test Methods are Inadequate Measurements of Product Quality

The sterility test is currently mandated within 211.165 of the Code of Federal Regulations (CFRs) and therefore presently constitutes one of the requisite tests implicit in a sterility assurance program. There remains a common belief or perhaps a convenient reliance on the sterility test as a definitive test of product sterility for an aseptically processed product. The purpose of the sterility test is to provide proof of absence of microorganisms (sterility); however, the absence of evidence of microorganisms by a sterility test is not adequate evidence of absence of microorganisms. A credible risk to the patient population likely accompanies our reliance on the sterility test; the risk of permitting the release of a microbially contaminated product may be described as a Type II error (false negatives). Type II errors in any testing or analysis of sterility has the potential for the disposition of nonsterile items into the market.

Fallibility of the sterility test to identify the presence of microorganisms within a product can be attributed to numerous factors, all potentially contributing to a Type II error. Generally, these factors contribute to two main categories, which are (i) the inability of the sterility test to grow microorganisms and (ii) the small sample size of this end product consuming test.

A microorganism may not grow or replicate by virtue of (a) debilitating injury (lethal or sublethal), (b) physiological prerogative (microbial dormancy), or (c) the mere fact that the nutritional and physicochemical conditions are not conducive. It has been estimated that only 1–5% of all microbial species have been successfully cultured [52,53]. With many aseptic processes involving human manipulation and intervention, the "culturability" of the human microflora to product risk assessment is of paramount importance. Disappointingly, utilizing common culturing technology and culture media only 10–50% of the human microflora (depending upon anatomical location) is recoverable by growth [43,44]. In contrast to nonculturability (because of suboptimal growth conditions), microbial dormancy is part of an essentially ordered developmental program such as sporulation [54,55]. In gram-positive microorganisms, there is evidence that the coordinated flux in metabolic and biochemical functions suggests dormancy is active and programmed [56]. Recent data have illustrated that this phenomenon is a characteristic of clinically relevant species of microorganisms. Sachidanandham and Gin [57] revealed that dormancy is a mechanistically ordered process permitting *Klebsiella pneumonia*, *Escherichia coli*, and *Enterobacter sp.* to endure adverse environmental conditions. In addition to human-borne microorganisms, the recovery and growth of microorganisms from clean room and aseptic processing environments are crucial considerations in the evaluation and quantification of risk. La Duc [45] has illustrated that the magnitude of the microbial hazard with the potential to contaminate a product processed within clean rooms and aseptic conditions may be far larger than previously considered. Indeed, when we consider that dormancy is a preferentially opted physiological state, adopted by microorganisms to endure suboptimal conditions, common to species of the human micro flora, and species causing infections, we must question the adequacy of our reliance on growth-based technology platforms for sterility testing and environmental monitoring, and risk evaluation.

Bryce [58] recognized the fundamental and unavoidable constraint of sample size that limits the sterility testing of finished articles to solely a determination of batches "sterility." Furthermore, for any batch contamination event, a relationship exists between the mean number of microorganisms per unit and the frequency of units within a batch containing at least one microorganism—see Tables 10.7 and 10.8 [59]. The relative frequency can be calculated using the following equation:

$$Q = 100\left(1 - e^{-m}\right)$$

where:

Q = percentage of units containing at least one microorganism
e = 2.7182818
m = the average number of microorganisms per unit

Applying this calculation, if one batch of a product contains on average one microorganism per individual unit (i.e., vial, container, or device), then only 63% of all units within that batch are likely to contain at least one microorganism. With sampling a finite ($n = 20$) number of units used in any testing strategy, there remains a significant risk of failing to identify a nonsterile unit; no refinement or optimization of sample size, sample choice, or frequency provides a satisfactory or adequate level of sterility assurance by end product sterility testing. The risk of reporting a false negative (Type 2 error), that is, the failure to identify a nonsterile unit by finished product testing is calculable from the following equation [60]:

TABLE 10.7 The Relationship Between Mean Microbial Load in a Batch (Described as Microorganisms/Unit) and the Frequency of Actual Contaminated Unit [59]

Contamination Distribution Mean Microorganisms/Unit	Frequency of Contaminated Units
1.0	63.2%
0.1	9.5%
0.01	1%
0.001	0.1%

TABLE 10.8 The Relationship Between the Frequency of Contaminated Units Within a Batch and the Number of Units Required to be Tested to Identify Contamination [59]

Units Needed to Test Positive Frequency of Contaminated Units	Number of Units Needed to be Test
0.1 (10%)	44
0.01 (1%)	458
0.001 (0.1%)	4603

$$P = 1 - e^{-\lambda}$$

where:
- P = probability of failing the sterility test
- e = 2.7182818
- λ = likelihood of a contaminated unit

Odlaug [61] used this equation to report the magnitude of insufficiency of final product sterility testing and to define the superiority of parametric release (for terminally sterilized product). Tables 10.9 and 10.10 summarize the probability of passing a sterility test (i.e., reporting a type 2 error) with different percentages of a batch contaminated with at least one microorganism. If 0.1% of a batch is contaminated, the risk (probability) of failing to identify nonsterility is 0.98 or 98%. Consider this statistical likelihood of assuring product quality when applying the mandated sterility test in counterpoise to CFR 211.165, which also states "Acceptance criteria for the sampling and testing conducted by the quality control unit shall be adequate to assure that batches of drug products meet each appropriate specification and appropriate statistical quality control criteria as a condition for their approval and release." The aim of the CFR for establishing statistical quality control criteria for sterile products might indeed appear contradictory when we fully appreciate the statistical constraints and culturability limitations of the sterility test.

TABLE 10.9 The Probabilities of Detecting a Nonsterile Unit with Varying Percentages of Unit Contamination Frequencies [61]

Frequency of Contaminated Units	Probability of Detection Probability of Positive Test
1.0 (100%)	1.0 (100%)
0.1 (10%)	0.86 (86%)
0.01 (1%)	0.18 (18%)
0.001 (0.1%)	0.02 (2%)

10.8.3 Superior Assurance of Product Quality is Achieved by Quantitative Risk Assessment

The previously described practical and technical constraints of sterility testing provide a profound justification for applying more innovative, quantitative risk assessment and risk-based approaches to assuring sterility of an aseptically manufactured product. The paradigm combining QbD, quantitative risk assessment integrated into pharmaceutical quality systems as defined in ICH Q8, Q9, and Q10 offers a realistic, practicable means of assuring product sterility to a far superior statistical basis.

Despite the currently accepted adequacy of the sum contribution of all technologies, practices, in-process tests, monitoring systems (qualified on a regular basis by process simulations), and final product testing, the disposition of an aseptically processed product can be greatly improved in terms of rigor and objectivity using QbD and quantitative risk assessment. For a considerable time, the food industry has successfully applied microbial risk models that model, calculate, and quantify the risk to the end consumer from microbial hazards derived from sources along a logic chain or risk pathway [62,63]. The fundamental logic path and associated microbial dynamics described by these quantitative microbial risk assessment methodologies is clearly mirrored by the application of risk hierarchies fundamental to aseptic processing risk assessments. Generating a risk pathway (essentially equivalent to a process flow) by applying risk hierarchies

TABLE 10.10 The Probabilities of Passing a Sterility Test with Varying Percentages of Unit Contamination Frequencies [61]

Frequency of Contaminated Units	Probability of Passing Test Probability of Passing Test ($n = 20$)
1.0 (100%)	0
0.1 (10%)	0.14 (14%)
0.01 (1%)	0.82 (82%)

at each process step and populating the associated risk components and risk factors with data that truly quantitatively describes the probability of microbial ingress realize the Q8-Q9-Q10 paradigm. Tidswell and McGarvy [19,27] innovated, developed, and successfully applied this strategy to quantitatively describe parenteral product sterility during aseptic manufacturing. Successful exploitation of this strategy is based on three fundamental tenets. First, that a single microorganism accessing an aseptically manufactured or manipulated product is unacceptable. The notion that species type per se is not imperative to patient infection and therefore salient to nonsterility is legitimate; fundamentally, any microorganism has the potential to cause patient harm and is salient to assessment of patient risk. Although a recent compelling discourse has pragmatically evaluated the risk from a limited number of microorganisms present within aseptically manufactured products [64], this contradicts the fundamental aim of aseptic manufacture—the generation of products devoid of microorganisms. Ironically, we continually seed our own bloodstream with microorganisms originating from our own inherent microflora and accessing at distinct locations. For example, the oral and buccal cavity contains large numbers of microorganisms accessing our bloodstream via small lesions and imperfections in our gums and oral mucosa. Any attempt to justify a low level of microorganisms within a parenteral product must be founded upon a proven clinical rationale, highly impractical and uneconomical to attempt. As previously described in terms of assessing risk to an aseptically manufactured product from the ingress of microorganisms, the fundamental risk is described by the probability of ingress and does not include evaluation of severity as a consequence of the realization of risk (Eq. 10.2).

The second fundamental tenet is that measurement uncertainty exists in the enumeration of microorganisms; this is especially the case when growth-based technologies are implicit in the sterility assurance program. It is appropriate, however, to account for this uncertainty and variability by the description of the magnitude of the microbial hazard by probabilistic population distributions. What is important to know is the maximum and minimum values possible, and the likely value. These describe the spread, shape, and likely bounds of any quantity of microbial hazard present. Traditionally, uncertainties implicit in the logic chain of quantitative microbial risk assessment methodologies have been accounted for using Monte Carlo simulations [65]; however, Bayesian belief networks may represent more sophisticated means of accounting for stochasticity [66]. These tools permit a systematic and repeated generation of random values for the microbial risk factor variables to populate probability distributions. In this manner, all possible permutations are evaluated to permit a thorough evaluation of risk, rather than rely on a single value that is likely inaccurate, somewhat subjective, and may bias the determination of risk. Simultaneous interpolation of all contributing permutations far exceeds our own mental acuity and does demand the use of software to generate the quantitative assessment of risk.

Finally, the third fundamental tenet is that the transfer of microorganisms from a source or location and any consequential ingress into an aseptically manufactured product can be defined, measured, and validated. Any microbial hazard is subject to the same forces and vectors acting upon inanimate particles; concomitantly, the physical disposition and movement must be predictable. Quite simply, under defined environmental conditions (including consideration of air flow dynamics, temperatures, humidity), the probability of a microorganism accessing a product, resulting in nonsterility, is dependent upon the number of microbial hazards and the time duration the product is vulnerable to ingress. The subsequent algorithm or transfer coefficient permits the determination of a statistically sound evaluation of product contamination; knowing the amount of a microbial hazard within an environment and the duration for which a product is vulnerable may thus be converted into a likelihood of contamination. Determination of such a transfer coefficient is highly dependent upon a well-defined testing environment with proven air flow dynamics and microbial distributions. Tidswell and McGarvey [19] achieved this relying on a quantification of microorganism using growth-based recovery and quantification, qualifying the methodology by direct comparison to process simulation data [27]. Within the risk assessment, transfer coefficients may also be assigned probabilistic distributions to build into the quantification an appreciation for uncertainty and variability. Augmentation with the application of real-time rapid microbial monitoring by the likes of cytometry [67,68] facilitates the real-time quantification of risk and a statistical determination of product sterility far beyond that feasible by the combination of current aseptic technologies, testing product disposition, and processes. Integration of real-time microbial monitoring data with an automated software platform is as yet unrealized but would provide a continuous automated risk calculation (ARC) for real-time deterministic assessment of product sterility, imparting a far superior assurance of sterility than that achievable by sterility testing.

10.9 FUTURE TRENDS AND THEMES

Healthcare will continue to experience a growing population of more aged patients and the continued burden of microbial infections linked to clinical administration of a diversifying range of therapies. It is foreseeable that where feasible healthcare practices, procedures, and treatments will be devolved out into community, clinical, and outpatient settings, diminishing the risk of hospital-associated infections and complications. This will further accentuate our need to understand and design out risk to product sterility in the context of out-of-hospital environments and settings. Economic and regulatory pressure will additionally encourage the automation and refinement of risk assessment processes and techniques that are not so heavily reliant on human input and where the uncertainty and subjectivity of content is designed out. Adequate assurance of the absence of microorganisms (sterility) from a product will likely never be achievable by end product testing. Universal acknowledgement of this fact will necessitate alternative strategies assuring sterility; automated risk calculation (ARC) for real-time, as described earlier, is one strategy likely to succeed. It is clear that risk analysis and risk evaluation of aseptic manufacture and aseptic processing will remain an integral part of the product and process life cycle. Risk assessment through the life cycle demands the diligent administration and management of "living" documents in the contemporary context and necessarily employing multiple tools. This chapter has provided some direction for prudent choice of the most pertinent tools for expedient, economical, and efficient execution of aseptic risk assessments.

REFERENCES

1. PDA. Technical Report No. 1, Validation of Moist Heat Sterilization Processes: Cycle Design, Development, Qualification and Ongoing Control. PDA; J Pharm Sci Technol 61(supplement S-1).
2. Agalloco JP, Akers JE. A critical look at sterility assurance. Eur J Pharm Sci 2002;7(4):97–103.
3. Agalloco JP, Akers JE. The myth called sterility. Pharm Technol 2010;34 Available at http://pharmtech.findpharma.com/pharmtech/Analytics/The-Myth-Called-Sterility/ArticleStandard/Article/detail/660544.
4. Tidswell. Risk assessment in parenteral manufacture. In: Williams KL, editor. Endotoxins, Pyrogens, LAL Testing and Depyrogenation. 3rd ed. New York, USA: Informa Healthcare; 2007.
5. Whyte W. A cleanroom contamination control system. Eur J Parenter Sci 2002;7(2):55–61.
6. McCormick D. Poor OOS review leads causes FDA citations. Pharm Technol 2006;30Available at http://pharmtech.findpharma.com/pharmtech/In+the+Field/December-2006/ArticleStandard/Article/detail/390980.
7. Jimenez L. Microbial diversity in pharmaceutical product recalls and environments. PDA J Pharm Sci Technol 2007;61(5):383–399.
8. Tellez S, Casimiro R, Vela AI, Fernandez-Garayzabal JF, Ezquerra R, Latre MV, Briones V, Goyache J, Bullido R, Arboix M, Dominguez L. Unexpected inefficiency of the European pharmacopoeia sterility test for detecting contamination in clostridial vaccines. Vaccine 2006;24:1710–1715.
9. Pittet D. Nosocomial bloodstream infections. In: Wenzel RP, editor. Prevention and Control of Nosocomial Infections. 3rd ed. Baltimore (MD): Lippincott Williams and Wilkins; 1997. p. 711.
10. Soufir L, Timsit JF, Mahe C, Carlet J, Regnier B, Chevret S. Attributable morbidity and mortality of catheter-related septicemia in critically ill patients: a matched, risk-adjusted, cohort study. Infect Control Hosp Epidemiol 1999;20(6):396–401.
11. Öncü S, Sakarya S. Central venous catheter-related infections: an overview with special emphasis on diagnosis, prevention and management. Internet J Anesthesiol 2003;7(1) Available at http://www.ispub.com/ostia/index.php?xmlFilePath=journals/ija/vol7n1/cvc.xml.
12. Safdar N, Maki DG. The pathogenesis of catheter-related bloodstream infection with noncuffed short-term central venous catheters. Intensive Care Med 2004;30(1):62–67.
13. Tidswell EC, Rockwell J, Wright M-O. Reducing hospital acquired infection by quantitative risk modeling of intravenous (IV) bag preparation. PDA J Pharm Sci Technol 2010;64:82–91.

14. A risk management standard. Available at http://www.theirm.org/. 2002.
15. ISO 31000. Guidelines on principles and implementation of risk management. 2009.
16. ICH Q9 Quality risk management. Available at http://EMEA/INS/GMP/157614/2005-ICH, http://www.emea.eu.int/pdfs/human/ich/15761405en.pdf#search='ich%20q9. 2005.
17. Guidance for Industry. Sterile products produced by aseptic processing–current good manufacturing practice, US FDA, September. 2004.
18. FDA risk pharmaceutical cGMPs for the 21st century: a risk-based Approach, US FDA. 2004.
19. Tidswell EC, McGarvey B. Quantitative risk modeling in aseptic manufacture. PDA J Pharm Sci Technol 2006;60(5):267–283.
20. Langille S, Ensor L, Hussong D. Quality by design for pharmaceutical microbiology. Am Pharm Rev 2009;12(6):80–85.
21. Hillson D. Business uncertainty. Threat or opportunity. Ethos 1999;13:14–17.
22. Sandal T. An approach for the reporting of microbiological results from water systems. PDA J Pharm Sci Technol 2004;58(4):231–237.
23. Akers J, Agalloco J. Risk analysis for aseptic processing: the Ackers-Agalloco method. Pharm Technol 2005;29(11):74–88.
24. Whyte W, Eaton T. Microbial risk assessment in pharmaceutical cleanrooms. Eur J Parenter Pharm Sci 2004;9(1):16–23.
25. Tidswell EC. Bacterial adhesion: considerations within a risk-based approach to cleaning validation. PDA J Pharm Sci Technol 2005;59(1):10–32.
26. Agalloco J, Akers J. Validation of aseptic processing. In: Agalloco J, Carleton FJ, editors. Validation of Pharmaceutical Processes. 3rd ed., InformaUSA, New York, 2007:317–326.
27. Tidswell EC, McGarvey B. Quantitative risk modeling assists parenteral batch disposition. Eur J Parenter Pharm Sci 2007;12(2):3–7.
28. WHO expert committee on specifications for pharmaceutical preparations. WHO Technical Report Series 908, World Health Organization, 99–112. 2003.
29. ISO 14971:2007(E). Medical devices – application of risk management to medical devices. 2007.
30. Tidswell. Risk-based approaches facilitate expedient validations for control of microorganisms during equipment cleaning and hold. Am Pharm Rev 2005;8(6):28–33.
31. Akers J, Agalloco J. The simplified Akers-Agalloco method for aseptic processing risk analysis. Pharm Technol 2006;30(7):60–76.
32. Ljungqvist B, Reinmuller B. Risk assessment with the LR-Method. Eur J Parenter Pharm Sci 2002;7(4):105–109.
33. Ackers J. Development of Highly Automated PAT-Compatible Aseptic Processing System. Biaritz, France: A3P Congress; 2004.
34. United States Pharmacopoeia. Available at http://www.uspnf.com/uspnf/login?bypass=false. 2010.
35. European Pharmacopoeia. Available at http://online6.edqm.eu/ep607/. 2010.
36. Pharmacopoeia of the people's Republic of China. 2005.
37. Japanese Pharmacopoeia. Available at http://jpdb.nihs.go.jp/jp14e/. 2001.

38. EudraLex The Rules Governing Medicinal Products in the European Union, Volume 4 EU Guidelines to Good Manufacturing Practice Medicinal Products for Human and Veterinary Use Annex 1 Manufacture of Sterile Medicinal Products. Available at http://ec.europa.eu/enterprise/sectors/pharmaceuticals/documents/eudralex/vol-4/index_en.htm#h2-annexes. 2008.
39. Costello EK, Lauber CL, Hamady M, Fierer N, Gordon JI, Knight R. Bacterial community variation in human body habitats across space and time. Science 2009;326:1694–1697.
40. Anderson T-H. Microbial eco-physiological indicators to asses soil quality. Agric Ecosyst Environ 2003;98:285–293.
41. Torsvik V, Golsor J, Daae F. High diversity in DNA of soil bacteria. Appl Environ Microbiol 1990;56:782–787.
42. Atlas RM, Bartha R. Microbial Ecology: Fundamentals and Applications. Redwood (CA): Benjamin/Cummings; 1998.
43. Gao Z, Tseng C-H, Pei Z, Blaser MJ. Molecular analysis of human forearm superficial skin bacterial biota. Proc Nat Acad Sci 2007;40(8):2927–2932.
44. Wade W. Unculturable bacteria-the uncharacterized organisms that cause oral infections. J R Soc Med 2002;95:81–83.
45. La Duc MD. Isolation and characterization of bacteria capable of tolerating the extreme conditions of clean room environments. Appl Environ Microbiol 2007;73(8):2600–2611.
46. De Kimpe SJ, Kengatharan M, Thiemermann C, Vane R. The cell wall components of peptidoglycan and lipoteichoic acid from Staphylococcus aureus act in synergy to cause shock and multiple organ failure. Proc Nat Acad Sci USA 1995;92:10359–10363.
47. Wray GM, Foster SJ, Hinds CJ, Thiemermann C. A cell wall component from pathogenic and non-pathogenic gram-positive bacteria (peptidoglycan) synergizes with endotoxin to cause the release of tumour necrosis factor-alpha, nitric oxide production, shock and multiple organ injury/dysfunction in the rat. Shock 2001;15:135–142.
48. Eaton T. Microbial risk assessment for aseptically prepared products. Am Pharm Rev 2005;8(5):46–51.
49. Tidswell EC, Bellinger M, McCullough D, Allexander A. Consecutive replicate contact plate sampling assists investigative characterization of surface-borne bioburden. Eur J Parenter Pharm Sci 2005;10(4):93–96.
50. Sutton S. Is real-time release through PAT compatible with the Ideal of "Science-Based Regulation?". Pharm Technol 2007 February. 2, 2007, www.pharmtech.com.
51. Whyte W. Collection efficiency of microbial methods used to monitor cleanrooms. Eur J Parenter Pharm Sci 2005;10(2):43–50.
52. Anderson T-H. Microbial eco-physiological indicators to asses soil quality. Agric Ecosyst Environ 2003;98:285–293.
53. Torsvik VG. High diversity in DNA of soil bacteria. Appl Environ Microbiol 1990;56:782–787.
54. Errington J. Determination of cell fate in *Bacillus subtilis*. Trends Genet 1996;12:31–34.

55. Losick R, Dworkin J. Linking asymmetric division to cell fate: teaching an old microbe new tricks. Genes Dev 1999;13:377–381.
56. Kell DB, Young. Bacterial dormancy and culturability: the role of autocrine growth factors. Curr Opin Microbiol 2000;3:238–243.
57. Sachidanandham R, Gin KY-H. A dormancy state in nonspore-forming bacteria. Appl Microbiol Biotechnol 2009;81:927–941.
58. Bryce D. Tests for the sterility of pharmaceutical preparations. J Pharm Pharmacol 1956;8:561–572.
59. Spicher G. Die mikrobiologie zwischen anspruch und wirklichkeit. Zentralbl Hyg. Unmweltmed 1993;3:223–235.
60. Moldenhauer JS. Towards and improved sterility test. PDA J Pharm Sci Technol 2004;58(6):284–286.
61. Odlaug TO. Sterility assurance for terminally sterilized products without end-product sterility testing. J Parenter Drug Assoc 1984;38(4):141–147.
62. Nauta MJ. Separation of uncertainty and variability in quantitative microbial risk assessment models. Int J Food Microbiol 2000;57:9–18.
63. Thompson KM, Burmaster DE, Crouch EAC. Monte Carlo techniques for quantitative uncertainty analysis in public health risk assessments. Risk Anal 1992;12:53–63.
64. Whyte W, Eaton T. Assessing microbial risk to patients from aseptically manufactured pharmaceuticals. Eur J Parenter Pharm Sci 2004;9(3):71–77.
65. Vose D. Quantitative Risk Analysis: A Guide to Monte Carlo Simulation Modeling. New York: John Wiley & Sons. Inc.; 1996.
66. Smid JH, Verloo D, Barker GC, Havelaar AH. Strengths and weaknesses of Monte Carlo simulation models and Bayesian belief networks in microbial risk assessment. Int J Food Microbiol 2010;139(1):S57–S63.
67. Miller MJ, Lindsay H, Valverde-Ventura R, O'Connor MJ. Evaluation of the BioVigilant IMD-A, a novel optical spectroscopy technology for the continuous and real-time environmental monitoring of viable and nonviable particles. Part I: Review of the technology and comparative studies with conventional methods. PDA J Pharm Sci Technol 2009;63(3):244–257.
68. Miller MJ, Walsh MR, Shrake JL, Dukes RE, Hill DB. Evaluation of the BioVigilant IMD-A, a novel optical spectroscopy technology for the continuous and real-time environmental monitoring of viable and nonviable particles. Part II: Case studies in environmental monitoring during aseptic filling, intervention assessments and glove integrity testing in manufacturing isolators. PDA J Pharm Sci Technol 2009;63(3):258–282.

11

PHARMACEUTICAL PRODUCT MANUFACTURING

MARLENE RASCHIATORE

This chapter focuses on the areas of risk that a drug company may encounter in pharmaceutical manufacturing. The chapter specifically addresses oral solid and liquid formulations, and uses a case study to illustrate how to apply a risk management tool to identify and mitigate risks that could affect product quality and patient safety. The beneficial outcome of this process is that the manufacturer critically evaluates the operation and from it gains product and process understanding. This information can then be used to make more informed decisions on a day-to-day basis if and when quality issues arise.

All pharmaceutical product manufacturing processes have inherent risks that can impact the product quality and patient safety of a drug product. Therefore, it is important to assess the risks in each manufacturing process. The assessment starts by identifying and understanding the potential risks, then controlling them to an acceptable level to ensure that the product consistently meets approved quality standards for patient safety. The use of risk assessment plays an important role in the following:

- providing a systematic approach to identify potential risks;
- providing measures to mitigate their occurrence;
- increasing awareness of what will happen in the event of a failure;

Risk Management Applications in Pharmaceutical and Biopharmaceutical Manufacturing, First Edition. Edited by A. Hamid Mollah, Mike Long, and Harold S. Baseman.
© 2013 John Wiley & Sons, Inc. Published 2013 by John Wiley & Sons, Inc.

- determining what areas need process improvements; and
- increasing product and process understanding.

11.1 ROLE OF QUALITY RISK MANAGEMENT

Risk assessment as part of an effective risk management program directs the product and process knowledge gathering process [1–11]. Decisions are made about how to assess, control, monitor, and/or accept risk based on predicted benefits and possible consequences. It determines through a systematic approach the product attributes and process parameters that are truly critical to the final product quality and safety and which are not critical. The value of this approach is that time, money, and other resources are not wasted on controlling, monitoring, or validating parameters that do not ensure final product quality.

Regulators are encouraging the use of risk assessment in the beginning during product development where the expected aim is to build quality into the product by design [12–16]. ICH Q8 (R2), Pharmaceutical Development, recognizes risk assessment as a tool to be used early on and throughout the development process to determine at a minimum, those aspects of drug substances, excipients, container closure systems, and manufacturing processes critical to product quality and to define their control strategy. Through a series of planned and designed experiments, these different aspects and any interactions between them can be examined. The end result is the creation of a comprehensive product profile that not only provides confidence that the final product will be safe and efficacious but also provides a foundation on which the impact of future changes can be predicted.

Although the critical product attributes and process parameters and their control strategy are best determined during the drug development stage, the risk assessment process can be applied at any point in the product's life cycle. Regulators are expecting that the pharmaceutical manufacturer knows the risks associated with their product and manufacturing process, which could result in product quality or safety issues whether the drug product is in development or already on the market.

Pharmaceutical manufacturers have been practicing some form of risk management, formally or informally, for some time to improve process efficiencies or investigate product quality issues. Historically, the risk management practices have been deductive in nature and performed through an informal process. A formal process is not always appropriate or necessary and informal processes are acceptable as long as the rationale is logical, documented, and the process follows company policy. What should be avoided is an "I made the decision in the coffee room" type of process where decisions may not be documented. Otherwise, the risk decision-making process may be too subjective rather than empirical and may not be well documented. Inadequate documentation and justification of risk decisions could open the door for questions by regulators. Proper documentation of the decision-making process and decision justification are essential to provide regulators with greater assurance of a manufacturer's ability to deal with potential risks. The use of risk assessment tools provides a good structure for prompt and adequate documentation of the information gathered and justification for the decisions made during the risk assessment process.

11.2 RISK ASSESSMENT IN PHARMACEUTICAL MANUFACTURING

Pharmaceutical manufacturing has a broad scope that includes many different types of drug products that can be administered differently and manufactured under different conditions. Drug product formulations range from powders, liquids, solids, creams, to ointments that can be administered as oral, parenteral, topical, transdermal, or inhalant. Each dosage form can be manufactured using different process steps and under different environmental conditions ranging from aseptic to nonaseptic, with varying levels of control depending on the risk of microbiological contamination to product quality. Some processes may not be aseptic but still have a requirement for bioburden control. For example, antiacids, some liquids, respiratory therapy products, or even terminally sterilized products are not manufactured aseptically, but there may be microbiological contamination risks to product quality.

The different combinations make it difficult to discuss in one chapter the risks associated with all product types included in the scope of pharmaceutical manufacturing. Risks associated with product quality and safety for one product type may or may not apply to another. For example, the risks impacting product quality in the manufacture of parenteral products where sterility is required for patient safety would obviously differ from risks in the manufacture of solid oral dosage products where sterility is not required for patient safety. Likewise, within one product type, some of the risks impacting product quality may be the same, while other risks may differ because one or more process steps differ. For example, the risks associated with the drying and milling process steps for a solid oral tablet may differ from the same steps for another solid oral tablet. The versatility of the risk assessment process is that it can be applied to manufacturing process steps for any product type.

The exercise should be product- and process-specific so potential risks are not missed. Ideally, a risk assessment would be performed for each product, but products that use the same operations can be grouped into families so that knowledge from common or shared process steps may be leveraged. For a company that manufactures more than one product, the initial step would be to identify each of the products, the process steps for each, and any shared or common process steps between products to help leverage common process risks across the different products. It would be important to document the criteria used and justification to group products into families in case questions or issues arise later on. Each risk-based assessment should be documented, and the results and conclusions approved by the site quality and production authorities.

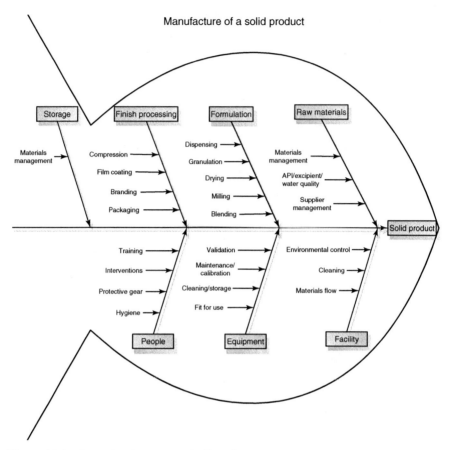

Figure 11.1 Example of a cause and effect diagram showing the manufacture of a solid product. (*See insert for color representation of the figure.*)

At first it may be best to start out by identifying all the manufacturing steps and inputs for a particular product. There are tools that can help. For example, one tool is a cause and effects diagram also known as a fishbone diagram [17]. Figure 11.1 is an example of a fishbone diagram where all the processes involved in the manufacture of a solid tablet are listed.

In this fishbone diagram, the top half of the fishbone contains the different major steps of the manufacturing process starting with the receipt of incoming raw materials in the warehouse (at the head of the fishbone) through the formulation steps (e.g., dispense, blend), the finish process steps (e.g., compression, packaging), and ending with the finished product storage in the warehouse (near the tail of the fishbone). The bottom half contains components that support the manufacturing process and would have an impact on the outgoing product quality: facility, equipment, and people.

The diagram lists the major steps or components at the end of the larger fishbone with substeps or components listed beneath on the smaller fishbones. For example, under the major step, "finish processing," the substeps are compressing, coating, filling, and packaging.

A multifunctional team of subject matter experts for the product, along with a skilled facilitator experienced in the risk assessment process, are gathered. As part of the analysis, the team would gather relative product and process knowledge already known such as basic scientific knowledge, historical test results, reject data, or stability data. From here, it can be determined what product or process knowledge is missing and if a plan is needed to acquire the knowledge. This information could be used to prioritize the assessments/studies.

The team would decide the scope of the assessment that will be undertaken and develop a process flow map. For example, Figure 11.2 illustrates the process flow map of a solid tablet starting with the incoming raw materials warehouse operation and ending with the finished product warehouse storage operation. The process flow diagram described in Figure 11.2 will also be used later in the chapter for the case study.

The team would identify the risks associated with each process step and look for those process steps and the process parameters critical to ensure final product quality. Any preventive measures in place to mitigate the risk are also identified and evaluated to determine if the measures are adequate or additional measures are needed.

It is important that the risk assessment is not done just sitting in a conference room. Walk the process. Understand the process looking at it out on the production floor and identify the hazards as they could occur there. Several process maps can be used to cover individual subprocess steps or components.

To help demonstrate the assessment exercise and provide some examples of potential risks to product quality in pharmaceutical manufacturing, the process

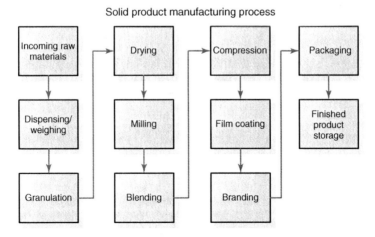

Figure 11.2 Example of a process flow chart.

steps listed in Figure 11.2 and others listed in Figure 11.1 were assessed and the potential risks identified are discussed later. In addition, a list of potential hazards and harms is included in Appendix I.

11.2.1 Raw Materials

11.2.1.1 Materials Management The risk assessment needs to start at the beginning of the manufacturing process. The first area considered here is incoming materials. The quality of the materials that will be used in the manufacture of a finished product needs to be assessed and protected starting at the receiving dock. There are at least three areas that require assessment.

One area is the physical examination of the materials when they are delivered to the receiving dock. What are the risks to quality under the following conditions?

- No one is checking to see that incoming materials are properly identified.
- Packaging is not intact to protect the contents from tampering or contamination.
- Packaging is not clean and dry.
- Proper storage conditions such as temperature or humidity were not maintained in transit as needed.

The other area is the physical protection of the materials once delivered. There are inherent risks to the quality of the materials if warehouse staging and storage areas are under the following conditions:

- not physically protected from ingress of environmental elements such as rain, wind, and possibly insects, birds, or rodents;
- have standing water that can invite the proliferation of mold; and
- have poorly defined material handling procedures that are not designed to protect the materials and their packaging from damage or contamination during warehouse storage operations.

Where materials have certain temperature or humidity requirements to maintain quality, such as a refrigerated temperature or controlled room temperature, the assessment will need to address whether there are adequate mechanisms in place or needed to maintain and monitor the storage conditions.

Lastly, there are risks involved if the materials are not properly handled according to their "release" status. Typically, the same storage facility is used to store both incoming materials and finished product in the quarantine, released, and rejected stage. For incoming materials, an apparent risk is the unintentional use of rejected materials to manufacture finished product. For finished products, the apparent risk is the inadvertent distribution of rejected products to the market. The risk assessment would include whether there are adequate systems in place to properly identify the status of materials (i.e., quarantine, released, rejected), segregate materials according to their status as needed, and ensure only released materials are sent to the production line or products to market.

11.2.1.2 Active Pharmaceutical Ingredients and Excipients Raw materials used to formulate the product need to be potent, safe, of a defined purity, and be of the correct composition, identity, and quality. If any of these aspects are compromised, product safety and efficacy are at risk. The risk assessment should address any of the chemical, physical, and microbial attributes of the material that are critical to product safety and be in relation to the specific use of the material. For example, water is used in the manufacture of both a liquid and a solid product. Poor microbial water quality would be a relatively greater risk for a liquid product where water is the main ingredient as compared to the manufacturing of a solid product where typically water plays a much smaller role and is eliminated by drying during the process.

In particular with raw materials, lot-to-lot variability of incoming lots poses a major risk to the quality of the finished product. Those characteristics of the material critical to consistently delivering the intended product quality need to be determined and controlled for incoming lots. For example, if low moisture content of a powdered material is critical to its use for a solid product, then moisture would be a characteristic required to be at a certain level in every lot received. Lots with high moisture content could adversely affect the granulation or drying processes. The bulk product may be too moist to properly compress to the proper hardness, affecting the final unit potency.

Particle size may be a characteristic identified to be critical to product quality. In a liquid formulation, a larger particle than expected could require more mixing time to allow for longer particle-to-water contact or greater agitation for full dissolution of the material. Likewise, in solid dose manufacturing, larger particles may not blend properly during the granulation process. Inadequate mixing or blending could produce a nonhomogeneous batch where pockets of the batch are superpotent, containing undissolved, concentrated amounts of active material, and other portions of the batch are subpotent, containing substandard amounts of active material. Depending on how and where the batch is sampled for testing, the discrepancy may not be uncovered during in-process or final product testing. The obvious safety risk is that the patient may either be overdosed (superpotent) or inadequately treated (subpotent).

Control of lot-to-lot variation in the microbial quality of incoming active and excipient ingredients is important for finished products that have an established microbiological limit. The control strategy would need to include monitoring of incoming lots for microbial content as well as controls to protect the material from microbial contamination during the sampling process from personnel, instruments, and the sampling environment.

11.2.1.3 Supplier Management In-house testing of the critical attributes in raw materials allows a manufacturer to know their incoming quality levels. Monitoring the incoming quality mitigates the risk of using materials that will not meet finished product specifications. It does not, however, mitigate the risk of receiving materials that do not meet incoming specifications. A drug manufacturer should have confidence that the quality of material received from the supplier will not vary from lot-to-lot and consistently will meet specifications. A good supplier management program, which selects a supplier based, at a minimum, on a good performance, validation status, and compliance with regulations, will improve the manufacturer's confidence that the materials received will consistently meet quality specifications. The supplier management program should be documented with rationales for supplier selection.

11.2.2 Formulation

The exact mixture and concentration of active pharmaceutical ingredient (API) and excipients are crucial for patient health and safety. Healthcare providers and patients inherently trust that the product delivers the intended potency and expected results if taken as listed in the product labeling. Formulation errors can impact product identity, potency, and stability, resulting in a product that fails to deliver on its intended purpose. Therefore, every formulation process step must be assessed for risks starting from selection of raw materials to completion of the in-process bulk.

In formulating a product, there are inherent risks with equipment so a specific reference to them will be made here (see Chapter 7 for equipment commissioning and qualification considerations). Knowing the risks and the potential outcome when equipment is substituted, incorrectly setup, or is used with the incorrect operating parameters will be helpful information when quality issues arise during routine operations. Several examples of how equipment can impact the potency, safety, or stability of the formulated bulk product are as follows:

- The use of equipment such as balances with less sensitivity than required may result in insufficient quantities of raw materials used (e.g., measures in whole numbers only when two decimal places are needed).
- A bulk tank of a different size or shape may result in insufficient space to allow for proper mixing or blending.
- A mixer of a different design, shaft length, or propeller size or speed capability may not be adequate to successfully complete the dissolution or blending step in the defined timeframe.
- Use of wrong mixer speed may not adequately dissolve materials or provide a uniform batch by the end of the defined mixing time.

11.2.2.1 Dispensing The steps involved in selecting the correct materials, weighing or measuring, and protecting them from contamination during the dispensing operation tend to be manual in nature. Therefore, when identifying the risks, close attention needs to be paid to the quality of operator instructions and managing operator performance. Procedures need to be sufficiently detailed and easy to follow, specifying the equipment or instruments to be used so that the operator has adequate instruction to execute each step precisely.

Some incoming materials are packaged in large quantities to be used in multiple batches. The risks of contamination from multiple operators, sampling equipment, and environmental exposure are greater for these materials than for the materials packaged for use in a single batch. The assessment would determine if controls are adequate to protect the material from chemical and environmental contamination during the weighing step. In addition, operators would need to employ procedural controls to protect the material's identity, making sure that the material is properly relabeled, reconciled, and returned to the proper storage area after use.

The obvious risk associated with incorrect weighing or measuring of materials is producing a batch that is subpotent or superpotent. Human errors such as operator technique, calculation, and labeling errors are common for weighing and measuring steps. Any of these errors can result in formulating a batch with the wrong amount of ingredients or the wrong ingredients. Errors can often be mitigated by a check by a second operator. A determination should be made if second checks are needed for calculations, measuring, or labeling and if they will adequately mitigate the risk. Operator technique can impact the final potency result if materials are not handled with care so that already weighed or measured material is completely transferred into the dispensing container without spillage. Controls should define what action is taken if accidental spillage occurs to ensure only the correct amounts of materials are used and to protect subsequently dispensed materials from potential cross-contamination.

Other factors play a role in this part of the manufacturing process including work flow. Operational steps that are handed-off from one department to another can lead to product error. For example, materials can be delivered to the incorrect formulation area, creating a risk that the bulk product will not be made in accordance with the correct batch instructions. Once materials are delivered, formulation operators should verify that the materials received are correct before use.

11.2.2.2 Granulation It is important to know what steps in the process of adding materials together to form a batch are critical in achieving the correct batch formulation that supports product potency and stability and include them in the risk assessment. For example, in some formulations the order of materials added, the amount of material per addition, or the timing between material additions is critical to product potency or stability. Once the critical steps are identified and defined, they should be fully described in written instructions such as SOPs or batch instruction records for personnel to follow.

Other common risks in the granulation process involve improper mixing of materials as a result of operator error. Installing the wrong chopper blades or applying the wrong mixing speed or time parameters can result in the risk that the granulation process is incomplete, leading to product instability or incorrect potency.

11.2.2.3 Drying If the granulation is a wet granulation process, a drying step is typically applied at the end of the process. The risks associated with this step include inadequate drying and uncontrolled or improper wet holding times.

Inadequate drying parameters (e.g., time and temperature) could result in a bulk that is too moist, which may ultimately impact product stability or its inability to be further processed. If the bulk is held too long in a wet state before drying, the wet conditions may allow chemical reactions between the materials to take place. These chemical reactions may change the final composition of the bulk such as forming a different product or by-products that could be detrimental to patient safety. There may be a need for wet holding times to be defined to avoid conditions that support chemical reactions.

11.2.2.4 Milling After drying, the bulk is typically processed through a mill or metal screens that break up clumps that may form as a result of the granulation and drying processes. In this process, it is important that the operating parameters and equipment setup are correct. Inadequate milling time or the wrong size milling screen can result in a nonhomogeneous batch, with clumps that will not be blended correctly. In addition, incorrectly fitted screens or warped screens can cause the metal screens to rub against each other, causing a risk of metal shearing and contaminating the batch with metal particulate.

11.2.2.5 Mixing/Blending The bulk granules are combined with a solution that will act as a lubricant to aid in the flow of the powder to the compression area, the next step in the process. In this process, there is a risk that the solution being added or that the amount of solution being added is incorrect. Material error will result in a product that is mislabeled and may be a risk to patient safety. Insufficient solution or insufficient blending time may result in a batch that will not compress properly, putting the patient at risk to receive a product that has a variable or an incorrect potency.

11.2.3 Finish Processing

11.2.3.1 Compression A powder bulk is compressed into metal cylinders or forms that determine the size, shape, and weight of a tablet. The tablet weight directly determines the delivered dose; and if the tablet weight varies so that some may be underweight, the patient is at risk of not receiving the intended treatment. Some of the common risks associated with weight variation are as follows:

- Inadequate or inconsistent powder flow as the result of equipment operating problems or inadequate humidity control causing the powder to cake. It is important that the powder flows freely at a consistent rate so that the forms are always correctly filled.
- Incorrect equipment setup using the wrong size or type dyes or punches.
- Incorrect pressure and speed operating parameters for the required weight or hardness. Incorrect hardness can result in chipping or flaking of the table or susceptibility to further cracking.

A common risk in the compression process is the improper setup of the metal punches or dyes. If the equipment is not setup so that the metal punches are in alignment, the metal parts will rub together, causing the metal to shear and emit metal particles that contaminate the product.

11.2.3.2 Film Coating The tablets may be film coated. In this step, the tablets are typically placed in a rotating drum, atomized with a coating solution, and air dried. In most cases the coating contains an inactive ingredient that protects the active ingredient inner core formed during the compression step. If the coating is not applied evenly and/or in the correct amount, there could be a risk that the dose is not properly delivered or that the exposure of the active ingredient to the environment may adversely impact the product stability. In addition, the tablet must be dried to ensure that the coating adheres to the core. If the tablets remain wet at the exit of the film coating equipment, they could stick together removing the coating when separated. Factors that can impact a proper film coating process are as follows:

- incorrect solution applied;
- incorrect solution preparation where the solution is too thick or thin, prohibiting even flow or appropriate flow rate;
- equipment malfunctions such as clogged spray nozzles; and
- incorrect operating parameters such as wrong rate and time of solution atomization, drum rotation, or air drying temperature.

11.2.3.3 Branding The tablets may be imprinted with an identifier such as the manufacturer's name. This identifier helps the patients to know that they are receiving the correct drug. An ink solution is typically injected onto the tablets. Similar risks are involved as those described for the film coating:

- incorrect ink solution preparation where the ink is too thick or thin and the flow is not even or consistent;
- equipment malfunctions such as clogged ink jets; and
- incorrect operating parameters such as inadequate pressure to apply ink.

11.2.3.4 Packaging In this step, the product is filled into its final container, which provides protection over its shelf life and provides the patient with information for its safe use. Some of the common risks are as follows:

- wrong product, product mix;
- incorrect label, expiration date, or patient information provided;
- missing or overfilled drug product; and
- inadequate or compromised package integrity.

Of maximum risk in the packaging operation is the potential for product mix-up. Providing the wrong product or the wrong dose can result in serious harm to the patient and send regulatory agencies to your door for investigation. A product mix-up can occur in any facility that manufactures more than one product or product dose. A few situations where product mix can likely occur are as follows:

- on packaging lines that process multiple products and is typically the result of inadequate cleaning of the line between different product batches;
- on packaging lines that are dedicated to one product but are adjacent to another line dedicated to a different product; and
- where the final package contains a combination of more than one dose, (usually two or more distinct dosage forms, typically identified by different product colors or shapes).

In all cases, key to avoiding product mix-up is separation or isolation of different products or product doses. The risk assessment should focus on adequate separation between packaging areas and for combination packaging, separation between the different doses. Here are some areas to consider when doing the risk assessment:

- product flow;
- packaging line clearances (e.g., proper bulk, labels, containers before startup);
- personnel process intervention, interaction with other lines, and work flow (e.g., operators handle more than one packaging line, one operator collects samples or conducts in-process challenges on multiple packaging lines);
- in-process sampling and/or routine challenge test sample handling and reconciliation (e.g., samples used to challenge automated vision color or shape detector);
- packaging line equipment cleaning; and
- facility design.

Tablets may be packaged in a variety of ways, such as in a bottle, a blister pack, or a sample packet. In each case, the equipment must be properly set up so that the tablets are aligned to fall into the package opening in the correct amount. Missing or inadequately filled product containers may result in the risk that the patient will not be properly treated. On the other hand, the patient may become overdosed if the container is overfilled.

The package contains information needed by the patient that identifies the drug and how it should be safely taken. The label on the package includes description of the product, its approved indication, correct dosing regimen, warnings, and expiration date. Incorrect product information is a risk to the patient who may take the wrong product or take a harmful dose.

Package integrity keeps the drug product protected from contamination and helps maintain its stability or potency over its shelf life. Having the proper package integrity requires that the packaging components fit together to form a proper seal. In the case of blister packs and sample packets, the packaging materials (e.g., foil thickness, plastic or paper stock, or glue) and sealing parameters (e.g., temperature, pressure) are critical to the seal formation. Any shortage in the amount of sealant applied or any mistake in setting the temperature, pressure, or timing parameters may result in an inadequate seal. For example, if the sealing temperature was set too low, the adhesive may fail to liquefy and could not be activated to form a seal. Likewise, if the amount of sealant is inadequate or misaligned to the sealing area, a seal would not be properly formed and the patient would be at risk of receiving a contaminated or degraded drug product.

11.2.4 Facility

11.2.4.1 Material Flow Facilities should be designed to allow for ease and efficiency of material and personnel flow, to allow for easy cleaning, and to safeguard product quality by avoiding the possibility of product contamination or cross-contamination. The risk of cross-contamination of one product with another is of particular concern where product areas or facilities are not dedicated to the manufacture of just one product. The risk increases if superpotent or highly allergenic products are being manufactured. There are inherent risks for product cross-contamination associated with the movement of personnel, equipment, and materials in shared production hallways, staging or preparation areas, and in nonproduction areas such as lavatories or cafeterias where employees interact. It may be necessary to modify the facility design to provide dedicated pathways or rooms for each product manufacture, control the use of a common cafeteria to avoid employee-to-employee contact, and have a qualified cleaning program that ensures removal of residual product for multiproduct production rooms or equipment.

11.2.4.2 Cleaning Cleaning agents, technique, and equipment and the required level of microbial bioburden control are factors to be considered in the risk assessment to support the cleanliness level required for the product. General area cleaning, product–contact surface cleaning, and, where applicable, surface sanitization practices should be assessed to ensure that, at a minimum, they adequately reduce the likelihood of environmental and product-to-product cross-contamination. The cleaning process, similar to dispensing, tends to be manual in nature. Therefore, the details of the cleaning program should be clearly defined in procedures that can be easily followed.

The assessment of the cleaning program should include the following:

- the type of cleaning agents;
- cleaning technique;
- cleaning equipment; and
- frequency of cleaning.

Unremoved product residue or other contaminants can pose a risk of introducing the contamination to the next batch when left on room surfaces or equipment. Dried product residue that builds up on surfaces may eventually flake off during subsequent manufacturing operations. Cleaning agents that are appropriate for pharmaceutical use, safe for the environment, and are able to dissolve any residual product or other soil generated during manufacturing operations should be selected. Likewise, where microbial bioburden control is important, sanitizing agents that are effective against the types of microbes found in the manufacturing environment should be chosen. Otherwise, the product is potentially at risk of microbial contamination from uncontrolled microbial levels in the manufacturing environment. For both cleaning agents and sanitizing agents, it is important to demonstrate that the agents selected at their use-concentration are effective.

In addition, there are risks that the agents could be rendered ineffective if not handled or stored properly. Handling and storage practices should be assessed and designed to ensure that the agents will be effective when used. For example:

- Are the handling practices designed to avoid contaminating the agents during preparation, use, or in storage?
- Are the agents stored under conditions that support their stability and efficacy at the time of use?
- Are there self-life expiration dates needed for in-use containers?

Just as there may be a risk with product contamination if cleaning and sanitizing agents are not used, using them may pose another risk. In all cases, risks introduced by taking measures to mitigate other risks need to be considered. In this case, agents are being used to reduce the risk of product contamination. However, some agents have been known to leave a residue after use that could potentially contaminate product if not controlled.

Just as important as the agents chosen is the techniques used to clean. The risk is introducing contamination to the manufacturing environment. These techniques should be designed to avoid redepositing contaminants onto already cleaned surfaces. Some examples that the risk evaluation would consider are the following:

- Order of rooms to be cleaned. What is the risk of introducing contamination if the most-soiled areas would be cleaned first followed by the least soiled ones or vice versa?
- Order and direction of surfaces within a room. What is the risk of introducing contamination if the floors rather than the walls are cleaned first or if they are cleaned starting at the back of the room rather than the front, from top-to-bottom or left-to-right?
- Number of passes of the mop, cloth, or rinse water before it is refreshed or replaced. What is the risk of introducing contamination if the mops are not rinsed, changed between rooms, or changed after a defined number of passes?

RISK ASSESSMENT IN PHARMACEUTICAL MANUFACTURING

Cleaning and sanitizing equipment are also part of the risk assessment. Some other factors that the assessment should consider are as follows:

- equipment type, e.g., mops, sponges, or buckets;
- materials of construction, e.g., cloth or sponge mops, cloth wipes or sponges, plastic or stainless steel buckets;
- cleanability of the equipment;
- size or ease of handling; and
- durability or life expectancy of the equipment.

If not readily cleanable, buckets, mops, and cloths can hold soil or contaminants from previous uses and redeposit or introduce these contaminants during subsequent uses. If the mops or cloths are not made from nonshedding materials, they could emit particles and thus contaminate surfaces or the environment.

The frequency of cleaning or sanitization can impact the cleanliness level. An insufficient frequency to keep contaminants at a controllable level poses the risk of product contamination.

11.2.4.3 Environmental Controls Each area and manufacturing room should be assessed to determine the level of environmental control needed to protect the product from airborne contaminants such as chemicals, nonviable contaminants, and microbes. The level of control needed will depend on the use of the area or room, the type and level of activities, and the extent to which the product will be openly exposed to the manufacturing environment. Airborne contamination is typically controlled by air filtration and may include air pressurization and temperature and humidity controls. Inadequate air filtration poses the risk that the environment will contain levels of contaminants sufficient to result in product contamination.

Air pressure differentials can be used to control the risk of contamination between adjacent areas with different air quality requirements. A room requiring a higher air quality would have a greater air pressure than that of the adjacent room. In practice, the higher pressure keeps the less-quality air of the adjacent room from entering. For example, a filling room, where a product is likely to be openly exposed to the environment during the process, would require a higher level of air cleanliness than that of an adjacent hallway where product exposure is not likely. To maintain its level of cleanliness, the filling room would have a greater air pressure to hold back the less clean hallway air from entering. If the level of air filtration, level of air pressure, or the air flow pattern is incorrect, the product could be at risk of contamination.

Room air pressurization is often used to protect a product from the risk of cross-contamination where highly potent or hypoallergenic chemical materials are part of the manufacturing process. The air flow pattern should be designed to contain the materials to a given area and keep them from traveling airborne to adjacent areas. Using the filling room example, the air pressure would be greater in the hallway adjacent to the filling room. The higher hallway air pressure would keep the potentially contaminated filling room air from traveling airborne to contaminate adjacent areas.

There may be other environmental conditions that pose a risk to product quality if not controlled. Therefore, product-specific requirements should be assessed to determine if environmental protective measures are required. These may include temperature and humidity or light sensitivity protection. Considerations for determining if temperature and humidity controls are critical to product quality should include the following:

- product hygroscopicity;
- product sensitivity to temperature and humidity;
- product specifications; and
- preclusion of condensate on room and equipment surfaces.

Additional considerations for temperature and humidity control would include operator comfort and equipment operation and exhaust.

Automated systems used to control air filtration, differential pressures, or temperature and humidity should be equipped with indicators (e.g. alarms) that allow operators to know if these systems are functioning or failing to maintain the critical parameters.

The exposure of product-contact packaging materials or clean product-contact equipment to an environment is another area of risk to product quality that should be considered in the risk assessment. It may mean that product-contact packaging materials (e.g., drum liners, super-sacks) or product-contact equipment need to be protected. (e.g., keeping them covered, use of continuous drum liners or use of laminar flow booths).

11.2.5 Equipment and Instruments

In general, any instrument or equipment used to sample, move, store, formulate, hold, dispense, test, clean, sanitize/sterilize, control, or monitor needs to be considered when performing the risk assessment. This includes the more simple manual equipment or instruments, e.g., scoops, container liners, or pH meters, to the more sophisticated electronically operated and controlled equipment or instruments. Different aspects of equipment selection and use should be considered; for example, some aspects to consider are as follows:

- appropriateness for the intended purpose;
- cleanability;
- maintenance and calibration schedule and routine; and
- validation.

11.2.5.1 Appropriateness for Use Each piece of equipment or instrument should be evaluated with regard to its functional role in the process step. As mentioned earlier in Section 11.2.2, with regard to formulation operations and Section 11.2.4 with regard to cleaning the facility, there are inherent risks in using the wrong equipment, equipment not fit for the purpose intended, or using the equipment incorrectly. Specifically, the risks, if any, associated with the equipment or instruments being used in the performance of a particular task are identified.

The assessment should consider if the equipment being used is the appropriate equipment for the job. For example, something as simple as a scoop, a tool typically with a short handle used to sample or extract materials from containers, would be assessed for its use in a given situation. For taking materials from the top of containers, a scoop may be an appropriate tool. However, its use for taking a sample from the bottom of a deep container would not be appropriate because it may require that the operator reach arm's length into the container exposing the material to potential contamination from the operator's clothing or bare skin.

11.2.5.2 Cleaning and Storage Equipment design should be assessed for risks associated with its inability to be completely cleaned, surfaces and especially internal parts that may come in contact with the product. Small internal crevices or open spaces within product-contact equipment can pose a risk of entrapping materials in hidden or impossible to clean locations. A tablet, for example, may be small enough to fall into the internal parts of a filling machine. In such situations, there is an inherent risk that these materials can later be dislodged during the manufacture of a subsequent batch, resulting in a product mix situation. Likewise, powder, liquid, or semi-solid materials can collect in internal pockets, may dry and harden, even possibly degrade to a different material or support microbial growth (e.g., mold) and later become dislodged, contaminating a subsequent batch if not removed.

Once cleaned, storage conditions should be assessed to ensure that they support maintaining the integrity of the cleaned equipment. For example, what is the risk to product quality if a piece of equipment (e.g., formulation tank or dispensing pump) that is cleaned and rinsed with water is stored wet? What if the equipment is not fully drainable and the residual water is allowed to remain? Will the wet condition pose a risk that residual moisture will support microbial growth and contaminate a subsequent batch? In this case, a drying step before storage may be needed to mitigate such a risk. Or, consider whether the storage room air quality or the type and extent of wrapping are adequate, as needed, to protect the equipment from environmental contamination.

11.2.5.3 Maintenance and Calibration Routine equipment maintenance, emergency repairs, and calibration practices should be assessed to ensure that:

- the methods are appropriate and adequate in scope;
- they are performed with sufficient frequency;
- they are performed at appropriate locations and time; and
- they are not performed at a time or in a way that would allow for product contamination.

Agents, if any, used to help equipment operate more smoothly or efficiently, such as lubricants, should be part of the risk assessment to determine their proper selection and use including the safeguards required to avoid their coming in contact with the product or product packaging components. Lubricants that are appropriate for pharmaceutical use and safe for the environment should be selected. The methods used to dispense them should avoid their being expelled into the environment (e.g., sprays) or coming in contact with surfaces that could lead to product contamination.

11.2.5.4 Validation A crucial control component of any process is validation. Once the proper critical steps are identified and it has been determined that these steps are properly controlled, the processes are validated. Validation activities are designed to provide data that confirms that the chosen parameters deliver the product as intended. The question is not whether or not to validate but whether or not the validation activities are adequate for confirmation.

11.2.6 People

Risk associated with personnel should be considered in either of two ways:

1. The risk to product quality resulting from microbial and other particulate contamination from personnel.
2. The risk to personnel who may be exposed to harmful materials during the manufacturing process.

Personnel have been credited with contributing the greatest risk particulate and microbial product contamination [18]. They routinely shed non-viable particulate from skin and hair and naturally harbor microorganisms as part of their biological makeup on their skin and in their nose and throat. To reduce the risk of introducing these contaminants to the product, it is important to assess whether or not adequate controls are in place.

11.2.6.1 Personal Hygiene Starting with the basics, personnel should be required to practice good personal hygiene. Depending on the level of control required, it may be necessary that personnel shower before entry to a manufacturing area or in the case of exposure to harmful materials, afterwards. Written procedures should instruct personnel on the proper cautionary measures to take when sick, sneezing, or coughing in the workplace to avoid the risk of contaminating the environment and product.

11.2.6.2 Protective Clothing and Gear Protective clothing or gear is a major player in both minimizing the introduction of contamination and minimizing personnel exposure to harmful materials. Protective clothing acts as a barrier between a person and the surrounding environment. In one role, its purpose is to contain particulate that would otherwise normally be shed by personnel. In its other role, protective clothing is needed to protect personnel from harmful materials such as highly toxic chemicals to which they may be exposed in the working environment. The assessment should take into consideration the type and the level of protection required for each manufacturing process. This information will help decide the types (e.g., body suit, hair bonnets, gloves, mask, booties, and respirators) of protective gear and the right materials of construction (e.g., latex, Tyvek®, woven cotton, polyester). The goal would be to have the maximum protection needed and yet provide comfort and allow for maneuverability.

It is important to consider whether the gear needs to be protected from environmental contamination before use, while being donned or while in use. The need to define the incoming packaging, storage area conditions, as well as proper donning and operating techniques while being used are worth assessing. To maintain control, personnel must be trained in the proper donning and operating techniques and these practices routinely monitored.

11.2.6.3 People Interactions Every step where personnel are interacting with in-process materials or products must be assessed to determine the risk of introducing contamination through these interactions. Every effort must be made to mitigate the risk through a number of options such as substitution by automation or barriers or strictly enforced technique and practices.

11.2.6.4 Training Regulatory agencies want to know if manufacturers have the right people doing the right jobs and are well trained to perform their job tasks [19]. They want manufacturers to take training seriously because product quality is at risk. Often, failure to train comes up as an issue in investigations or deviations. The question arises whether the personnel had the appropriate training so they can perform their functions properly or whether the training was effective. Employees must be adequately trained, their training managed, recorded, and monitored; and their training records evaluated all in an effort to determine if the training program is producing the intended results.

When an operator error occurs, it is of utmost importance to first find the root cause of the problem. Retraining an operator on a procedure may not substitute as a fix for other deep-rooted manufacturing issues. It must be determined whether the root cause is the employee, the procedure, or the process as it is being performed. Retraining on a procedure that may not work may not fix operator error. Having said that, good training and training programs are key to operational success and minimizing the risk to product quality.

11.3 CASE STUDY: MICROBIOLOGICAL CONTROL IN NONSTERILE MANUFACTURING

As a general observation, the risk of microbiological contamination during the manufacturing of nonsterile products has been an area that has not been well defined [20]. Typically, environmental control systems and monitoring data have been used to control and indicate the microbiological quality of nonsterile manufacturing. However, as a result of a lack of set standards, the application of environmental control and monitoring of nonsterile processes ranges from nonexistent to programs parallel to aseptic processing. In some cases, the type and frequency of data generated from some programs may be of little value in determining the microbiological quality of environments in which the product is manufactured. In addition, uncertainties exist on how the data will be used and interpreted and its significance regarding product quality and safety to the patient.

The use of environmental control and monitoring data has often been misapplied as a means of microbial and process control. Hence, the question is how microbial control can be applied effectively in the manufacture of nonsterile products? The answer lies in the use of a risk-based approach to understand the manufacturing process, in defining where microbial contamination could occur and in effectively determining the best level of control and applying monitoring methods to minimize microbial contamination of the final product.

Does the microbial count in the air provide meaningful information for determining the state of control of a nonsterile manufacturing process or will the critical manufacturing control points provide more protection from microbial contamination?

In this case, where the risk factors are microbial in nature, the critical manufacturing control points provide more protection. The reason is that the microbial quality of the air would be more indicative of the level of general cleanliness provided by routine surface cleaning and the efficiency of the air filtration systems rather than the level of microbial contamination control in the manufacturing process. Samples for microbial counts of the air are only taken typically at designated times (e.g., quarterly) and at designated locations (e.g., taken in two to three locations). This information is limited. It reflects only the snap shot in time when the samples are taken. Therefore, it is not relevant to the immediate activity happening on a daily basis. Further, samples may require incubation for several days to enumerate microbes on growth media. Results are not known until days later and this prevents immediate correction of an uncovered problem. For these reasons, there is greater value in determining the critical manufacturing control points to protect the product against microbial contamination.

The process steps where microbial contamination is likely to be introduced or occur are determined. Controls are put in place during the process steps to eliminate or reduce the likelihood of contamination. The controls provide a means for real-time or near-time performance feedback, so corrections or adjustments can be made if needed in a timely manner.

To minimize the risk of microbial contamination, the application of technical and scientific principles for control of the microbial hazards because of the facility, equipment, and production process needs to occur. One risk management tool for assessing microbial hazards is hazard analysis and critical control points [21–34].

11.3.1 Risk-Based Approach Management Tool: What is HACCP?

Hazard analysis and critical control points (HACCP) is a worldwide-recognized systematic and inductive tool that can address biological, chemical, or physical hazards through anticipation and prevention, rather than through end-product inspection and testing. The approach is to control identified critical processing parameters as the product is being manufactured [14]. It can be used as a way of identifying and controlling microbial risk for the production of pharmaceuticals that are not required to meet the test for sterility.

11.3.2 History of HACCP

HACCP has its roots in food safety and was first developed and used by the Pillsbury Company in the late 1960s to provide safe food for the United States space program, NASA. The program focused on applying science-based controls to prevent hazards that could cause food-borne illnesses. In 1973, the United States Food and Drug Administration (FDA) instituted mandatory HACCP programs for juice, for low acid canned food processing and in 1997 for seafood processors. The United States Department of Agriculture (USDA) instituted mandatory HACCP programs for meat. EU member states have also adopted HACCP for food safety. ICH Q9, "Quality Risk Management" listed HACCP as a primary tool for risk management.

11.3.3 Benefits of HACCP for Microbiological Control

In the past, periodic inspection and environmental testing were used to ensure the quality and safety of the product. However, they provide information about the product that is only relevant for the specific time of the inspection or testing, and microbiological test results would not be available in real time. HACCP provides a system to control safety as the product is manufactured rather than trying to detect problems by testing or inspection. HACCP identifies critical control points (CCPs) within the process in which a microbial hazard could have an impact on final product quality.

A drug product is defined by physical, chemical, or microbiological characteristics or attributes. When these characteristics or attributes impact product quality or safety, they are referred to as *critical quality attributes* (*CQAs*). These product attributes may be impacted by the processing steps during the manufacturing process. Those process parameters, whose variability may impact a critical quality attribute, are referred to as *CPPs*. By definition in ICH Q8 (R2), *Pharmaceutical Development*, the critical quality product attributes should be within established specifications to ensure product quality and the critical process parameters should be controlled. By applying "control" at the CCP, the amount of total process monitoring should be minimized and the value of the activities maximized. It provides just in time control rather than after the fact.

11.3.4 HACCP Process

The HACCP process consists of seven principles:

- Principle 1: Conduct a hazard analysis.
- Principle 2: Determine the CCPs.
- Principle 3: Establish critical levels at each CCP.
- Principle 4: Establish a system to monitor control at each CCP.
- Principle 5: Establish the corrective actions to be taken when monitoring indicates that a particular CCP is not under control.
- Principle 6: Establish procedures for verification.
- Principle 7: Establish documentation.

A case study is used to work through the seven HACCP principles to show how a pharmaceutical manufacturer could apply the HACCP approach to review operations. The case study involves the manufacture of a solid tablet as illustrated in Figure 11.2 where incoming materials are granulated, milled, blended, compressed into the final tablet, and packaged. The tablet has a bioburden requirement to safeguard its preservative system and ensure product quality, but it is manufactured in a nonsterile manufacturing environment. The objective in the case study is to perform a risk-based assessment of the nonsterile manufacturing process to identify the microbial hazards and ensure that they are adequately controlled so the product meets its bioburden requirements.

Before the application of the HACCP principles to a specific product or process, there are preliminary to be accomplished. These tasks are as follows:

- Assemble the HACCP team.
- Describe the product and raw materials.
- Construct a process flow diagram.
- Verify the process flow diagram.

11.3.4.1 Assemble the HACCP Team The team selected in the case study consisted of four to six individuals with specific knowledge and expertise relative to the product and process. The members were from multiple disciplines:

- facilitator (knowledgeable in HACCP process);
- production personnel (including local operators who know the variations and limitations of the operations);
- microbiologist(s);
- quality control/assurance personnel;
- engineering; and
- other expertise (external experts knowledgeable in microbial hazards).

The team members received training in the HACCP process to ensure that they first understood what HACCP is and to learn the skills necessary to make it function properly.

They met for 2–3 h at least once a week. The length of time to complete the HACCP analysis depends on a number of variables including the complexity of the subject matter and the dedication of resources. In this case, it took the team about three months to complete the HACCP analysis.

It is important in undertaking the project that the team does due diligence in working through the steps of the HACCP process, taking the responsibility seriously, and not using the process as an exercise to justify a predetermined decision.

11.3.4.2 Describe the Product and Raw Materials Part of the analysis was to describe the product and raw materials. The team started by defining such product attributes as follows:

- route of administration;
- type of packaging;
- intrinsic factors (e.g., pH and water activity [A_w]);
- regulatory or compendia requirements;
- storage conditions; and
- list of raw materials.

11.3.4.3 Construct a Process Flow Diagram Another part of the analysis was for the team to construct a process flow of the operations. The process flow diagram provides a clear, simple outline of the steps involved in the process within the scope of the assessment. The HACCP team members walked the process in constructing the diagram to understand firsthand and in detail the personnel and material flow, interconnections, and potential hazards in the processes. The team constructed a process flow diagram as shown in Figure 11.2.

The types of data that the HACCP team considered and included in the process flow diagram were as follows:

- all raw materials;
- packaging components;
- storage conditions;
- microbiological data;
- sequence of all process operations (including raw material addition;
- holding times and temperature;
- equipment design features, cleaning, and storage; and
- details of any product rework or recycling.

Site information was integrated into the process flow diagram and included the following:

- details of segregated areas and personnel flow;
- routes of potential cross-contamination;
- flow of raw materials and packaging materials;
- facility HVAC and utilities; and
- personnel hygiene and uniform requirements;

11.3.4.4 Verify the Process Flow Diagram After the process flow diagram was constructed, the HACCP team performed an on-site review of the operation to verify the accuracy and completeness of the flow diagram. Any deficiencies or modifications were documented in the process flow diagram.

11.3.4.5 Principle 1: Conduct a Hazard Analysis Principle 1 consists of the following:

- Conduct a hazard analysis.
- Identify preventive measures.

11.3.4.6 Conduct a Hazard Analysis Hazard analysis is a two-phase exercise. The first phase is to develop a list of steps in the process where hazards, which could result in injury or illness if not controlled, could occur. In the second phase, the hazards are evaluated to determine which must be part of an HACCP control plan.

The HACCP team reviewed the ingredients used in the product, the activities conducted at each step in the process, the equipment used, and the final product.

On the basis of this review, the team listed all of the microbial hazards that may be reasonably expected to be introduced, increased, or controlled at each of the steps identified in Figure 11.2, and listed in the following:

- raw materials (API and excipients);
- pre- and post-manufacturing storage;
- dispensing;
- granulation;
- drying;
- milling;
- blending;
- compression;
- film coating;
- branding; and
- packaging.

In addition, the following areas were also considered for potential microbial hazards:

- sampling;

- equipment (cleaning and storage);
- utilities;
- personnel; and
- facility (material flow, cleaning and sanitation, environmental controls).

Not all of these processes mentioned are covered in the example. The example will cover an analysis of the compression step to show how the HACCP process works. The compression step occurs after the raw materials are granulated, milled, and blended. In the compression step, a tablet press compresses the granulation material into the size and shape of the finished tablet. The process flow map created for the compression step is illustrated in Figure 11.3.

Each processing step was assigned a sequential number on the process flow diagram. Next, the team conducted a hazard analysis of the identified hazards. The challenge that the team faced was to prepare a list of steps in the process where significant microbiological hazards could occur and determine what are truly significant versus insignificant hazards. Significant microbial hazards are of such a nature that their elimination or reduction to acceptable levels is essential to meeting the bioburden requirement for product quality.

Each potential hazard was assessed on the basis of the severity of the consequences of the exposure to the potential hazard (e.g., magnitude and duration of illness) and its likely occurrence. It is helpful to consider the likelihood of exposure and severity of the potential consequences if the hazard is not properly controlled.

11.3.4.7 Identify Preventive Measures If a microbial hazard exists, preventive measures must be identified. Preventive measures would include the following:

- effective cleaning and sanitization procedures;
- temperature control;

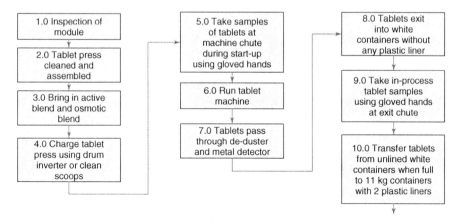

Figure 11.3 Process flow diagram example. Process flow diagram for tablet compression process step.

- moisture control;
- bioburden testing of raw materials;
- in-process microbial reduction steps (e.g., pasteurization);
- control of storage conditions; and
- in-process filtration.

The team identified any preventive measures required. Preventive measures already in place and those that need to be put in place were considered. It is important to remember that more than one preventive measure may be required for each hazard or one preventive measure may address more than a single hazard.

The team initiated a hazard analysis chart to record the results of the hazard analysis. Table 11.1 provides an example. The team recorded the following information:

- each step, with an assigned a reference number;
- the determination whether or not there could be a microbial hazard associated with the step along with the rationale for the determination; and
- any preventive measure(s).

For process steps 1.0, 3.0, and 4.2, no microbial hazards were identified. For process step 2.0, purified water is used to clean the equipment. The water quality is tested to ensure it meets quality specifications before use. This measure ensures that the microbial burden in the water is being controlled, which reduces the risk that high levels of microbial contamination is being introduced during the cleaning process. For process step 4.0, the room air is filtered by high efficiency particulate air (HEPA) filters and the personnel are gowned. These measures help reduce the risk of microbial contamination being introduced into the product container during its exposure to the manufacturing environment and personnel. For process step 4.1, the equipment is set to sound an alarm when the correct air temperature (300°F) for drying is reached. Once the alarm is sounded, the operator can start the 30-min drying step. This measure helps reduce the risk that the air temperature is not sufficient to dry the equipment in 30 min, resulting in wet equipment. In addition, procedural controls are in place and personnel are trained to ensure that the equipment is dry because moisture supports the growth of microorganisms. For process step 5.0, contamination of the equipment or finished product from operator contact during the sampling process is controlled with the use of gloves. Operators are trained in the proper sampling techniques to don the gloves properly, keep them clean and dry, and avoid product contact.

Going forward, not all of the process steps included in Table 11.1 will be covered. Steps 4.0 and 4.1 will be used to demonstrate how the HACCP approach continues.

11.3.4.8 Principle 2: Determine the Critical Control Points (CCPs) Although other approaches may be used, an effective tool to help determine where the CCPs should be in the process is a decision tree. The CCP decision tree is based

CASE STUDY: MICROBIOLOGICAL CONTROL IN NONSTERILE MANUFACTURING   301

TABLE 11.1   Hazard Analysis Chart Example

Hazard Analysis Chart
Microbiological Control for Tablet Compression in Non-Sterile Manufacturing

(1) Process Step	(2) Microbial Hazard? (Y/N)	(3) Justification for Decision in (2)	(4) Preventive Measures	(5) CCP Y/N
1.0 Inspection of module	No	Step occurs before cleaning	N/A	No
2.0 Tablet press cleaned and assembled	Yes	If microbial quality of water used for cleaning is poor	Purified grade water used for cleaning, tested on daily basis before use to meet microbial limits	Yes
3.0 Bring in active blend and osmotic blend	No	Product is not exposed to the environment	Dispensing containers are sealed	No
4.0 Open container and use drum inverter to charge mill	Yes	Product open to environment	Room air is HEPA filtered. Operators are properly gowned	No
4.1 Turn on equipment with heated air for 30 min and visually check for dryness	Yes	If tablet press is stored wet	Equipment is alarmed when air reaches 300°F to start 30 min drying step. SOPs are in place and operators have been trained to verify dryness	Yes
4.2 Start up tablet press	No	Operator does not have product contact	N/A	No
5.0 Take samples of tablet at machine chute during start-up using gloved hands	Yes	If gloves are not clean, dry, and fully covered hand	SOPs are in place operators trained in proper gowning and sampling techniques	No

Product:
Date:
Signature(s):

on answering five questions at each step that has an identified microbial hazard. The use of the decision tree tool promotes structured thinking and a consistent approach at each process step. The five-question decision tree is illustrated in Figure 11.4.

Figure 11.4 graphically illustrates the CCP decision tree.

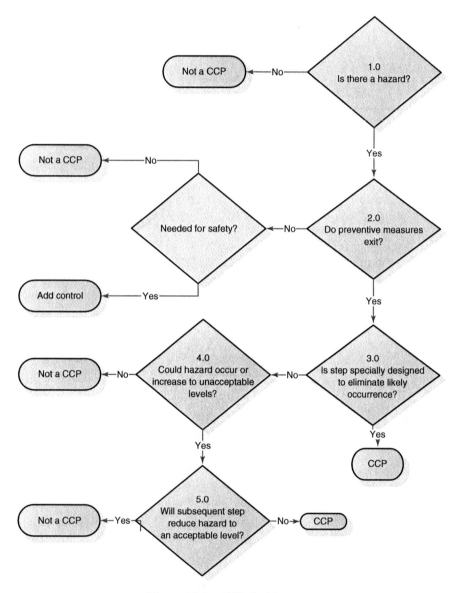

Figure 11.4 CCP decision tree.

The CCP decision tree analysis process is described as follows:

1. Is there a microbial hazard at this step? If the answer is "No," then the process step is not a CCP. If the answer is "Yes," then proceed to question #2.
2. Do preventive measures exist for the microbial hazard? Consider the measures in place in your hazard analysis chart. If the answer to this question is "Yes," then move to question #3. If the answer is "No" and control measures are not in place, then consider whether control is necessary for product safety. If control is not necessary, then the process step is not a CCP. If control is necessary and no process control measures exist, then the product or process should be modified at this step or some earlier or later stage to include a control measure. If a control measure is not possible, then consider the introduction of a new procedure to build in control.
3. Is the step specifically designed to eliminate or reduce the likely occurrence of the hazard to an acceptable level? The question refers to the process step and is really asking whether the step itself controls the microbial hazard. If the answer is "Yes," the process step is a CCP. If the answer is "No," then proceed to question #4.
4. Could contamination occur or increase to unacceptable levels? The answer would be based on the hazard analysis information, the team's expertise, and literature references. It would be important to consider such factors as the effect of immediate environmental or facility design; possible cross-contamination from personnel, other products, or raw materials; possible contamination from equipment or utilities; physical conditions, e.g., hold times, temperature; product or raw material buildup in dead leg spaces; or any other factors or conditions that could cause bioburden to increase in the decision. In addition, any additive effects during the process that may increase bioburden should be considered. If the answer is "No," then the step is not a CCP. If the answer is "Yes," move to question #5.
5. Will a subsequent step or action eliminate or reduce the hazard to an acceptable level? This question is designed to allow the presence of a microbial hazard at a particular step if it will be controlled later in the process. If the answer is "Yes," then the step is not a CCP. If the answer is "No," then the process step is a CPP.

Each question in the decision tree is addressed for each process step until the determination is made that the step is a CCP or not a CCP. Figure 11.5 illustrates the decision tree analysis for process step 4.0, "Open container and use drum inverter to charge mill." The results of the decision tree analysis concluded that the process step was not a CCP.

Figure 11.6 illustrates the decision tree analysis for process step 4.1, "Turn on equipment with heated air for 30 min and visually check for dryness." The decision tree analysis concluded that this process step is a CCP.

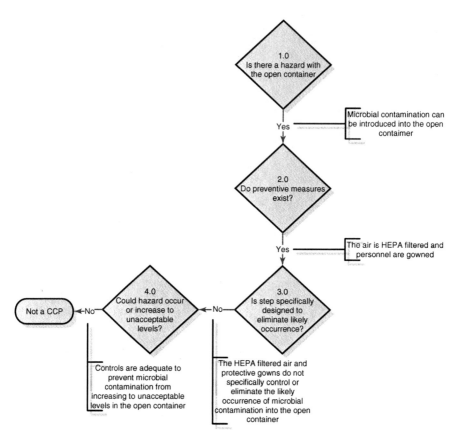

Figure 11.5 CCP decision tree analysis for process step 4.0, charge mill.

The information is documented. Table 11.2 is an example of a CCP decision chart used to document the results of the decision tree analysis. The responses will indicate if the identified hazard is a CCP. The table includes the response to each question, the conclusion of the analysis whether the step is a CCP or not a CCP, and a place to record any notes. The CCP decisions can also be recorded on the hazard analysis chart illustrated in Table 11.1.

As shown in the chart for process step 4.0, the answer to question 1, is there a hazard with the open container, is recorded as "Yes." There is a hazard that microbial contamination can enter the open container. If the answer to question 1.0 is "Yes," the step could be a CCP and question 2.0 is addressed. The answer to question 2.0, do preventive measures exist, is also recorded as "Yes." Filtered room air and gowned personnel are preventive measures taken in an effort to reduce the likelihood of microbial contamination. If the answer to 2.0 is "Yes," the step could be a CCP and question 3.0 is addressed. The answer to question 3.0, is the step specifically designed to eliminate the likely occurrence, is recorded as "No." Filtered air and protective gowns do not specifically control the likely

CASE STUDY: MICROBIOLOGICAL CONTROL IN NONSTERILE MANUFACTURING 305

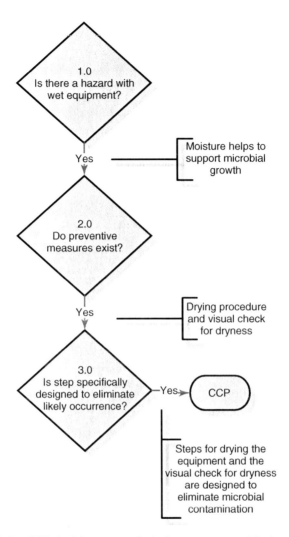

Figure 11.6 CCP decision tree analysis for process step 4.1, dry equipment.

occurrence of microbial contamination. If the answer to question 3.0 is "No," the step could be a CCP and question 4.0 is addressed. The answer to question 4.0, could the hazard occur or increase to unacceptable levels, is recorded as "No." Filtered room air and protective clothing do not directly control the likelihood that microbial contamination would not occur, but they would prevent the microbial levels from increasing to unacceptable levels. If the answer to question 4.0 is "No," the step is not a CCP and the decision process is complete. The answer to question 5.0 is recorded as "not applicable (N/A)" and "No" is recorded in the CCP column.

TABLE 11.2 CCP Decision Chart Example

	CCP Decision Chart						
	Microbiological Control for Tablet Compression in Non-Sterile Manufacturing						
Process Step-Hazard	Q1	Q2	Q3	Q4	Q5	CCP	Notes
4.0 Open container and use drum inverter to charge mill	Yes	Yes	No	No	N/A	No	None
4.1 Turn on equipment with heated air for 30 min and visually check for dryness	Yes	Yes	Yes	N/A	N/A	Yes	None
Product:							
Date:							
Signature(s):							

For process step 4.1, the answer to question 1.0, is there a hazard, is recorded as "Yes." Moisture trapped in the equipment could help support the growth of microorganisms. If the answer to question 1.0 is "Yes," the step could be a CCP and question 2.0 is addressed. The answer to question 2.0, do preventive measures exist, is recorded as "Yes." A drying step and a check for dryness are preventive measures taken to control the moisture in the equipment. If the answer to question 2.0 is "Yes," then the step could be a CCP and question 3.0 is addressed. The answer to question 3.0, is the step(s) specifically designed to eliminate the likely occurrence of the hazard, is recorded as "Yes." Drying the equipment and ensuring it is dry are steps specifically designed to eliminate the occurrence of moisture that would support microbial growth. If the answer to question 3.0 is "Yes," the step is a CCP and the decision process is complete. "N/A" is recorded for questions 4.0 and 5.0 and "Yes" is recorded in the CCP column.

On the basis of the responses to the decision tree analyses, the process step of opening the container and using the drum inverter to charge the mill is not considered to be a CCP, whereas the process step of turning on the equipment with heated air for 30 min to dry the equipment is a CCP. Controlling the moisture in the process will control the potential growth of microbial contamination.

If a hazard has been identified at a step where no control exists but control is necessary for product safety, then the process should be modified to include a control measure.

11.3.4.9 Principle 3: Establish Critical Levels at each CCP Critical levels must be specified and validated if possible for each CCP. The team ensured that critical levels were established for preventive measures associated with each identified CCP and served as the boundaries for each CCP. Levels are associated with a measurable factor, e.g., time, temperature, pH, moisture level, and sensory parameters such as visual observation.* In the CCP example of drying the equipment, the critical level is the minimum temperature requirement of 300°F before the start of the 30-min drying step.

11.3.4.10 Principle 4: Establish A System to Monitor Control at each CCP
CCP monitoring is the act of conducting a planned sequence of observations or measurements of control parameters to assess whether a CCP is under control and to produce an accurate record for future use in verification. Monitoring is best performed as the process occurs with results received at the actual time the control occurs (i.e. real-time data) to rather than with results that would be produced days later as would be the case with microbiological laboratory testing. The first steps to establish monitoring procedures for each CCP would be to identify the best atline, online, and/or offline systems and observational procedures. Continuous monitoring is ideal, but when continuous monitoring is not feasible a monitoring frequency must be determined. The next steps would be to establish procedures for the operating personnel at each CCP to define the tasks for performing and recording the monitoring activities and who is responsible for these activities.

General examples of monitoring activities include visual observations or measurements of temperature or moisture levels. The monitoring frequency, technique, and documentation are defined in procedures. For the specific CCP example of drying the equipment, the temperature of the air used to dry the equipment is monitored. An alarm sounds to indicate that the proper temperature has been achieved before starting the 30-min drying step. Shortly after the 30-min equipment drying step, the operator monitors the success of the control step by visually checking for dryness. Monitoring is best performed immediately following the 30-min drying step to confirm that the allotted drying time produced the desired results of total dryness and if not, to address the process failure or any other issues that may have arisen before moving on to the next process step.

11.3.4.11 Principle 5: Establish the Corrective Actions to be Taken when Monitoring Indicates that A Particular CCP is Not Under Control In manufacturing, deviations will occur. Specific corrective actions must be developed for each CCP to address the deviation when it occurs and must ensure that the CCP has been brought under control. A deviation from an established critical level is only a symptom of a process failure and a thorough investigation into the root cause of the process failure is required. Regardless of when or how the deviation is detected, it is important to promptly notify the proper quality and/or production authorities and to decide whether to temporarily halt operations until the relative risk of the failure on product quality can be determined and appropriate corrective measures are taken.

---

*Limits may be set on the basis of results from the design of experiments, historical trends, and regulatory guidance or compendia standards.

The investigation activities need to begin immediately. All operators and line management involved in the manufacturing process should participate in the investigation and subject matter experts should be consulted for their input.

A good starting place to investigate a process failure is to perform a risk assessment.[†] A risk assessment works just as well when conducted on a reactive basis as on a proactive one to gather information and determine the relative risk of failure on product quality. It is a good problem-solving tool because it organizes the fact finding process by laying out a road map of every process step, each control measure, and monitoring method in place. Each of these aspects can then be reviewed in a systematic manner. The underlying questions may change the focus from what could go wrong and why in the proactive setting to what did go wrong and why in the reactive setting but the overall process is still the same. Some of the questions to ask the operators should also include the following:

- Were procedures adequate to describe the activity and were procedures followed?
- Was the equipment involved in proper working order?
- Were there any changes implemented recently?
- What are the recommendations for correction?

If laboratory testing was involved, a similar investigation of the laboratory work should also be conducted.

The review may uncover that a CCP was not properly identified, not properly controlled or monitored, or a change was implemented whose impact on the process was not adequately assessed. Corrective actions, designed to prevent reoccurrence of the failure, need to be implemented. Corrective actions may include new controls, new monitoring methods or criteria, and revised or new procedures to ensure that the new actions are carried out as planned. One additional aspect is to determine if and how the changes implemented in the corrective actions may have impacted other CCPs. Corrective action may also need to address the impact of the failure on product disposition. All fact finding results and actions taken must be documented. Deviation and product disposition procedures must be in place and documented in the HACCP record keeping. Finally, it must be determined that the manufacturing process is again under control, which can be shown through another risk assessment.

11.3.4.12 Principle 6: Establish Procedures for Verification Any HACCP plan must be verified initially and on an ongoing basis to ensure that the process is being maintained in a state of control, that all hazards have been identified, and identified hazards are either being eliminated or controlled at acceptable levels. Verification involves activities other than routine monitoring ones and may

---

[†]A risk assessment may be performed by using any one of several risk assessment tools such as failure mode and effect analysis (FMEA) or fault tree analysis (FTA).

CASE STUDY: MICROBIOLOGICAL CONTROL IN NONSTERILE MANUFACTURING 309

include sampling and analytically testing products, reviewing monitoring records and deviations, or auditing monitoring procedures or manufacturing operations. Verification is an ongoing process and the HACCP team must continue to update the HACCP program if there are any changes in the processes or materials.

Verification activities should be established through procedures. Procedures should define the types of verification activities that will be performed and describe a plan when to conduct these activities. An example of a verification procedure that describes the verification plan is documented in Table 11.3. The table defines for each verification activity (initial, subsequent, or comprehensive verification or record review), the frequency or conditions under which a verification activity needs to occur, the person or persons responsible to perform the verification activities, and the reviewer of the completed verification results. For example, the chart defines that there will be a monthly review of the monitoring records, corrective action reports, and internal or external audit or regulatory inspection findings by the Quality Department and reviewed by the Operations Manager. This review verifies that the controls already in place and the corrective actions taken in response to issues were effective and adequate to control the hazard. If not, further action should be taken.

11.3.4.13 Principle 7: Establish HACCP Program Documentation and Record Keeping It is essential that the HACCP program is documented. Documentation of the HACCP program provides an organized plan of procedures and the rationale for maintaining control of a manufacturing process. In addition, a summary of the team deliberations and the rationale developed during the hazard analysis should be kept. This information will be helpful for future reviews and could be helpful during regulatory inspections.

Examples of documentation are as follows:

- hazard analysis;
- CCP determination;
- critical limit determination;
- SOPs; and
- team meeting minutes.

Record keeping of the process monitoring activities is essential to the application of HACCP and should be designed to be easy to complete. Record keeping should include information that provides feedback that critical limits are consistently being met using the current procedures. The information obtained from process monitoring lends support to product testing, can be used to justify reduced final product testing or justify a decision not to perform noncritical process monitoring, e.g., environmental monitoring.

Examples of records are as follows:

- CCP monitoring activities; and
- deviations and corrective actions.

TABLE 11.3   Verification Plan Example

Activity	Frequency	Responsibility	Reviewer
\multicolumn{4}{c}{Verification Plan}			

Activity	Frequency	Responsibility	Reviewer
Initial verification	Before plan implementation	Operations supervisor	Operations manager, technical operations manager
Subsequent verification	When a change in critical level, process, or equipment occurs or after system failure	Operations supervisor	Operations manager, technical operations manager
Review of monitoring, corrective action records, internal and external audit, or regulatory inspection findings	Monthly	Quality assurance supervisor	Quality assurance manager, operations manager
Comprehensive verification	Yearly	Operations manager, technical operations manager	Operations director, technical operations director

Verification Plan — Microbiological Control for Tablet Compression in Non-Sterile Manufacturing

Product:
Date:
Signature(s):

Table 11.4 presents an example of an HACCP control chart used to document the HACCP program. The chart holds in one place all essential details about the steps in the process where there are CCPs. For each CCP, the chart defines the process step, the potential hazard that may result if the process step fails or the rationale for performing the process step, the measures that are in place to ensure that the process step will be executed properly, and critical levels required to be met to ensure the process step is under control. In addition, the chart provides the type of monitoring, the frequency of monitoring, and who is responsible to perform the step. The chart also defines the steps to be taken if the control levels are not met.

11.4   CONCLUSION

It is important to realize that risk management is not simply about "checking off the box" by completing an assessment for each process step, one at a time

TABLE 11.4 HACCP Control Chart Example

HACCP Control Chart
Microbiological Control for Tablet Compression in Non-Sterile Manufacturing

(1) CCP	(2) Microbial Hazard	(3) Preventive Measure	(4) Critical Levels	(5) What	(6) How	(7) Frequency	(8) Who	(9) Corrective Action	(10) Record
CCP #4.1 dry step	Tablet press is wet	Heated air 30 min	30 min	Timer	Check timer SOP #101	Start of each drying step	Online staff	Repeat dry step	Batch record
		Alarm sounds at air temp. of 300°F	≥300°F heated air	Temp. chart, alarm	Check chart, SOP #102 test alarm SOP #105	Start of each batch	Online staff	PM on alarm, temp. device, heat source	Batch record temp. chart
		Operator verifies dryness	Dry	Tablet press	Visual SOP #104	End of drying step	Online staff	Repeat dry step.	Batch reccrd

Product:
Date:
Signature(s):

in isolation, but requires stepping back and gaining a full picture, linking the risks from the different process steps to the full manufacturing process, and then looking at all the manufacturing processes together. This offers a snapshot of all the risks, the overall control, and residual risk the company is faced with that either cannot be further controlled or mitigated or that the company decided it cannot mitigate. Further, it is important to realize that the risk analysis is not a one-time task but a continual process that requires a link to change control and the investigation systems so that the risk analysis is updated when changes are required.

What could be the benefit of risk management in pharmaceutical manufacturing? The benefits could include the following:

- fewer compliance issues such as recalled product, reduction in customer complaints;
- a well-trained and informed staff;
- increased process efficiencies resulting in cost savings;
- building trust with regulatory agencies; and
- improved assurance of product quality that reduces the risk to the patient.

The use of risk management principles and tools has a clear sense of purpose in pharmaceutical manufacturing. It fosters product and process knowledge and it sets direction and priorities during product development, problem solving, and process improvement activities. It provides for a proactive strategic approach to determine how best to engineer the manufacturing process with the appropriate process controls and monitoring so that product quality is achieved reliably and consistently from batch to batch. We used HACCP as the tool of choice in the example for assessing microbial control in solid manufacturing; however, other types of risk assessment tools could have been used for this example .Likewise, the HACCP tool could be applied to other types of manufacturing processes. The benefit of HACCP is that the tool fits well into the life cycle approach. With HACCP, the hazards are always being looked at, critical parameters and limits monitored, and adjustments made to improve the process and control strategy.

Appendix I: List of Potential Risks Associated with Drug Product Manufacture

Major Area	Subarea(s)		Potential Hazard	Potential Harm
Raw Materials	Material Management	Materials on Receipt:	Inadequate Identification	Use of Wrong Material or Wrong Grade of Material
			Packaging is not intact and not tamper evident	Contaminated or tainted/adulterated materials
			Packaging not dry	Moisture damage, contamination of materials
			Packaging not clean	Contamination of materials
			Required environmental conditions not maintained	Loss of material's potency/stability
			Procedures do not address material disposition and management notification for deviations	Use of inappropriate materials. Continued problems with quality of received materials
		Design of receiving areas	Ingress of environmental elements such as rain, wind, animals, or insects	Contamination of materials and/or infestation of staging and storage areas
			Standing water or excess moisture	Potential for microbial (mold) proliferation, or contamination of materials
		Material handling procedures	Practices not designed to prevent damage to packaging, contamination	Materials can be exposed to environmental conditions and contamination
		Environmental storage conditions	Required environment conditions are not maintained. Adequacy of temperature and humidity control depends on season of year.	Loss of material's potency/stability. Lots do not meet incoming specifications for critical quality attributes
			Procedures address material disposition and management notification for deviations	Use of substandard materials, failed product quality
		Material status	Adequate identification and/or separation of different material status (i.e., quarantine, release, rejected)	Use of rejected or quarantined materials in a released batch

(continued)

Appendix I: (*Continued*)

Major Area	Subarea(s)		Potential Hazard	Potential Harm
Raw Materials	Material Management	Materials on Receipt:	Inadequate Identification	Use of Wrong Material or Wrong Grade of Material
	API/ Excipient/ Water Quality	Incoming lot quality	Lot-to-lot variability in quality characteristics such as moisture content and microbial bioburden	Lots do not meet incoming specifications for critical quality attributes. Use of materials impact the outgoing quality of the product
			Incoming lot monitoring program is inadequate to monitor incoming quality	Use of substandard materials, failed product quality
	Supplier management	Supplier management program	Supplier management program does not exist or is inadequate	Inconsistent incoming lot quality. Use of substandard materials, failed product quality
Formulation		Equipment	Substituted equipment: incorrect design, size, or incapable to run required operating parameters	Inadequate or incorrect granulation, milling, or blending. Product quality compromised
	Dispensing	Materials	Incorrect raw materials weighed or measured	Incorrect batch composition; mislabeled product; patient safety
		Material labeling	Materials incorrectly labeled	Incorrect batch composition, not delivering on intended use, patient safety
		Measurement/ weighing	Inaccurate measurement	Mislabeled product, subpotent or superpotent product
		Material delivery	Incorrect raw materials delivered to formulation area	Mislabeled product, not delivering on intended use; product safety
	Granulation	Material addition	Materials added in incorrect sequence, amounts per addition, or timing between additions	Improper formulation, product instability, or incorrect potency
		Operating procedures	Inadequate (e.g., every step not defined), incorrect written instructions	Improper formulation; product instability or incorrect potency
		Equipment	Incorrect chopper or mixing blades	Improper product formulation

Appendix I: (*Continued*)

Major Area	Subarea(s)		Potential Hazard	Potential Harm
Raw Materials	Material Management	Materials on Receipt:	Inadequate Identification	Use of Wrong Material or Wrong Grade of Material
		Compressed gases	Inappropriate venting or use of compressed gases	Impacts environment, product contamination
	Drying	Operating parameters	Inadequate drying	May impact product stability, unable to further process
		Wet holding time	Inadequate control of wet holding times	Wet conditions could allow chemical reactions to take place, degrading or changing bulk composition
	Milling	Operating parameters	Inadequate sieving	Nonhomogenous granules, product quality compromised
		Equipment setup	Wrong screens	Clumps not eliminated as needed
		Equipment	Warped screens	Contamination with metal particulate
	Mixing/ blending	Blending operating parameters	Inadequate mixing/blending	Nonhomogeneous batch; variable or incorrect finished product potency
		Materials	Wrong material	Mislabeled product; patient safety
Finish Processing	Compression	Bulk flow	Wrong amount of material added	Bulk quality, affects compression process
			Inaccurate amount measured to be compressed	Mislabeled product, subpotent or superpotent product
		Environmental conditions	Humidity is not adequately controlled	Powder absorbs moisture and cakes preventing adequate flow
		Hardness– friability	Incorrect compression parameters	Chipping or flaking off, susceptible to further cracking
		Equipment setup	Wrong punches or dyes	Inadequate product quantity, weight variation
			Punches are out of alignment	Parts shed metal contaminating product
	Film Coating	Coating material	Incorrect solution preparation (e.g., too thick or thin)	Inadequate or uneven application; product degradation or instability
		Equipment	Spray nozzles are clogged	Inadequate coating; product degradation or instability

(*continued*)

Appendix I: (*Continued*)

Major Area	Subarea(s)		Potential Hazard	Potential Harm
Raw Materials	Material Management	Materials on Receipt:	Inadequate Identification	Use of Wrong Material or Wrong Grade of Material
		Operating parameters	Wrong rate of spin, drum rotation, wrong air drying temperature	Inadequate coating; product degradation or instability
	Branding	Ink solution, preparation	Wrong ink material (e.g., ink too thick or thin)	Inadequate or smeared imprint, product not properly identified
		Equipment setup	Incorrect pressure parameters	Inadequate imprint, product not properly identified
		Equipment	Clogged ink applicator	Inadequate imprint, product not properly identified
	Packaging	Filling	Product mix-up. Incorrect product in package	Adulterated product
			Incorrect number or amount in package	Mislabeled product, patient treatment impacted
		Patient information	Incorrect information	Patient safety
		Seal integrity	Inadequate seal (e.g., incorrect sealant or sealant amount, application misaligned, incorrect sealing temperature or pressure)	Product contamination, product stability may be compromised
Facility	Material flow	Material and personnel flow patterns (e.g., undedicated/ multi-product facility)	Shared production and non-production areas	Product cross-contamination
	Cleaning	Cleaning agents	Inadequate to dissolve dried, residual product	Product-to-product contamination
		Sanitizing agents	Inadequate to reduce environmental microbial bioburden, as needed	High microbial bioburden levels, microbial contamination
		Cleaning technique	Practice, as designed, may re-introduce contamination	High levels of residual product and microbial bioburden; microbial contamination
		Cleaning equipment	Inadequate design to support technique/practices	Inadequate cleaning, e.g., mop handle too short to reach upper wall
			Not readily cleanable	May re-introduce contamination or cross-contaminate another area

Appendix I: (*Continued*)

Major Area	Subarea(s)		Potential Hazard	Potential Harm
Raw Materials	Material Management	Materials on Receipt:	Inadequate Identification	Use of Wrong Material or Wrong Grade of Material
			Inappropriate materials of construction, e.g., produces lint	Product contaminated with particulate, failed product quality
		Cleaning frequency	Insufficient frequency to keep contaminants at controllable level	High levels of contamination, product contamination
	Environmental control	Room air quality requirements	Inappropriate air quality requirements/classification for room's intended purpose	Inadequate controls in place, product contamination
		Air filtration system	Inadequate air filtration to keep environment clean of air-borne contaminants	High levels of air-borne contaminants, product contamination
		Air pressure	Inadequate air pressure or incorrect air flow design	Introduction of contaminate into room; product contamination
		Light protection	Inadequate light protection for light-sensitive product	Product degradation, instability
		Temperature/ humidity	Inadequate temperature or humidity control	Condensate in room, increased microbial bioburden levels, hygroscopic or temperature sensitive product degradation/instability
			Insufficient monitoring of room environment (air particles, temperature, humidity)	Product contamination/inability to correct issue in timely manner
Equipment/ Instruments	Fit for purpose	Correct instrument or equipment needed for job	Inappropriate sensitivity used for application (e.g., balance measuring to only one decimal when three decimals are needed)	Inadequate or inaccurate measurements, product quality, potency, stability compromised
			Incapable of meeting required operating parameters	Incomplete process step
		Size	Different size than required, e.g., smaller bulk formulation tank	Improper product formulation, e.g., mixing, nonhomogeneous bulk
		Model or design	Different model or design than required	Improper product formulation or filling operation

(*continued*)

Appendix I: (*Continued*)

Major Area	Subarea(s)		Potential Hazard	Potential Harm
Raw Materials	Material Management	Materials on Receipt:	Inadequate Identification	Use of Wrong Material or Wrong Grade of Material
		Use parameters	Incorrect use parameters, e.g., mixing speed, drying temperatures, or timeframes	Improper formulation; inadequate material dissolution or mixing resulting in nonhomogeneous batch; product instability
		Materials of construction	Wrong materials of construction, e.g., wrong filter substrate	Introduce foreign matter into formulation such as particulate or leachable material, product degradation
	Cleanability	Designed for easy cleanability	Design allows for internal spaces that harbor product	Product cross-contamination; foreign matter contamination
	Storage	Maintains integrity of cleaned equipment	Incomplete or inadequate practices	Product contamination
	Lubricants, other operating aids	Use	Uncontrolled use of lubricants; inappropriate applicators (e.g., spray applicator)	Product contamination from contaminated environment or product-contact surfaces
	Maintenance	Frequency	Inadequate to ensure consistent performance (e.g., infrequent air filter changes)	Compromise product quality (e.g., clogged filters do not provide adequate air filtration to support air quality levels)
		Parts inventory	Incorrect parts in inventory, like-to-like parts not available or used	Inappropriate equipment operation, parts may become warn earlier than expected, use of inappropriate parts may cause friction and particulate generation
		Timing (e.g., for stationary equipment)	Inappropriate time (e.g., during operations)	Contaminate product or product components
		Methods	Inappropriate practices that may not be aligned with maintenance needs	Inadequate or inconsistent equipment performance, product quality not ensured
		Emergency repairs	Inappropriate practices (e.g., use of unclean tools on product-contact parts)	Product contamination

Appendix I: (*Continued*)

Major Area	Subarea(s)		Potential Hazard	Potential Harm
Raw Materials	Material Management	Materials on Receipt:	Inadequate Identification	Use of Wrong Material or Wrong Grade of Material
	Calibration	Frequency	Inadequate to ensure proper performance (e.g., temperature recorder)	Incorrect results; failure to accurately measure or perform
		Methods	Insufficient points used in calibration to cover entire use range	Inaccurate measurements outside of calibrated range, compromise of product quality
	Validation	Extent of validation activities	Inadequate to ensure equipment is operating consistently as intended	Inconsistent product quality
People	Hygiene	Health habits	Inadequate personal hygiene	Contribute to environmental contamination
		Illness or injury	Sneezing, coughing, open wounds	Microbial contamination of environment or product
	Protective clothing or gear	Type	Inappropriate type or insufficient protective garb	Inadequate environmental or personnel protection
		Material of construction	Improper material of construction (e.g., loose weave bodysuit or materials shed particulate)	Unable to provide adequate barrier; particulate contamination of environment and product
		Size	Improper size (e.g., too large or too small)	May not completely cover and expose skin; may cause rubbing of skin leading to excessive shedding; product contamination
		Incoming packaging	Inappropriate and inadequate packaging to protect garb during transit	Use of garb contaminated from outside sources, introduce contamination to manufacturing areas
		Internal handling practices and storage conditions	Inappropriate practices and conditions to maintain integrity of packaging	Environmental contamination
		Disposal or replacement practices	No replacement schedule or standards for nondisposable garb	Inadequate protection leading to contamination
		Donning practices	Inappropriate practices	May introduce environmental contamination

(*continued*)

Appendix I: (*Continued*)

Major Area	Subarea(s)		Potential Hazard	Potential Harm
Raw Materials	Material Management	Materials on Receipt:	Inadequate Identification	Use of Wrong Material or Wrong Grade of Material
	Intervention/ interaction with product, components	Operator techniques	Procedures are not designed to avoid operator contact with product (e.g., operator uses hands to reach into or leans over open container	Product contamination
		Operator behaviors	Procedures do not define operator behaviors that avoid product contamination (e.g., operator leave doors open between rooms of different air pressures)	Product contamination
	Training	Appropriate training	Training not associated with tasks performed	Untrained personnel incorrectly performs step, product quality impacted
		Effective training	Inappropriate training technique	Training incomplete, incorrectly performs step impacting product quality
		Training records archive and retrieval	Poor documentation of training and uncontrolled storage. Inability to review training to assess training needs	May assign untrained person to perform task; step incorrectly performed impacting product quality
		Performance monitoring	Inadequate monitoring, training needs not assessed	Incorrect step performance impacts product quality

GLOSSARY

The following definitions are provided to aid understanding:

Critical quality attribute (CQA): A physical, chemical, biological, or microbiological property or characteristic that should be within an appropriate limit, range, or distribution to ensure the desired product quality [14].

Critical control point (CCP): A process parameter whose variability has an impact on a critical quality attribute and therefore should be monitored or controlled to ensure the process produces the desired quality [14].

Corrective action: Action to eliminate the cause of a detected nonconformity or other undesirable situation. NOTE: Corrective action is taken to prevent recurrence, whereas preventive action is taken to prevent occurrence [15].

Microbiological hazard: Any circumstance in the process flow of a nonsterile pharmaceutical that could cause a microbial contamination event affecting the quality of the product.

Preventive action: Action to eliminate the cause of a potential nonconformity or other undesirable potential situation. NOTE: Preventive action is taken to prevent occurrence, whereas corrective action is taken to prevent recurrence [15].

State of control: A condition in which the set of controls consistently provides assurance of continued process performance and product quality [15].

Critical levels: A criterion that separates acceptability from unacceptability.

Process flow diagram: A system representation of the sequence of steps or operations used in the production, control, and distribution of a pharmaceutical product.

Monitoring: The act of conducting scheduled measurement or observation of a CCP relative to its critical limits. The monitoring procedure must be able to detect loss of control at the CCP [24].

REFERENCES

1. Global Harmonization Task Force, July 2005, Implementation of Risk Management Principles and Activities Within a Quality Management System, SG3/N15RB/2005. Available at http://www.ghtf.org/documents/sg3/sg3n15r82005.pdf. Accessed 2011 Jul 18.
2. Gunter F., Hideki A., March 2008, R8 Implementation of Risk Management Principles and Activities Within a Quality Management System GHTF.SG3.N15 N15. Available at http://www.ghtf.org/meetings/conferences/4thapec/5_APEC_SG3_RM% 20Principles%20within%20a%20QMS%20KL%202008-Gunter%20Frey-Hideki%20 Asai.pdf. Accessed 2011 Jul 18.
3. Potter C. PQLI application of science and risk-based approaches-(Q8, Q9, and Q10) to existing products. J Pharm Inn April 2009;4:4–23.
4. Bush L., September 2009, Quality Risk Management Demystified at CMC Strategy Forum. Available at http://biopharminternational.findpharma.com/biopharm/GMPs% 2fValidation/Quality-Risk-Management-Demystified-at-CMC-Strateg/ArticleStandard /Article/detail/624456. Accessed 2011 Jul 18.
5. Kozlowski S., October 2006, Risk Management for Complex Pharmaceuticals. Available at http://www.fda.gov/ohrms/dockets/ac/06/slides/2006-4241s2_6.ppt. Accessed 2011 Jul 18.
6. European Medicines Agency (EMA), January 2007, Report of the CHMP Working Group on Benefit-Risk Assessment Models and Methods. EMEA/CHMP/1504/2007. Available at http://www.ema.europa.eu/docs/en_GB/document_library/Regulatory_ and_procedural_guideline/2010/01/WC500069668.pdf. Accessed 2011 Jul 18.

7. Coburn J., Levinson S.H., Weddle G., July 2008, A Precedent for Risk-Based Regulation. Available at http://www.johnsoncontrols.com/publish/etc./medialib/jci/be/commercial/capabilities/my_building__vertical/life_sciences/articles_and_white.Par.27473.File.tmp/A%20Precedent%20for%20Risk-Based%20Regulation.pdf. Accessed 2100 Jul 18.
8. Therapeutic Goods Agency (TGA), May 2011, TGA's Risk Management Approach to Regulation of Therapeutic Goods, Available at http://www.tga.gov.au/pdf/basics-regulation-risk-management.pdf. Accessed 2011 Jul 18.
9. Ahmed R et al. PDA Survey of Quality Risk Management Practices in the Pharmaceutical, Devices, & Biotechnology Industries. PDA J Pharm Sci Technol January 2008;62(1):1–21.
10. Cox B., The Gold Sheet©, December 2009, Roche Builds Quality Risk Management Program Response to Viracept Crisis.
11. The Chartered Quality Institute, 2010, A Guide to Supply Chain Risk Management for the Pharmaceutical and Medical Device Industries and their Suppliers, Volume 1.0.
12. International Conference on Harmonisation (ICH), November 2005, Quality Risk Management Q9. Available at http://www.ich.org/fileadmin/Public_Web_Site/ICH_Products/Guidelines/Quality/Q9/Step4/Q9_Guideline.pdf. Accessed 2011 Jul 18.
13. PIC/S, Quality Risk Management, Implementation of ICH Q9 in the Pharmaceutical Field, An Example of Methodology from PIC/S, 2010 January. Available at http://www.picscheme.org/bo/commun/upload/document/psinf012010exampleofqrmimplementation.pdf. Accessed 2011 Jul 18.
14. International Conference on Harmonisation (ICH), November 2009, Q8(R2) Pharmaceutical Development. Available at http://www.ich.org/fileadmin/Public_Web_Site/ICH_Products/Guidelines/Quality/Q8_R1/Step4/Q8_R2_Guideline.pdf. Accessed 2011 Jul 18.
15. International Conference on Harmonisation, June 2008, Pharmaceutical Quality System Q10. Available at http://www.ich.org/fileadmin/Public_Web_Site/ICH_Products/Guidelines/Quality/Q10/Step4/Q10_Guideline.pdf. Accessed 2011 Jul 18.
16. United States Food and Drug Administration (FDA), Center for Drug Evaluation and Research (CDER), Applying ICH Q8(R2), Q9, and Q10 Principles to CMC Review (MAPP 5016.1), February 2011. Available at http://www.fda.gov/downloads/AboutFDA/CentersOffices/CDER/ManualofPoliciesProcedures/UCM242665.pdf. Accessed 2011 Jul 18.
17. Introduction to Quality Control by Kaoru Ishikawa, Jan 1, 1990).
18. Ahmed R., et al., March 2008, PDA Technical Report No. 44 (TR 44), Quality Risk Management for Aseptic Processes, PDA Journal of Pharmaceutical Science and Technology, Supplement Volume 62, No. S-1.
19. 21 CFR 211.
20. Miele, W., March 2001, PDA Annual Meeting, A Science and Risk Based Approach to Microbiological Control of Manufacturing Environments Applicable to Multiple Dosage Forms Produced in a Non-Sterile Setting.
21. United States Department of Agriculture (USDA), July 1996, Federal Registry,9 CFR Part 304, et al., Pathogen reduction, Hazard Analysis and Critical Control Point (HACCP) Systems Final Rule. Available at http://www.fsis.usda.gov/Frame

REFERENCES

/FrameRedirect.asp?main=http://www.fsis.usda.gov/OPPDE/NIS/HIMP/HACCP_Final_Rule.pdf. Accessed 2011 Jul 18.

22. Ohio State University, February 2007, HACCP Information, HACCP Models. Available at http://meatsci.osu.edu/HACCPsupport.html. Accessed 2011 Jul 18.
23. United States Department of Agriculture (USDA), September 1999, HACCP-1;Guidebook for the Preparation of HACCP Plans. Available at http://www.fsis.usda.gov/OPPDE/nis/outreach/models/HACCP-1.pdf. Accessed 2011 Jul 18.
24. Annex to Codex Alimentarius Commission (CAC)/RCP 1–1969, Rev. 3 (1997), Hazard Analysis and Critical Control Point (HACCP) System and Guidelines for Its Application (Food and Agriculture of the United Nations). Available at http://www.fao.org/docrep/005/Y1579E/y1579e03.htm. Accessed 2011 Jul 18.
25. Quality and Standards of Ethiopia (QSAE), Basic Concepts and Implementation of Hazard Analysis Critical Control Points (HACCP). Available at http://www.qsae.org/web_en/pdf/HACCPImpGuide.pdf. Accessed 2011 Jul 18.
26. Burson D., University of Nebraska, HACCP Principle 3: Establish Critical Limits. Available at http://foodsafety.unl.edu/haccp/principles/Principle%203%20Establishing%20Critical%20Limits.pdf. Accessed 2011 Jul 18.
27. World Health Organization (WHO), 1997, HACCP: Introducing the Hazard Analysis Critical Control Point System,WHO/FSF/FOS/97.2 (1997). Available at http://www.who.int/foodsafety/fs_management/en/intro_haccp.pdf. Accessed 2011 Jul 18.
28. Industry Council for Development (ICD), WHO/ICD (ICD): Course on Training in Hazard Analysis and Critical Control Point System, Module 1: Introducing the HACCP System, Module 3: HACCP Application, Module 4: HACCP Plans and Their Implementation. Available at http://www.icd-online.org/an/html/courseshaccp.html. Accessed 2011 Jul 18.
29. United States Department of Agriculture (USDA), August 1997, Hazard Analysis and Critical Control Point Principles and Application Guidelines. Available at http://www.fda.gov/Food/FoodSafety/HazardAnalysisCriticalControlPointsHACCP/HACCPPrinciplesApplicationGuidelines/default.htm. Accessed 2011 Jul 18.
30. United States Department of Agriculture (USDA), April 1997, Guidebook for the Preparation of HACCP Plans. Available at http://haccpalliance.org/alliance/haccpmodels/guidebook.pdf. Accessed 2011 Jul 18.
31. United States Food and Drug Administration (FDA), October 2001, HACCP: A State-of-the-Art Approach to Food Safety.
32. Rushing J. E., Ward D.R., North Carolina State University, December 1995, HACCP Principles. Available at http://www.ces.ncsu.edu/depts/foodsci/ext/pubs/haccpprinciples.html. Accessed 2011 Jul 18.
33. Savell J.W, Texas A&M University, September 2000, Introduction to HACCP Principles in Meat Plants. Available at http://meat.tamu.edu/HACCPintro.pdf. Accessed 2011 Jul 18.
34. Schwartz A.R., October 1998, HACCP Inspectional Approach for Medical Device Industry Manufacturers; Should Industry be Accepting?, Journal of cGMP Compliance, Volume 3, Number 1. Available at http://www.mdiconsultants.com/publishing/haccparticle1.htm. Accessed 2011 Jul 18.

12

BIOPHARMACEUTICAL MANUFACTURING

Ruhi Ahmed and Thomas Genova

12.1 A GENERAL APPROACH TO IMPLEMENTING QRM IN BIOPHARMACEUTICAL MANUFACTURING

Risks are associated throughout the biopharmaceutical manufacturing process, from raw material supply through manufacturing and filling operations to final distribution via a controlled cold chain process. Although many of these risks apply equally to small molecules and biologics, assessing relevant attributes and risks for biotechnology-derived products is more complicated—as stated by the FDA "a much greater challenge for complex pharmaceuticals" [1]. Control often is more challenging with biotechnology products because they are difficult to characterize and the manufacturing processes are complex and variable. However, the application of risk management practices enables manufacturers to design processes that can proactively identify, mitigate, and/or control risks in a manner which ensures high-quality biotechnology drug products to patients.

As noted in ICH Q9 "Quality Risk Management," risk management is integral to product development because it enables manufacturers to "design a quality product and its manufacturing process to consistently deliver the intended performance of the product" [2]. For manufacturing process development, the implementation of risk management can begin with the identification of the quality target product profile and an analysis of user requirements. Process knowledge is gained through a comprehensive process development route, which carefully

---

Risk Management Applications in Pharmaceutical and Biopharmaceutical Manufacturing,
First Edition. Edited by A. Hamid Mollah, Mike Long, and Harold S. Baseman.
© 2013 John Wiley & Sons, Inc. Published 2013 by John Wiley & Sons, Inc.

identifies variables (e.g., raw material variability, process parameters) that may influence product quality. All of these can be explored via designed experiments to elaborate the effects of the variables and any parameter interactions. Risk assessment and management during process development helps companies identify the process parameters and attributes that will impact the quality of the product. It also allows companies to confidently validate only the parameters and attributes that are critical to the manufacturing process. This enhanced understanding of the process parameters and quality attributes eventually leads to a more thorough understanding of the product and its quality and safety profile.

In recent years, the trade-offs between achieving optimal supply chain efficiencies and management of supply chain risk has created some major problems for companies because they have jeopardized patient safety. Global supply chains are even more risky than domestic supply chains because of numerous links interconnecting a wide network of firms and communication and cultural gaps. Therefore, supply chain risk assessment is absolutely essential for biotech products and should begin with incoming raw materials and excipients and extend through the distribution chain to the patient. Information gained from experience and product development can be used to identify potential risks and establish mitigations to minimize their occurrence and increase the ability to detect failures should they occur.

Finally, risk assessments should become a living part of the product life cycle. Specifically, as the product/process design matures, technology content improves, and patient needs and safety concerns are better defined or regulatory requirements change, the risk assessment plan should be maintained and updated as new knowledge is gained about the product and process.

This chapter briefly discusses the process steps traditionally used for manufacturing biopharmaceutical products and identifies areas of risk associated with this process step. The reader should note that, while this chapter provides a list of risks that are associated with the process steps outlined, it is not exhaustive or all inclusive. This chapter also does not address risks associated with the preparation of finished dosage forms, for example, prefilled syringes. These risks are addressed elsewhere in this book (see Chapter 9 for a complete discussion of aseptic processing risks). A detailed case study is provided at the end of the chapter, which illustrates how a risk assessment tool can be applied in the manufacturing process to identify, mitigate, and control risk. Finally, this text is based on the opinion and experience of the authors and is not meant to replace regulatory guidances or regulations.

12.2 UNDERSTANDING CRITICAL QUALITY ATTRIBUTES AND CRITICAL PROCESS PARAMETERS

All products have a set of characteristics or attributes by which they may be defined. Frequently, such attributes are included in specifications and product release tests and are thought of as quality attributes. When a quality attribute

UNDERSTANDING CRITICAL QUALITY ATTRIBUTES

affects product safety, efficacy, and purity, it is referred to as a *critical quality attribute* (CQA) [3]. By definition, all specifications are considered to be CQAs [4].

CQAs should address those aspects of the product that are indicative of product safety and efficacy. Historically, CQAs have been synonymous with product release tests. In recent years, however, health authorities have acknowledged that not all release tests are necessarily CQAs. Companies have been encouraged to utilize risk management techniques to analyze products and identify those attributes that are truly critical to safety and efficacy. While many risk management tools may facilitate such analysis, a failure mode and effects (FMEA) type approach is frequently used. A simple model for identifying CQAs is shown in Figure 12.1.

In such a model, the severity of harm resulting from a failed attribute and the likelihood of occurrence of such a failure are used to identify potential CQAs. The data needed to perform the assessments noted in Figure 12.1 may be drawn from a variety of sources. Data from preclinical and clinical studies may be used to assess severity and likelihood of specific compounds. In addition, analysts may utilize historical information from similar compounds to establish a preliminary assessment of risk.

Quality attributes are influenced by process parameters. Process parameters with a high potential to influence CQAs are defined as critical process parameters (CPPs) because they have the potential to impact product safety and efficacy. Again, the identification of CPPs can be facilitated using risk management approaches such as fish bone diagrams and FMEAs. A simple approach to CPP identification is shown in Figure 12.2.

The analysis is undertaken by first creating a process map that establishes the process steps, parameters, and target values for the entire manufacturing operation. While separate process maps may be prepared for individual process steps as described in this chapter, they should include all phases of the process having the potential to impact product quality.

Once a process map has been prepared, the individual process parameters may be evaluated to assess their potential impact on final product quality. The

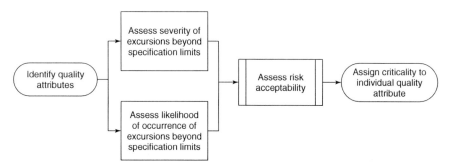

Figure 12.1 Identification of critical quality attributes.

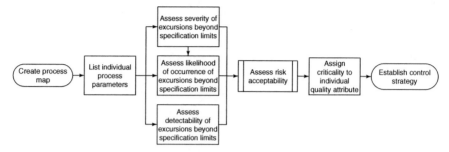

Figure 12.2 Identification of critical process parameters.

evaluation can be based on retrospective information such as general scientific knowledge, experimental data, and nonconformance results. Prospective evaluations can also be performed using the results of designed experiments, and validation studies. Using this information, the evaluators may identify a subset of parameters that have the potential to impact product safety and efficacy.

The next step in the process is to assign criticality to the parameters identified. Criticality is assigned on the basis of three risk factors: severity, occurrence, and detectability [5].

Severity is evaluated by assessing the impact of a process parameter on product quality and patient safety. Determine how the parameter would impact a CQA if it were to exceed its maximum operating range, for example, batch record limits. In some instances, the maximum operating range for a parameter is well within the range proved to be acceptable to product quality. In these instances, an out-of-tolerance parameter would be expected to have minimal impact on its associated CQA. In other instances, a parameter's maximum operating range may be very close to its proven acceptable range and excursions have a greater potential to impact product quality. Thus, its potential impact on product quality may be very severe.

The next step in assigning criticality to a parameter is to assess the likelihood that a parameter will exceed its maximum operating range. The likelihood of exceeding an operating range can be identified through both experimental studies and practical experience with the same or similar processes. Appropriate engineering controls should be established to insure that processes operate within specified ranges, thereby reducing the likelihood of an excursion.

Detectability is a factor that describes one's ability to identify an excursion before it affects patient safety or product quality. Detectability evaluations should consider both the ability to detect a process excursion and the ability to detect the failure of a quality attribute. Detectability is a mitigating factor in that it facilitates the identification of potential problems and permits corrective actions to be performed to preclude the product and/or release of a nonconforming product.

Taken together, the three factors, severity, likelihood, and detectability define the criticality of a process parameter. In an FMEA analysis, the product of the three factors would be compared to a preestablished value to assign criticality. Parameters identified as CPPs might be subject to additional engineering controls to ensure compliance with specified limits. Alternatively, the process might be redesigned to reduce likelihood or additional control strategies to improve process performance. Finally, in-process monitoring might be established to improve detectability, thereby reducing risk and, consequently, criticality.

An overall understanding of a product's CQAs and their underlying CPPs facilitates the design of processes and control strategies to ensure that products comply with their performance requirements. CQA and CPP identification strategies rely heavily on risk analysis techniques. Risk analysis not only facilitates the identification of these important elements but also facilitates the implementation of appropriate mitigations, thereby creating robust processes that are less likely to result in a nonconforming product.

12.2.1 Quality Risk Management in the Identification of CQAS and CPPS

Two risks are associated with CQA/CPP identification: (1) the risk of not identifying an attribute or parameter as critical when, in fact, it is, and (2) the risk of identifying an attribute or parameter as critical when it is not.

The first scenario, not properly identifying an attribute or parameter may result in directly impacting product safety and efficacy. This risk can be addressed by having robust procedures in place to guide the identification process and by using a cross-functional team that is familiar with all aspects of the product and process.

The second risk, identifying an attribute or parameter as critical, when it is not, may result in over-testing products and/or over-engineering or controlling processes. Both these would require resource expenditures, which may not be necessary to achieve adequate control.

12.3 INDIVIDUAL PROCESS STAGES AND ASSOCIATED RISKS

12.3.1 Raw Materials

12.3.1.1 Process Overview Establishing and maintaining a consistent supply of raw materials is critical to all manufacturing processes. However, biopharmaceutical processes are not only subject to the same issues as pharmaceutical processes but are also impacted by added challenges, for example, microbiological contamination, arising from the use of materials derived from animals, yeast, humans, etc., and the need for those materials to be compatible with, and supportive of, biological systems. These challenges have led manufacturers to substitute animal-derived materials with others derived from vegetable or recombinant sources. Not all products or materials are, however, amenable to substitution because of uncharacterized growth factors contained therein, and thus some products continue to utilize animal-derived materials.

Raw material control contributes to lot-to-lot variability [6]. Thus, manufacturers must be aware of and control raw materials and their variability. Raw material control also contributes to product safety because of impurities in the material. Manufacturers must be able to test and verify the identity of raw materials. Vendor-supplied information, in the form of a certificate of analysis, may not be sufficient to adequately control raw material. Vendors should be certified to ensure a continuous supply of safe, consistent material.

The origin of materials may be an important factor in the supply chain. Historically, raw materials used in biopharmaceutical manufacturing have been naturally derived and have included animal serum, albumin, enzymes, etc. These materials are used to prepare both active substances and excipients, (e.g., polysorbates, collagen, and gelatin), and may contain microbial contamination and/or support microbial growth. Contamination by microorganisms including bacteria, fungi, mycoplasma, viruses, and transmissible spongiform encephalopathies (TSEs) and/or proteolytic enzymes is not unusual. Bacteria and fungi can express extracellular proteins and directly influence the material itself and/or the cell cultures from which the biopharmaceutical is derived. Mycoplasma, viruses, and TSEs can interact with and influence the cell cultures, potentially impacting yields, purity, and the efficacy of downstream processes. Perhaps the greatest challenge provided by TSEs is the inherent difficulty that they present to removal from manufacturing processes. Not only is their removal difficult but the ability to validate the removal process is itself difficult to demonstrate.

Contamination is controlled and/or minimized by a series of actions that span the entire process from collection through processing and testing. The collection process is controlled through strict requirements regarding geographic sourcing of materials, definition and documentation of donor animal herds, the collection of serum according to protocols and by trained personnel.

Current regulatory requirements include provisions to test animal-derived raw materials for adventitious contaminants before use [7–9]. The tests include assays to determine bacterial and fungal levels, mycoplasma assays, and a series of tests to detect both viral antibodies and viral particles. These tests are to be performed before the implementation of any inactivation steps to assess total contaminant load as the total load may impact the efficacy of removal/inactivation steps.

Gamma irradiation is currently the preferred means of microorganism control. The irradiation process sterilizes the raw materials and renders them safe for use. After gamma irradiation, manufacturing processes that physically remove contaminants, for example, filtration or precipitation, are the preferred means of microorganism control; however, chemical inactivation of contaminating microorganisms may be acceptable in some situations. In instances where physical removal or inactivation of TSE forms the basis of a regulatory claim concerning product safety relative to TSE, all removal and inactivation process must be adequately validated to demonstrate efficacy for their intended purpose.

In addition, many companies impart a degree of control over their raw materials by maintaining an active supplier quality program through which they verify the continued compliance of raw material suppliers to applicable health authority and company standards. Suppliers in such programs are included on the basis of the criticality of the individual raw material and its associated product. A relative risk ranking tool can be used to establish the priorities with which suppliers are audited for compliance to standards [2]. Thus, animal-derived raw materials being used in parenteral products would be expected to assume a higher priority than non-animal-derived raw materials or materials intended for use in topical applications. Similarly, suppliers having a spotty compliance history may be subject to closer scrutiny than those who have a "cleaner" compliance record. Prioritization for follow-up activities may be established using relative risk ranking tools or checklists.

Supplier quality programs typically assess a supplier's quality system including the compliance with GMPs, nonconformances, process changes, validation issues (method and process), etc. These assessments provide a window into supplier processes and can provide an early warning of problems associated with raw material quality. Given this early warning, a company may be able to work with their suppliers to establish mitigations and alleviate potential problems before they occur.

Managing suppliers and cold chain issues are important to assure the quality of raw materials. Vendor experience often dictates the frequency of vendor audits. In selecting vendors it is important to consider their financial status, their experience with a particular raw material, the geographic proximity to the manufacturing sites and to whom the supplier provides materials, multiple industries, or just the pharmaceutical industry. Risk may be minimized by qualifying multiple suppliers to avoid reliance on a single source of materials.

Supplier agreements should include provisions for oversight of change control processes such that manufacturers are notified of changes to a supplier's manufacturing processes.

Including impact assessments in such agreements provides a ready form for making decisions relative to the importance of changes to material quality.

Risk assessments for suppliers should focus on material quality and availability. A nine-block assessment tool can be used to integrate information about supplier performance and material risk into a single risk ranking. The ranking can then be used to establish supplier controls and audit frequencies. An example of a nine-block assessment tool is shown in Figure 12.3.

As seen in the nine-block tool in Figure 12.3, the degree of control exerted over a supplier is directly related to the risk associated with an individual material and the overall performance of the supplier. Thus, a marginal supplier who provides a high risk material would require much greater control than a best-in-class supplier providing a material having a similar risk characterization.

Raw materials must be traced by suppliers to their country of origin. Risk assessments should be reviewed periodically and should include lessons learned from past experiences. It is highly recommended that they be updated to reflect current knowledge and changing conditions relative to the supplier.

12.3.1.2 Quality Risk Management for Raw Materials The major risks associated with the raw materials management are discussed in the following section.

		Material risk		
		Low	Medium	High
Supplier performance	Best in class	Audit every five years Vendor certified for CofA acceptance	Audit every five years Vendor certified for CofA acceptance	Audit every three years Vendor certified for minimal testing
	Acceptable	Audit every two years Vendor certified for minimal testing	Audit every two years Skip-lot testing of incoming lots	Annual Audit Skip-lot testing of incoming lots
	Marginal	Annual audit	Annual audit Test all lots received	Annual audit Test all lots received Review all vendor non-conformances

Figure 12.3 Example of a 9-block assessment tool.

Several risks are associated with raw materials. Chief among these for biopharmaceuticals is the use of animal-derived materials. This risk is addressed through adherence to sound animal husbandry and procedural controls, as well as, through the use of screening tests to assess material quality before use. The use of animal-derived materials may be linked to microbiological contamination, by bacteria, fungi, mycoplasmas, viruses, and/or prions. This risk is also addressed through procedural controls and then verified through testing. Manufacturers should also be concerned with lot-to-lot variation in incoming materials. The effects of lot-to-lot variation were discussed earlier. They can be assessed by using multiple lots during material qualification and through supplier qualification, as well as through the use of monitoring programs that periodically test and assess the quality of incoming materials.

12.3.2 Cell Banking

12.3.2.1 Process Overview Cell banking assures an adequate supply of homogeneous and well-characterized cells for manufacturing over the expected lifetime of a biopharmaceutical product. Cell substrates that are banked must be stable, reproducible, and available in sufficient quantities. Specifically, cell substrate stability is important in terms of safety and quality because it ensures that only the desired and specific biopharmaceutical product is manufactured without the introduction of any mutations and with minimal likelihood of contamination by infectious agents. Cell substrate reproducibility is important over time because the biopharmaceutical product manufactured during the first manufacturing campaign has to match the product that is produced in subsequent campaigns. Finally, cell substrate availability is important to ensure that there is adequate long-term inventory for commercial production. Typically, manufacturers prepare a master and a working cell bank to ensure supply of stable and reproducible genetic source materials [10].

A cell bank construction strategy is usually in two stages. The first stage is the construction of a master cell bank (MCB). The MCB is produced directly from the R&D cell bank constructed for early production. Recombinant production cell lines are often initially constructed by the introduction of a plasmid (containing the nucleotide sequence that expresses the protein of interest) into a selected cell line. For the MCBs, the cell line of interest is expanded and cultured for a defined number of cell passages. The resultant product producing cells are then aliquoted in small amounts into ampoules or vials and frozen and stored under defined conditions. Thus, the cells with the desired genetic source material are effectively cryopreserved for an indefinite amount of time. MCB ampoules are then used to generate a working cell bank (WCB). The generation of a WCB typically entails the thawing of a single MCB ampoule, culturing and propagating the cells contained in the ampoule, and subsequently aliquoting these cells into multiple hundreds of ampoules. This new batch of ampoules is then cryopreserved to form the WCB. When a single batch of new product is desired, one ampoule of the WCB is thawed and used to seed that batch. When all the ampoules of the WCB have been utilized, another ampoule of the MCB is thawed and used to generate another WCB.

Contaminated cell lines pose a risk to product quality and safety. As such, the characterization and testing of cell banks is critical to confirm the identity, purity, and genetic stability of the cell line. The potential risk of introducing adventitious agents such as bacteria, mycoplasma, fungi, or viruses to the biopharmaceutical must be eliminated or minimized during the cell banking process. Regulatory agencies (e.g., FDA, EMEA, and World Health Organization) require an appropriate and sufficient characterization of the MCB and WCB because the quality and safety of the cell banks impact the quality and safety of the product. And since there are a diverse number of banks in use, regulatory agencies such as the FDA and EMEA have also provided specific guidance on the characterization tests needed for the types of banks used for biopharmaceutical production [11–13]. In accordance with the guidances cited earlier, some common tests recommended for characterization of bacterial/mammalian cell banks are outlined in Tables 12.1 and 12.2.

These comprehensive characterization tests ensure that a contaminated bank is not used for production, and they confirm the identity, purity, and suitability of the bank for manufacturing use. In addition, it is also critical to check that the cellular material at the end-of-production (EOP) run remains essentially the same as that in the MCB or the WCB. That is, EOP cells should be evaluated once for those endogenous viruses, which may not have been detected in the MCB and WCB. Also, by conducting adventitious virus assays once, it shows that the production process is not prone to such contamination.

12.3.2.2 Quality Risk Management for Cell Banking The major risks associated with the cell banking process are discussed in the following section.

Presence of Infective Agents (Viruses, TSE/BSE, Bacterial Endotoxin, Mycoplasma)
Manufacturers of biopharmaceutical products, products derived from animal or human tissues, blood products, and some medical devices are required to assess

TABLE 12.1 Tests for Master Cell Banks (MCBs)

Attribute	Test Methods
Microbiological contamination	Bacteriostasis and fungistasis Sterility Mycoplasma (cultivable and noncultivable strains) Endotoxin
Cell line identification	Isoenzyme analysis/DNA fingerprinting or karyology Copy number determination Restriction map analysis DNA/RNA sequencing
Retroviruses	Reverse transcriptase, infectivity assays, PCR-based assays
Adventitious viruses	*In vitro* and *in vivo* assays for viral contaminants
Specific virus assays	Antibody production assays, for example, mouse antibody production (MAP) assay or hamster antibody (HAP) production assay; PCR-based assays
Other virus assays	Specific bovine or porcine virus assays

TABLE 12.2 Tests for Working Cells Banks (WCBs)

Attribute	Test Methods
Microbial contamination	Bacteriostasis and fungistasis Sterility Mycoplasma (cultivable and noncultivable strains) Endotoxin
Adventitious virus testing	*In vitro* and *in vivo* assays for viral contaminants
Cell line identification	DNA fingerprinting or karyology or isoenzyme analysis

[a] Limited testing is required because WCBs generally originate from a well-characterized MCB.

the ability of their purification and manufacturing processes to produce a product that is safe for use in humans. Incoming raw materials (such as animal-derived serum) for cell banking should be assessed for risk of contamination, for example, transmissible spongiform encephalopathy (TSE) or bovine spongiform encephalopathy (BSE). Moreover, the type of testing required needs to be host dependent. Elimination of the use of animal-derived raw materials/excipients, selection of cell substrates with low inherent viral (or phage) risk, and the implementation of appropriate in-process and lot release detection tests are all important strategies to mitigate the risk of viral and TSE/BSE contamination. The documentation regarding viral and TSE/BSE clearance studies are required by regulatory

INDIVIDUAL PROCESS STAGES AND ASSOCIATED RISKS 335

authorities as an integral part of Biologic License Applications (BLAs) before approval. For more information, refer to EP 5.1.7 Viral Safety [14], FDA's 1993 Points to Consider in the Characterization of Cell Lines Used to Produce Biologicals [15], and ICH Q5A (R1) [11].

The presence of bacterial endotoxins should also be monitored and levels controlled throughout the production process to ensure that the downstream purification process is not overloaded with this contaminant. *In vitro* bacterial endotoxin testing and *in vivo* pyrogenicity testing are critical for demonstrating that the manufacturing appropriately reduces, removes, or inactivates these hazards. Mycoplasma are the simplest, smallest, self-replicating organisms. Production culture contamination usually originates from components of cell culture medium (e.g., serum) or from an infected person working in production. Tests such as direct fluorescent assay or PCR can be employed to readily detect mycoplasma contamination.

Facilities, Equipment, and Personnel   For all areas in which cell banking operations are performed, precautions should be taken to prevent contamination. Generally, GMP banks need to be prepared in well-controlled facilities where rooms are appropriately classified to demonstrate control in quality of air, movement of equipment (clean and dirty), and segregation of activities. In addition, thorough personnel training and control of work practices are necessary to ensure that the preparation and maintenance of cell banks are free from cross-contamination. Inappropriate preparation and handling of the cells for banking is one of the common reasons for contaminated cell lines.

Documentation   Inappropriate or inadequate documentation of proper control procedures used during the assembling, preparation, and storage of the cell banks is a risk because it does not provide assurance regarding the true identity of the components and the absence of introduced contamination. The results of all test performed on the MCB, WCB, and EOP cells have to be appropriately documented and filed with regulatory agencies to demonstrate the safety and quality of the genetic source material for the biopharmaceutical product. Inadequate documentation can be a significant problem for companies because according to the FDA, per 21 CFR 601.2(a), the Agency can refuse to file a submitted BLA/NDA because of "...insufficient description of source material (including characterization of relevant cell banking systems)..." [16].

Storage   Finally, the appropriate and reliable storage of cell banks is also critical for ensuring the availability and the quality of the genetic source material. Cell banks have to be maintained in a suitable and controlled environment (typically in the vapor phase of liquid nitrogen) to ensure viability, stability, absence of contamination, and their availability. Therefore, the selection of the appropriate cell banking conditions and facilities is important in minimizing risk to the integrity of the MCB and WCB banks.

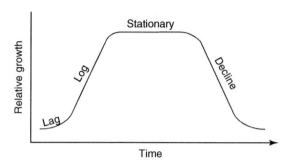

Figure 12.4 Cellular growth curve for microbial cells.

12.3.3 Fermentation/Cell Culture

12.3.3.1 Process Overview

Fermentation In biopharmaceutical manufacturing, the term "fermentation" describes any process for the production of a product by the large-scale cultivation of a microorganism (occurring with or without air). The growth of microbial cultures can be divided into a number of stages as shown in Figure 12.4.

Initially, the microbial cells of interest are inoculated into a selected growth medium, a period during which growth does not appear to occur (adaptation phase or lag phase). The next phase is characterized by a period where the growth rate of cells gradually increases as the cells grow at a constant, maximum rate (log or exponential phase). During this phase, the microbial cells take up nutrients from the fermentation broth and release products, byproducts, and waste metabolites. Eventually, the growth of cells ceases, or the number of new cells formed equals that of others dying (stationary phase), owing to the continuously falling concentrations of nutrients and/or a continuously increasing (accumulating) concentrations of toxic substances. Finally, after a further period of time, the viable cell number declines and the cells die (death phase). It is important to keep the cells in growth phase because if they are allowed to reach the lag or stationary phase, the culture may cease or lag depending on the cell line and growth medium used. Once acceptable microbial cell growth is achieved, that microbial culture can be used to inoculate the production fermentor. Microbial fermentations typically do not require elaborate cell accumulation and expansion steps (as required for mammalian cell culture processes) because of the short doubling time for the microbes (e.g., 15–20 min for *Escherichia coli* in the laboratory).

Cell Culture Tissue/cell culture is the general term for the removal of cells, tissues, or organs from an animal or plant and its subsequent placement into an artificial environment conducive to growth. This environment usually consists of a suitable glass or plastic culture vessel (e.g., shake flasks) containing a liquid or semi-solid medium that supplies the nutrients essential for survival and growth.

As cells are placed in a suitable culture environment, they will divide and grow. This is called the "primary culture." When the cells in the primary cell culture have grown, they are subcultured to allow continued growth with a fresh medium. For mammalian cells, cell subcultures are normally used to inoculate intermediate bioreactors to generate enough cell mass or cell density ($>1.5 \times 10^6$ cells/ml) so that they can be used to inoculate the large production bioreactors.

Process Overview The schematic in Figure 12.5 represents a typical fermentation/cell culture process. In brief, the fermentation/cell culture process can be described as follows: the first steps involve the removal and thawing of a vial from the WCB or MCB into an appropriate medium in the presence or absence of selective agents; the next steps are the production of an active, pure culture in sufficient quantity to inoculate the production vessel, followed by growth of the organism in the production fermenter/bioreactor under optimum conditions for product formation. The final steps involve the collection of the product containing harvest fluid and the disposal of the effluent waste.

12.3.3.2 Quality Risk Management for Fermentation/Cell Cultures The major risks associated with the fermentation/cell culture process are discussed in the following section.

Contamination One of the primary risks in cell culturing and fermentation is contamination. There are two main types of contamination: chemical and biological. Chemical contamination is usually harder to detect and can be caused by agents such as endotoxins and extractables/leachables from plastic storage vessels and tubing, metal ions, or minute traces of chemical disinfectants. Biological contamination is usually easier to detect and is caused by fast-growing yeast, bacteria, and fungi that usually have a visible effect on the cell culture (e.g., phage in a fermenter can destroy a culture in an hour with mass cell lysis; the first hints are the production of large amounts of foam in the culture vessel or increased oxygen consumption). However, two other sources of biological contamination, mycoplasmas and viruses, are not easily detectable and usually require special detection methods, such as sensitive PCR-based methods and immunocytochemical procedures.

Therefore, in order to minimize risks of both chemical and biological contamination, it is essential that personnel engaged in cell culture are appropriately trained in aseptic techniques; and that properly designed, maintained, and sterilized equipment is used during the process.

Suitable Environment There is a significant risk to production if microorganisms/cells are not provided quality materials, selection agents, and a suitable environment in which to grow, proliferate, and carry out their desired physiological and biochemical functions. Therefore, appropriate control and monitoring of equipment, culture media, environment, etc. is required to ensure provision of a good suspension/substrate for attachment (i.e., for anchorage-dependent cells),

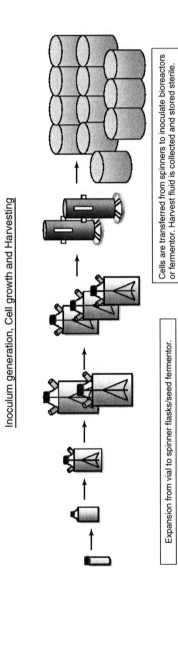

Figure 12.5 Schematic representation of typical fermentation/cell culture process. Detailed information on fermentation and cell culture can be found in P.F. Stanbury et al. [18] and J.A. Ryan [34].

the supply of proper culture medium (with the appropriate growth factors, pH, osmolality, etc.), and the control/monitoring of appropriate temperature, CO_2, and DO parameters.

Selection of Stable Strains The ability of the producing strain to maintain its high productivity during both culture maintenance and fermentation is a very important quality. Therefore, certain characteristics of the producing organism that affect the process are critical for its selection; otherwise, there is a risk to its commercial success. Some examples of these characteristics are strain stability, resistance to phage infections, response to dissolved oxygen, tolerance of medium components, the production of foam, the production of undesirable byproducts, and the morphological form of the organism.

Culture Media The quality and composition of culture media play an important role in the production process. Development studies are typically necessary for selecting the most suitable media for fermentation/cell culture processes; otherwise, there is a significant risk that process and product quality will be compromised. Generally, the media selected should produce maximum yields and concentration of product per gram of substrate used. Selected media should also permit the maximum rate of product formation with a minimum yield of undesired products. In addition, media quality should be consistent, be readily available, and should cause minimum problems during preparation and sterilization. Media should also cause minimal problems in other aspects of the production process, such as aeration and agitation, extraction, purification, and waste treatment. Another consideration to be kept in mind while selecting appropriate media is that the selection of media can sometimes affect the design of the fermenter or bioreactor to be used.

Oxygen Requirements The majority of fermentation processes is aerobic and, therefore, requires the provision of oxygen enrichment. The oxygen demand of industrial fermentation and cell culturing processes is normally supplied by aerating and agitating the fermentation broth or culture medium. However, there is a risk that cellular metabolism of many processes can be limited if there is not enough oxygen availability. It is critical that factors affecting the fermenter's or bioreactor's efficiency in supplying microbial or animal cells with oxygen are considered and carefully monitored to ensure that there is enough oxygen in the reactor.

Sterilization A biotechnology product is produced by the culture of a specific organism or organisms in a nutrient medium, and foreign microorganism contamination can have serious consequences. For example, a foreign microorganism can contaminate the reactor and compete with the desired culture organism for nutrients, thereby ruining the fermentation and causing costly delays in production. Therefore, procedures should be employed to avoid the risk of contamination. Examples of such procedures are using a pure inoculum to start fermentation/cell culture, sterilizing the medium to be employed, sterilizing the fermenter/biorecator vessel, sterilizing all materials to be added during the process, and maintaining aseptic conditions during fermentation and cell culture process. Clean-in-place (CIP) and sterilization-in-place (SIP) processes should be qualified and/or validated where required. The extent to which these procedures need to be adopted is determined by the risk of contamination and the nature of the consequences.

Facilities and Equipment The maintenance of aseptic conditions during fermentation/cell culture and the design of suitable fermentation/bioreactor equipment are critical for the fermentation/cell culture process. The main function of a fermenter/bioreactor is to provide a controlled environment for the growth of microorganisms or mammalian cells used for obtaining the desired product. As such, the fermentation/bioreactor vessel should be capable of being operated aseptically for a number of days and should be reliable in long-term operations. Significantly, it should also provide adequate aeration and agitation to meet the metabolic needs of the microorganisms and cells and should not cause damage to the microorganisms/cells during mixing and agitation.

Scale-Up The challenge of developing a process from a laboratory to the pilot scale and subsequently to the industrial scale is also an important consideration during fermentation/cell culture. Additional details on scale-up considerations are provided in Section 12.3.5.

12.3.4 Downstream Processing

12.3.4.1 Process Overview Downstream processing refers to the recovery and purification of biotechnology products from the recombinant culture or plant tissue or fermentation broth. The purification of biotechnology-derived products can be difficult, resource intensive, and expensive. Therefore, the main goal of the recovery and purification process is to obtain a sufficient quantity of good quality product and to do that as quickly and efficiently as possible. In general, downstream processing steps can be broadly categorized into four main groups:

- *Removal of insolubles* involves the recovery of the product as a solute and removal of large particulates and/or cell debris.
- *Product isolation* involves the removal of those components whose properties vary markedly from that of the desired product, for example, viral inactivation.
- *Product purification* is done to separate those contaminants that resemble the product very closely in physical and chemical properties.
- *Product polishing* involves the final processing steps to obtain a concentrated and pure product and ends with packaging/storing of the product in a form that is stable and transportable.

The choice of a recovery/purification process is typically based on process/product-specific criteria such as the intracellular or extracellular location of the product, the concentration of the product in the harvest fluid, the minimal acceptable standard of purity, facility limitations, etc. Detailed information on downstream processing can be found in R. Bates [17].

Microbial Products For fermentation or microbial products (i.e., obtained from bacteria, yeast, and fungi) that are usually intracellular, the recovery and purification process can be summarized as follows: the first step is the removal of large solid particles and microbial cells. The method commonly employed for this purpose is either filtration or centrifugation. In the next step, the cells are lysed either by chemical or mechanical methods to release the product. The cell lysate is then fractionated (primary isolation of the product) using techniques such as ultrafiltration, reverse osmosis, adsorption/ion exchange/gel filtration, or affinity chromatography. After this step, the product-containing fraction is further purified by fractional precipitation, or more precise chromatography techniques (intermediate purification/concentration) to obtain a product that is highly concentrated and essentially free from impurities and ready for filtration and fill [18]. The flowchart in Figure 12.6 depicts the typical stages of extraction and purification of fermentation products.

Mammalian Products For therapeutic proteins (i.e., obtained from mammalian cells) that are usually extracellular and secreted, the recovery and purification process are briefly described as follows: the first step is the clarification and initial purification stage. The main goal of this step is to reduce the working volume of the process stream and to remove the cell debris and harmful contaminants. The speed of recovery is critical in this step to minimize degradation and product loss. Typical unit operations include the use of filtration (including depth filtration to reduce solids content and ultrafiltration and diafiltration to reduce volume and change buffering conditions), chromatography, precipitation, or centrifugation techniques. In the next step, once the whole cells are separated from the product-containing harvested cell culture fluid (HCCF) during clarification, the HCCF can often be applied directly onto a chromatographic column for initial purification. Typically, filtration, centrifugation, ion exchange column (IEC), or hydrophobic interaction columns (HIC) are used for this purpose. Ideally, this initial purification results in a substantial reduction in the level of proteases or other harmful components. In the next step, intermediate processing, most of the remaining contaminants are removed from the process stream (e.g., virus particles, nucleic acids, endotoxins, and other host cell proteins). Unit operations with a large capacity, high recovery, and good resolving power such as adsorption chromatography (e.g., ion exchange, hydrophobic interaction, affinity chromatography, etc.) and membrane chromatography are common methods for use in this stage. The final step of the process is concentration and polishing. This step removes impurities such as deamidated isoforms or aggregates using chromatographic techniques such as gel permeation chromatography (GPC). In

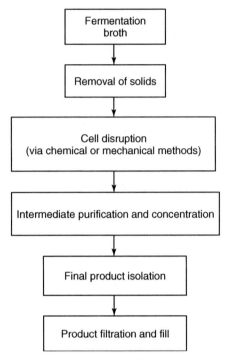

Figure 12.6 Stages in the extraction and purification of intracellular fermentation products.

general, the feedstock for the final polishing step is very clean, with few impurities, so this step typically involves only one unit operation having both high resolving power and high recovery. However, if the processing capability of the chromatographic step is limited, a concentration step may be required before column loading [19]. Drug substance after the final concentration and polishing step is ready for formulation and fill/storage. The schematic in Figure 12.7 depicts the typical stages of recovery and purification of therapeutic products and the methods employed during the processing steps.

12.3.4.2 Quality Risk Management for Purification Process The major risks associated with the extraction and purification process are discussed in the following section.

Inefficient Harvest/Recovery Conditions One major risk during the manufacturing process is inefficient harvesting of the product from cells. For example, low cell culture titers (10–100 mg/l) may require the collection of large volumes of product-containing HCCF that then requires rapid processing; or if the desludge time in the centrifuge is too long or too short, it can result in low yields of product, poor clearance of medium components from solids, or product contamination. Other risks to the product may include degradation risk from high

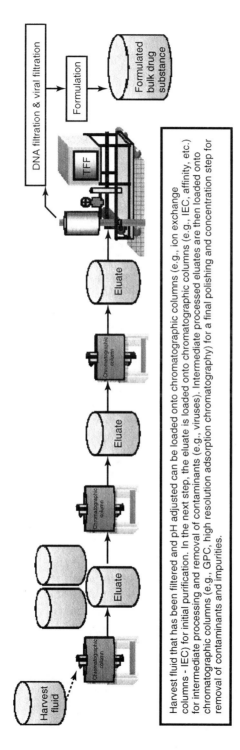

Figure 12.7 Typical stages in the recovery and purification of extracellular therapeutic products.

temperatures, long hold times, inappropriate processing times, and degradation due to shear. Given the fragility and the high value of the cell-derived products, cell separation must be accomplished with minimum cell damage, degradation, and loss of product activity during the recovery process. Therefore, thorough process/equipment risk assessments need to be performed for quick and efficient recovery and to mitigate product loss.

Inefficient Purification Conditions Downstream processing is used for the removal of a variety of process- and product-related impurities, typically via efficient, orthogonal, and robust purification processes. However, many downstream processes often consist of only two or three separation steps, and usually avoid conditioning steps such as ultrafiltration/diafiltration (UF/DF) for buffer exchange to reduce the total number of processing steps. This strategy can significantly increase the risk that contaminants are not completely removed during the purification process, especially if complex microbial feedstocks or *E. coli* lysates are used as starting materials. In addition, a typical chromatographic purification step has numerous operating parameters that can impact its performance. Therefore, it is important to apply risk analysis at the process characterization stage to identify key process parameters that may impact the purification process and product quality.

Viral Contamination The major concern when using mammalian cell lines for production of a biotechnology product is the risk of viral contamination. Such contamination could have serious clinical consequences and can arise from source cell lines from infected animals, virus endogenous to the cell line, use of contaminated reagents or equipment, and improper handling of the cell line.

There are three complementary approaches that are generally employed to reduce the risk of viral contamination. Specifically, these approaches are as follows:

- Selecting and testing cell lines and raw materials (including cell culture media and animal-derived raw materials) for the absence of undesirable viruses that may be infectious and/or pathogenic for humans
- Assessing the capacity of the manufacturing process to inactivate or remove infectious viruses.
- Testing the product at appropriate stages during production for the absence of contaminating infectious viruses.

The selection of a cell line with undetectable virus levels and the use of contaminant free raw materials are essential for the production of therapeutic proteins by cell culture. However, even with such a cell line, the absence of viral contamination is not guaranteed, so the downstream processing must demonstrate sufficient viral clearance. Viral inactivation (by acidic pH conditions and/or by addition of detergent) along with virus removal (by centrifugation or filtration) are methods generally used to minimize the risk of viral contamination. In addition, viral clearance can be demonstrated through viral spiking and challenge

INDIVIDUAL PROCESS STAGES AND ASSOCIATED RISKS

studies utilizing appropriate viral agents (e.g., generic as well as specific types of virus particles). The above-mentioned strategies along with testing for absence of infectious agents at appropriate stages during the purification process are typically used to minimize this critical risk to the product.

Endotoxins (Pyrogens) Endotoxins are lipopolysaccharides found in the cell membranes of gram-negative bacteria, which are released during cell division, cell growth, and cell death. Because of their high toxicity *in vivo* and *in vitro*, their removal is essential for a safe parenteral administration. Ideally, in cell culture supernatants the endotoxin level should be zero, which is not always the case because endotoxins can be found virtually everywhere (e.g., the process water used in cell culture). Methods used for decontamination of water, such as ultrafiltration, have little effect on endotoxin levels in protein solutions. Therefore, to reduce the risk of endotoxins, removal techniques tailored to meet specific product requirements must be built into the purification process. The standard protein purification process often includes an ion exchange step, a hydrophobic interaction step, and some sort of size-based separation to remove risk of endotoxin contamination. These three steps frequently are adequate for endotoxin removal, but the product should be tested at appropriate stages for their absence.

12.3.5 Scale-Up of Production Process

12.3.5.1 Process Overview Scale-up is a standard aspect of product development and/or life cycle management. Typically, material requirements increase significantly from clinical trial stage to product launch, and the increase in product demand drives the decision to increase the scale of production. A successful scale-up strategy requires identifying parameters that should either remain constant or need to be scaled appropriately when transferred to large-scale manufacturing. Therefore, an understanding of the cellular mechanisms that regulate cell physiology, the physicochemical characteristics of the product, and the basic engineering design principles of equipment and process parameters is essential. In addition, at the completion of a process scale-up operation, it is important that a comparison of the laboratory and large-scale operation should be performed and documented. The comparison report would typically focus on the process performance, process design, and product quality aspects between the two manufacturing scales, and is important to demonstrate comparability (or noncomparability) of a process/product.

This section presents a brief overview of some of the typical unit operations that are scaled up during upstream and downstream manufacturing operations of biotechnology products.

Scale-Up of Upstream Operations
MEDIUM PREPARATION While it may be feasible to use custom, completely liquid media for production runs at the laboratory scale and pilot scale, commercially available powdered media or liquid concentrate media is typically

used during large-scale production because of economic reasons (e.g., shelf life), storage, and shipment concerns. At the large scale, the culture medium is typically prepared by the addition of base powder or liquid concentrate mixtures to appropriate grade water. These base media mixtures usually contain amino acids, vitamins, cell membrane precursors, antioxidants, and growth factors. Additional components such as lipids and proteins may be added separately. Media may also contain poorly characterized ingredients such as yeast extract or protein hydrolysates. It is important that media be identical in terms of composition. One major issue that needs to be monitored when using powdered media is the issue of blend uniformity or homogeneity of powdered mediums from different drums [20].

INOCULUM EXPANSION   Inoculum expansion increases the number of cells available for inoculation of the large-scale production fermenters or bioreactors. Cells are cultured in successively larger flasks by adding fresh medium during their growth phase, so a vigorously growing culture can be used for start of commercial production. Inoculum expansions are typically carried out in T-flasks and shake flasks for smaller volumes at the beginning of expansion, and subsequently roller bottles or spinner flasks are used for larger volumes (10–20l). For volumes greater than 10–20l, bioreactors of successively large scale can be used for expansion of cells until the working volume of the production fermenter/bioreactor is reached [20].

BIOREACTOR OPERATIONS   The choice and scale of any fermenter or bioreactor will depend upon the needs of the process, product, and the market demand. At a minimum, at commercial scale, fermenter/bioreactor vessels should be capable of being operated aseptically for a number of days, reliable in long-term operations, and meet containment requirements. In addition, vessels should be capable of providing adequate mixing, agitation, and aeration to meet the metabolic requirements of the microorganisms or cells.

Cell culture and fermentation processes are often considered the most difficult to scale up, partly because causal links between culture conditions (e.g., aeration, mixing, dissolved oxygen and carbon dioxide, and nutrient concentrations) and product characteristics are often poorly understood, and also because not all critical process parameter values can be kept the same during scale-up.

HARVEST OPERATIONS   Biotechnology-derived products can be either intracellular or extracellular (or secreted). As mentioned in Section 12.3.4, in the initial stages of recovery and purification, a product needs to be purified from the accompanying cells, cell debris, and other large particles or contaminants. Typical unit operations employed for this clarification step include centrifugation, filtration, etc. To ensure a successful scale-up strategy, it is important to select a method that is robust, readily scalable, efficient, and can provide high processing rates.

Scale-Up of Downstream Operations

MEMBRANE/FILTRATION OPERATIONS Membranes and filters are an integral part of most protein purification schemes, and are used for a variety of operations, such as control of bioburden and particulate levels, medium exchange during cell growth, cell harvest, product concentration, diafiltration, formulation, and for removal of viruses and other agents. The important process parameters that should be kept in mind when selecting membranes/filters for scale-up operations are cross-sectional area, trans-membrane pressure, volume processed per unit area, filtration area, shear rate, operating time, temperature, protein concentration, solution viscosity, and, where applicable, number of uses [20]. A conservative approach to scale-up involves increasing flow rates and filter areas while keeping other variables constant. Additional details on this topic can be found in PDA Technical Report No. 15, *Validation of Tangential Flow Filtration in Biopharmaceutical Application* [21].

CENTRIFUGATION Centrifugation is used to separate or concentrate materials suspended in liquid medium, based on sedimentation rates in an increased gravitational field. That is, two particles of different size, density, and shape will settle in a tube/bowl at different rates in response to gravity. For biotechnology operations, batch centrifugation is often used at the laboratory scale, while continuous centrifugation is preferred at the production scale. Scale-up operations typically use centrifugation for separating whole cells from the supernatant. It is also used after precipitation steps to separate solid from liquid phases.

For cell culture operations that involve secreted proteins, filtration may be preferable because of the relatively mild operating conditions. Another advantage of filters is the relatively simple cleaning validation as compared to the cleaning validation required for centrifuges. However, as process volume increases, the economics of using filters decreases and makes them unsuitable for very large-scale operations [20].

CHROMATOGRAPHY The majority of processes currently used to manufacture biotechnology products employ column chromatography as the main method for product recovery and purification. It is capable of combining relatively high throughputs with high selectivity to either capture the product or purify it from accompanying impurities. Key process parameters for scaling up and using chromatographic columns include protein and product loading, linear velocity, buffer volume, bed height, temperature, cleaning capacity, and gel lifetime. Other parameters include buffer properties (e.g., pH, ionic strength) and measures of packed bed quality (such as number of theoretical plates or asymmetry in a pulse test).

The most commonly used types of chromatographic columns in commercial production are ion exchange columns (impurities are removed by manipulating the pH and conductivity of equilibration wash and elution buffers), hydrophobic interaction columns (contaminants and proteins are selectively removed on the basis of their hydrophobic interactions), and affinity chromatography (provides high specificity, selectivity, and volume reduction based on a highly specific biological interaction such as that between antigen and antibody, enzyme and substrate, or receptor and ligand). Additional details can be found in the PDA Technical Report No.14, *Validation of Column-Based Chromatography Processes for the Purification of Proteins* [22].

12.3.5.2 Quality Risk Management for Scale-Up Process The major risks associated with the scale-up process are discussed in the following section.

Inappropriate Purification Techniques/Process Design For purification, scale-up considerations are important from the earliest steps of product/process development because the risk of using purification techniques that have limited scale-up potential can be rate limiting for large-scale manufacturing. For example, designing an elaborate peak collection scheme for column chromatography may be unfeasible for implementation during large-scale manufacturing. In order to minimize this risk to large-scale manufacturing, the relationship between controlled purification parameters (e.g., temperature, pH, conductivity) and process performance parameters (purification factors, yield) should be defined early in development using laboratory systems to model the manufacturing equipment and scale. This strategy helps define the edges of failure through experimental analysis of the input parameters or output variables, and also helps identify the parameters necessary for reliable process performance. In addition, it is also important to verify that a scaled-down process is an accurate representation of the production process so that validation studies for issues such as viral clearance and column lifetimes are justified and can be performed at the laboratory scale.

Viral Contamination One of the major risks during scale-up is viral contamination. Viruses can be introduced into the process through the following routes: source cell lines from infected animals, viral establishment of the cell line, use of contaminated reagents or equipment, and improper handling of the cell line.

It is critical that the small-scale viral studies are accurately representative of the production process. In addition, it is recommended that viral studies include the use of typical critical operating parameters for each step of the process, as well as conditions that represent a worst case for viral removal. For example, for process validation of chromatography, the criticality of the viral inactivation step usually demands that the validation be done at the extremes: for example, at 90% of minimum time, at the highest pH (or lowest temperature), protein concentration, reduced height, or contact time, and total protein capacity. Finally, testing for absence of infectious agents should also be conducted at appropriate stages/steps during the entire process (e.g., routine testing should include the end-of-culture samples). Refer to Section 12.3.4 for additional information on viral contamination.

Inadequate Process Controls Owing to the difficulties involved in maintaining homogeneity during large-scale production, adequate monitoring/control of

INDIVIDUAL PROCESS STAGES AND ASSOCIATED RISKS   349

process parameters (e.g., temperature, oxygen dissolved concentration, pH etc.) is essential to ensure a successful operation. Inadequate or inefficient process monitoring and/or process controls pose a significant risk for large-scale production. Therefore, it is important to perform optimization experiments at a small scale, and identify relevant analytical techniques and issues during small-scale production so that they can be integrated into the large-scale manufacturing process. A deficiency in understanding and control of critical process parameters could significantly impact the validation campaign. For example, during inoculum development, it is important to know the oxygen requirements for the cell cultures because culture flasks used during large-scale operations may have oxygen transfer limitations, and a different aeration strategy may need to be employed than the one used during laboratory or pilot-scale production. Optimization experiments can significantly help in reducing this risk.

Inappropriate Facility Design The use of inappropriate facilities or equipment poses a considerable risk for biopharmaceutical production because biotechnology products are inherently more complex and susceptible to harm (e.g., degradation, aggregation, denaturation, or contamination). Scale-up primarily depends upon a product's specific characteristics and the market demand for the product. As it is not always feasible to design a new facility or to implement a new design of large-scale equipment, and retrofitting a facility can be cost prohibitive, the design of a process sometimes has to consider constraints imposed by existing production facilities. As such, it is critical to review scale-up calculations for the different process steps in an existing environment with minimal changes in equipment, for example, buffer preparation, bioreactor operations, etc. [20].

12.3.6 Excipients

12.3.6.1 Process Overview Excipient control also influences product quality and patient safety. Its impact is seen in microbial, pyrogen, and chemical contamination. Basic information about excipient quality can be obtained from the USP-NF if a monograph exists for that particular item, USP/NF 2009 and C. Moreton [23,24]. Compendia standards can provide some assurance of material quality; however, they may not always require tests for all relevant quality attributes and should not be a substitute for rigorous material qualification.

The primary interest for excipients is the components in them. Excipients are generally not pure compounds but rather mixtures of compounds, portions of which may be critical to performance. Thus, the critical concern for excipients is functionality, and purity may not be the major concern. Data from material Certificates of Analyses may not coincide with performance in a specific formulation.

Data from the manufacturer and previous lots can be analyzed to assess variability. Lot-to-lot variation in incoming lots can influence product variability. Thus, manufacturers need to understand excipient variability to establish robust formulations. The information needed for excipients varies with formulation and application. Oral, parenteral, and topical applications all differ. Excipient variability is a critical issue and one that should be studied as part of formulation development.

Use of multiple lots of a material during qualification reduces overall risk because it facilitates the definition of lot-to-lot material variability and the effects thereof on product quality. The influence of material degradants on product quality should also be studied. For example, Polysorbate-80 (PS-80) is a surfactant used to prevent aggregation in biotechnology products. However, PS-80 degrades over time to form peroxides and if lots are not monitored/tested appropriately before use, products manufactured using the old PS-80 material may not meet specifications. Therefore, manufacturers should be concerned about the stability of the excipients used in their formulations. Stability testing on excipients is difficult because of the large and varied range of container quantities in which they come, for example, small package through railcars. In addition, excipients are exposed to a wide range of environmental conditions, that is, temperature and humidity, while transiting from supplier to pharmaceutical manufacturer. General International Pharmaceutical Excipient Council (IPEC)-recommended storage conditions for excipients are 4–40 °C and 20–90% relative humidity [25]. Some excipients require more stringent storage conditions and should be so labeled. Suppliers should be able to provide data supporting their recommended storage conditions. Stability studies may be performed to generate data showing that the excipient continues to perform as expected from the time it is packaged until the time when the biopharmaceutical manufacturer uses it (i.e., shelf life).

Risk assessments should be performed to assess the risks associated with a specific excipient in a given application. The assessment should address several distinct attributes of the material including safety, efficacy, and availability to better understand the materials' impact on product quality and patient safety. They should identify the material attributes that are critical to the application in question. Adulteration may be an issue and should be addressed through ongoing assessments.

Risk assessments should be reviewed periodically and should include lessons learned from past experiences. They should be updated to reflect current knowledge about the materials such that new hazards, failure modes, and occurrences are included in the assessment. The assessments can then be reevaluated to insure that known risks continue to be acceptable. A model for performing raw material risk assessments may be found in Beck et al. [6].

12.3.6.2 Quality Risk Management and Excipients   The major risks associated with excipients are discussed in the following section.

Risks associated with excipients include, but are not necessarily limited to, the potential for microbiological contamination, lot-to-lot variability, a potential lack of stability, and the presence of additional chemical compounds that are critical to product performance. Many of these risks are monitored and controlled through supplier qualification and testing programs. Lot-to-lot variability may be addressed by including multiple lots in product qualification exercises and by periodically testing incoming lots of material.

12.3.7 Primary Packaging

12.3.7.1 Process Overview Packaging for biopharmaceuticals is a major area of concern primarily because of its impact on product quality. The FDA's requirement, as spelled out in the "Guidance Container Closure Systems for Packaging Human Drugs and Biologics," provides companies with the guidelines as to what is expected from them to demonstrate that the proposed package and its components are suitable for their intended use and will maintain the quality, safety, potency, and stability of the drug product over its intended shelf life. In addition to maintaining product quality, primary packaging for biotechnology drug substances also need to be sterile, scalable, disposable, and readily available. There are three important points to consider when choosing primary packaging:

1. Does the container-closure system protect the product from environmental challenges (e.g., moisture, light, oxygen, shipping)?
2. Is the dosage form identified in the FDA guidance as a high risk for packaging concerns (e.g., inhalation or injectable drug)?
3. Does the container-closure system play a functional role in the delivery of the drug product (e.g., inhalations and transdermals)?

An affirmative response to these questions indicates that the primary packaging could significantly impact the product and the information supplied to the agencies will undergo careful scrutiny in the license applications.

Drug Substance Typically, after the final purification step, bulk drug substance is formulated, filtered, and stored under defined conditions in either plastic bottles or bioprocessing containers (e.g., ethylene vinyl acetate or EVA bags), or glass containers (e.g., tetrafluoroethylene or Teflon glass bottles), or in stainless steel vessels. There is no single packaging configuration that is suitable for all products; they all have advantages and disadvantages (e.g., the glass bottle containers are inert but fragile), so the selection of the correct primary packaging depends upon the unique product/process characteristics and the business needs of the sponsor.

Drug Product For biotechnology-derived drug products, parenteral or intravenous (IV) injections of proteins often provide the most efficient route of delivery for protein-based formulations. Despite the significant advances in delivering therapeutic enzymes, peptides, and proteins through nontraditional means, injection remains the principal delivery system. The typical primary packaging presentations used for protein drug products are single-dose vials, IV bags, or prefilled syringes. The product is provided either as a solution, or, as a lyophilized cake, which is reconstituted and injected via syringe. Requirements for product purity, activity, and shelf life are critical and necessitate high standards for injectable drug packaging, particularly for highly active peptides and proteins. Therefore, packaging should not only protect the product from leakage, or contamination from foreign particles, at the same time it must also be fully compatible with the product [26].

12.3.7.2 Quality Risk Management for Primary Packaging Process The major risks associated with primary packaging process are discussed in the following section.

Excipient Interactions Excipients aid in protecting the drug substance, supporting and enhancing its stability, improving bioavailability, and in general contributing to the overall safety and effectiveness of the drug substance. However, there is a risk that excipients may impact product quality by interacting negatively with the primary packaging. For example, polysorbate 80 (PS80) is a surfactant that is typically used to prevent protein aggregation in drug substance, but PS80 can adsorb to glass surfaces and its levels can decline significantly during storage, thus reducing its ability to prevent protein aggregation. Therefore, it is important to evaluate the risk of potential excipient interactions with the primary container closure during development studies.

Environmental and Chemical Hazards Biopharmaceuticals are proteins and peptides molecules with unique chemical, physical, and mechanical properties. Proteins are sensitive to heat, light, and chemical contaminants. As proteins and peptides have a tendency to adsorb onto the surface of packaging containers and closures, there is a risk that minute concentrations of metals, plasticizers, and other materials from packaging may deactivate or denature therapeutic proteins. This can essentially remove all active materials from the drug formulation. For example, storing drug substances that contain sodium chloride in their formulation in stainless steel containers is a risk because metal oxidation can occur. In situations where the drug desorbs back into the solution, such interactions could cause the drug to lose potency.

Many biotechnology products are also sensitive to silicone oil, a material commonly used to lubricate elastomeric stoppers during fill/finish to facilitate insertion of the stopper into the vial. However, silicone oil poses a risk to product because it has been associated with protein inactivation through nucleation of proteins around oil droplets. Recently, fluoroelastomer coatings on stoppers have mitigated this risk and helped provide chemical inertness, barrier protection, and safety for the product [27]. Therefore, it is important to perform risk analysis studies to assess the interactions of environment and chemical hazards with the packaging and to minimize product risk.

Packaging Operations Another major risk to the product is inappropriate packaging/handling conditions. For example, most lyophilization cakes are sensitive to moisture, and an inadequate seal on the vial could cause water and other contaminants to enter the package and deactivate the drug. Therefore, vial-capping operations play an important role in ensuring product quality and as such should be validated and monitored throughout the production process to minimize product risk. Risk assessment on the packaging operations is critical for identifying equipment and process steps that need to be validated and/or monitored.

12.3.8 Extractable/Leachables

12.3.8.1 Process Overview The potential impact of extractables and leachables on biotechnology-derived drug products is significant, especially since the drug product may contain just micrograms of the active ingredient. Extractables are the most common source of leachables contamination. An extractable is any chemical species that can be released from a packaging component, and which has the potential to contaminate the pharmaceutical product. Extractables are typically generated by interaction between strong solvents and the package (including the glass vial and stopper) over time depending on temperature and extraction conditions.

A leachable is an extractable or chemical that actually migrates from packaging or other components into the drug product under normal storage conditions.

Evaluation of extractables from packaging materials should be performed as part of product development and qualification operations. Leachable tests should be carried out at the point of use, and in real-life situations in the presence of the actual drug product.

Section 501(a) [3] of the Federal Food, Drug, and Cosmetic Act defines a drug as adulterated "if its container is composed, in whole or in part, of any poisonous or deleterious substance which may render the contents injurious to health ..." [28]. The concept may easily be extended from containers to the processing aids used in the manufacture of biopharmaceuticals.

Risks associated with extractables and leachables include both product quality and regulatory risks. FDA requests for extractable and leachable data began with inhalation and nasal dosage forms and have now expanded to parenteral products. FDA provided guidance to industry in its 1999 document *Container Closure Systems for Packaging Human Drugs and Biologics*, wherein the concept of suitable for intended use [29] was defined. Suitable for intended use was defined as addressing product protection, compatibility, safety, and performance [29]. Elaborating on compatibility, the Agency noted that packages should not interact with the product such that unacceptable changes in product or package quality occur. Examples of such interactions include loss of potency due to absorption or adsorption to the package, degradation due to leachates, excipient absorption onto the package, pH and color changes due to leachates. The guidance notes that while some problems may be detected during qualification, others may not be manifested until the package is placed on stability.

The large amount of plastics used in processes provides a significant opportunity for process streams to interact with and extract materials from process components.

The primary reason for testing these materials is to address the issue of suitability for intended use. By testing extractables, one can generate information about the potential of a process material to leach contaminants into the process stream. This information can inform decisions regarding the need for further testing of the process stream.

Extractable testing can be based on model solvents and thus data provided by suppliers can be used to assess actual conditions. Adequate characterization of extractables will facilitate choosing containers that contribute minimally to the drug product over the product lifecycle. The initial materials used to formulate polymers and elastomers all degrade during processing and may potentially leach degradants into the drug product. Other contaminants in elastomers that may leach into the drug product include curing agents, additives, accelerators, plasticizers, processing aids, and reaction products.

Potential interactions between biologicals and containers are influenced by the characteristics of the biological and the container itself. While plastics and elastomers have the most potential for interaction, other materials, for example, glass, metallics, etc. may also interact with biological materials.

Leachates from the package may adversely impact the safety profile of products and should therefore be carefully characterized. Characterization processes include both the extraction and the identification of contaminants as well as an assessment of the toxicological impact of the contaminants. Establishing levels at which no or minimal toxic effect is observed provides important information in assessing the risks associated with extractables and leachables.

Extractable/leachable testing should address several risk questions including, but not limited to, those in Table 12.3.

Answers to these questions can be used to establish risk-based test schemes to characterize materials from the perspective of compatibility and safety [30,31]. Such characterizations will minimize adverse patient impact.

TABLE 12.3 Extractable/Leachable Considerations

Category	Consideration
Drug–Container Interaction	Container material of construction
	Container surface finish
	Drug solvent characteristics
	Drug/Container surface-volume ratio
	Drug and container manufacturing processes, molding, sterilization, etc.
	Product storage conditions, temperature, humidity
Drug administration	Oral
	Parenteral
Patient demographic	Age
	Sex
	Degree of illness, acute, chronic

12.3.8.2 Quality Risk Management and Extractable/Leachable Concerns

Extractables and leachables may interact with biopharmaceutical products causing degradation of the molecule, thereby impacting product performance. This may be addressed by careful selection and characterization of processing aids and packaging materials. Table 12.3 describes the factors that should be considered in terms of biopharmaceutical compatibility and product performance.

12.3.9 Distribution and Cold Chain Supply

12.3.9.1 Process Overview Distribution processes for biopharmaceutical products can be very complicated. They extend from the loading dock of the pharmaceutical manufacturer to the patient (Fig. 12.8).

From the manufacturer, products can be delivered either via air or local highway transport to distribution centers, or, in the case of local highway transport, directly to hospitals and pharmacies. From the distribution centers products may be further transported to hospitals and pharmacies or, in the case of specialty services providers, the product may be repackaged and delivered directly to a patient. Patients may also receive products from local hospitals and pharmacies in which case they themselves transport the product home via local transport, for example, automobile or public transport.

The distribution chain is subject to variations that manufacturers are unable to control and thus cannot be validated. However, manufacturers may qualify portions of the distribution chain, for example, shippers, warehouses, etc. A typical approach would be to qualify primary, secondary, and tertiary packaging to demonstrate its ability to protect the product during rough handling/condition. Tertiary packaging may include cooling units and be qualified to prove the adequacy of those units to ensure that the product is not exposed to temperature excursions during transport. Generally, distribution tests are executed according

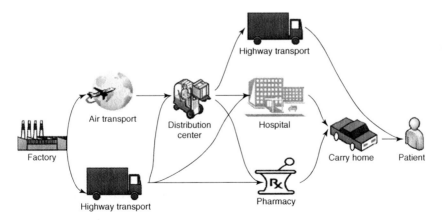

Figure 12.8 Typical distribution chain. (*See insert for color representation of the figure.*)

to international standards to demonstrate the ability of the final packaged, cartoned product to withstand the drops, vibrations, etc. that accompany routine transportation and handling.

The risks associated with the distribution of biopharmaceutical products are primarily associated with their temperature sensitivity [32]. Some products require refrigerated or freezer storage to maintain product quality. In addition, other products, although stable at ambient temperatures are degraded if exposed to temperatures exceeding 0 or 30 °C, which represent extreme temperatures for many biopharmaceutical products . Products such as these require controlled transport conditions to ensure that product degradation does not occur during the distribution process. Distribution environments are influenced by climate zone, season, and transportation modality.

A risk assessment of the distribution chain should begin at the manufacturer's loading dock. There, the product is palletized in its tertiary container and loaded onto trucks for transport. The risks associated with this step are associated with rough handling, which may result in crushed cartons and damaged product. Mitigations at this step demonstrate that primary and secondary packaging are able to protect the product.

Once loaded onto trucks, the product is transported to the next stage in the distribution chain, as shown in Figure 12.8. At this step, there are several risks associated with the transportation equipment. Transportation vehicles may have faulty suspensions or mechanical failures that may result in carton and product damage. Malfunctioning or uncontrolled cooling units may result in temperature excursions that result in a degraded product. Biopharmaceutical products are often shipped in qualified insulated containers with cold packs or refrigerated trucks to ensure that controlled temperature conditions are maintained during the shipping process.

If being transported to an airport, the product may need to pass through a Customs inspection, which may add additional time to the transport process and may exceed qualified times for product cooling containers, thus resulting in potential temperature excursions and degraded product. Once through Customs, the product is loaded onto an airplane where it may subjected to rough handling, with risks that are similar to being loaded onto a truck. While in transit on the airplane, the cartons can experience turbulence, engine vibrations, and pressure changes all of which have the potential to impact the cartons and result in product damage. Thorough testing of primary and secondary packaging is required to insure adequate protection of the product.

Failure modes similar to those already discussed are again present as the product is unloaded from the aircraft, loaded onto trucks, and transported to distribution centers. The same mitigations apply in these circumstances.

The final step in the distribution chain is the transport from a local hospital or pharmacy to the patient's home. This may occur via public transportation or personal automobile. At this step, the product is outside of the manufacturer's control, but must still be protected against rough handling and temperature excursions. Mitigations at this step rely on well-designed packages to protect the product and a robust product formulation that is able to withstand slight excursions in temperature. The former is demonstrated via validation exercises, while the latter is demonstrated through stability studies that challenge the product at temperature extremes.

12.3.9.2 Quality Risk Management in Distribution The major risks associated with distribution and supply chain are discussed in the following section.

While biopharmaceutical products share many of the same distribution risks as pharmaceutical products, their primary risk is associated with temperature excursions beyond acceptable ranges. This risk is controlled through careful qualification of packaging and shipping containers as well as a thorough characterization of the distribution chain to ensure that product temperatures are well maintained and excursions are minimized.

12.4 SUMMARY AND CONCLUSIONS

Risks are associated with biopharmaceutical manufacturing processes from the acquisition of raw materials and excipients through distribution of the final product (Table 12.4). The identification (Table 12.5) and analysis of risks at each step of the process facilitates better process knowledge, resulting in a more robust process and a product of more uniform quality. A variety of tools are available to help the practitioner complete this task. Examples of these tools have been provided in this chapter and elsewhere in this book.

The overall control of process risks begins with the identification of the product's critical quality attributes and the critical process parameters by which they are controlled. Given this information, appropriate control strategies can be established to mitigate the potential for process shifts and subsequent product quality issues.

Raw material and excipient supplies must be carefully assessed and controlled to assure consistency and safety of the incoming material. The microbiological attributes of raw materials are of particular importance and must be carefully controlled to preclude the possibility of contamination. Excipient variability and its impact on product quality must be well defined.

Cell banks must be adequately characterized and maintained. Cell banks carry a risk of introducing adventitious agents into the process stream, or changing genetically or biochemically, and thereby altering the characteristics of the product produced. These risks need to be addressed and mitigated through appropriate cell bank handling, characterization, and storage.

Biopharmaceutical manufacturing risks can be similar at several stages of the process, a fact that can be advantageous for manufacturers. For example, cell culture and downstream purification processes carry common risks including that of microbial contamination and unsuitability of growth media. Improper resins, cleaning failures, or changes in process parameters may also adversely impact product quality. Robust process characterization analyses through either

TABLE 12.4 Summary Table of Manufacturing Process Stages and Associated Risks

Process Stage	Risks Associated with Individual Process Stages
Raw materials	Lot-to-Lot variability; vendor qualification; origins of raw material; contamination; stability
Cell banking	Viral/TSE/BSE clearance; bacterial endotoxins; mycoplasma; facilities; equipment; personnel training; documentation; improper storage conditions
Fermentation/cell culture	Contamination; suitable environment; selection of stable strain; culture media; oxygen requirements; sterilization; facilities and equipment; scale-up considerations
Downstream processing	Inefficient harvest/recovery conditions; inefficient purification conditions; viral contamination; endotoxins (pyrogens)
Scale-up of production process	Inappropriate purification techniques/process design; viral contamination; inadequate of process controls; inappropriate facilities
Excipients	Lot-to-lot variability; vendor qualification; origins of raw material; contamination; stability
Primary packaging	Excipient interactions; environmental and chemical hazards; packaging operations
Extractable/leachables	Suitability of use of materials; technical risks; regulatory risks; contaminants in elastomers
Distribution and cold chain supply	Insufficient primary/secondary/tertiary packaging; inappropriate qualification of distribution and supply chain warehouses and shippers; inadequate testing; inappropriate storage and handling conditions; transit times

TABLE 12.5 Example of Risk Acceptability Definitions

Risk Level	Risk Acceptability
Broadly acceptable	These are acceptable risks. No further risk control measures needed. Minimal impact to product safety and efficacy
As low as reasonably practical	All practical risk mitigations have been undertaken. Further risk reduction is not practical and the benefits outweigh the residual risk
Intolerable	Unacceptable if no further risk reduction measures are feasible. Individual risks may be accepted on a case-by-case basis by proving that the risk/benefit ratio is favorable, once all reasonable risk reduction measures have been taken

FMEA analysis or other risk assessment methodologies are an effective way of identifying and mitigating these common risks at all stages. Designed experiments may also be used to identify critical parameters, their interactions, and their impact on process limits. Given this information, analysts can validate the process to establish reproducibility within identified ranges and demonstrate the adequacy of control strategies for identified risks.

In addition to the manufacturing process, primary packaging is critical as a biotechnology product may be sensitive to a variety of environmental factors such as surface of its primary package, temperature, moisture, and light. Primary packaging must also be characterized for extractables and leachables to ensure absence of adverse interactions between these components and the product.

Finally, the product is packaged and ready to begin its journey to a patient. Distribution processes must be carefully qualified and controlled to preclude adverse product interactions. Typical distribution tests address product impact by emulating transportation challenges such as vibration, time, dropping, etc. Through such testing, adequate secondary and tertiary packaging may be designed and distribution processes developed so that product quality is not compromised.

The use of risk management to identify, assess, and mitigate risks associated with biopharmaceutical manufacturing can reduce the impact of unplanned and uncontrolled events that periodically occur. Risk management activities result in more robust processes that are capable of handling a broader range of inputs while still yielding a product of consistent quality. Although risk assessments require somewhat more effort (time and resource) to complete, they ultimately benefit both the manufacturer and the final patient.

APPENDIX A: APPLICATION OF RISK MANAGEMENT TOOLS TO BIOPHARMACEUTICAL MANUFACTURING

A.1 Case Study: Raw Material Hazard Analysis

This example addresses the use of a hazard analysis to identify the potential hazards associated with the use of fetal bovine serum (FBS) to manufacture a master cell bank for a biological product. The example follows the general outline provided in PDA TR44 and utilizes the same scales [33].

A.2 Prerequisites

Initiation: Risk assessments should begin with careful planning to establish the scope and boundaries of the task at hand. A basic risk question that addresses the issues to be considered should be identified. For example, one might ask "what are the risks associated with the use of FBS obtained from ABC country for the production of a master cell bank for XYZ product line?"

A tool used to map the process at a high level and establish boundaries around the analysis is the SIPOC analysis. The acronym SIPOC stands for supplier,

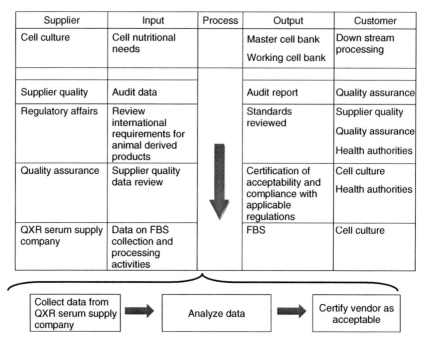

Figure A.1 SIPOC analysis: FBS utilization.

input, process, output, customer. It is a standard Six Sigma graphic that may be used to gather fundamental information into an organized graphic that will allow the team to focus the assessment. The SIPOC assessment shown in Figure A.1 indicates that the scope of this issue extends from the vendor, QXR Serum Supply Company, through to certifying the vendor as acceptable.

Team selection: Teams that include subject matter experts who have knowledge of the issue and can thoroughly identify and assess the associated risks should be chosen. For example, an assessment of FBS might include representatives with expertise in cell culture, formulation development, supplier quality, quality assurance, regulatory affairs, etc.

Product analysis: At this stage, the team should develop a high level understanding of the product and its usage. The serum itself will be used as a nutritional supplement in the production of cell banks for the XYZ product line. In this context, the team should identify the critical quality attributes of the product. For FBS, a critical quality attribute would be its ability to support the growth of cells.

A.3 Risk Assessment

Risk identification: The hazard analysis in the following, Table A.1, shows that five principle hazards were identified: contamination (microbiological),

TABLE A.1 Hazard Analysis: FBS Utilization

No.	Hazard	Cause	Harm	Severity	Likelihood	Risk	Mitigations
1	Contamination, microbiological	Serum collection does not comply with applicable protocols	Death	High	Low	As low as reasonably practical	Supplier quality audit indicates compliance with all applicable regulations
2		Serum collection by untrained personnel					Supplier quality audit indicates compliance with all applicable regulations
3		FBS processing does not comply with cGMPs					Certificate of Conformance indicates compliance with requirements
4		Irradiation dose too low to inactivate microorganisms					Irradiation process validated and monitored. Certificate of Irradiation provided with each lot
5		Testing does not comply with European Pharmacopeia requirements					Material tested in compliance with regulations after irradiation
6	Contamination, prions, for example, TSE	Serum collection does not comply with applicable protocols	Death	High	Low	As low as reasonably practical	Country of origin complies with GBR level 1 requirements as specified by the EFSA. A Certificate of Suitability is available for all lots

TABLE A.1 (*Continued*)

No.	Hazard	Cause	Harm	Severity	Likelihood	Risk	Mitigations
7		Serum collection by untrained personnel					Country of origin complies with GBR level 1 requirements as specified by the EFSA
8	Contamination, particle	Serum collected in contaminated vessels	Process failure	Low	Low	Broadly acceptable	FBS filtered to remove particulates prior to irradiation
9	FBS out of specification	Testing does not comply with European Pharmacopeia requirements	Medium does not support cell growth.	Low	Low	Broadly acceptable	Supplier Quality audit confirms compliance with regulations. Certificate of Conformance indicates compliance with requirements
10		FBS processing does not comply with cGMPs	Process failure	Low	Low	Broadly acceptable	
11	FBS degraded	Irradiation dose too high	Medium does not support cell growth.	Low	Low	Broadly acceptable	Supplier Quality audit confirms compliance with regulations

TABLE A.2 Example of Risk Analysis Scales

	Risk Factor	
Rank	Severity	Likelihood
High	Failure has a severe impact on product safety and efficacy	Frequently occurs
Medium	Failure has a medium impact on product safety and efficacy	May occur if not controlled
Low	Failure has a severe low on product safety and efficacy	Rarely occurs

Risk evaluation: Risk evaluation is the act of assessing the risks and determining their acceptability in terms of potential impact to product safety and efficacy. Risks are frequently rated as broadly acceptable, as low as reasonably practical, and intolerable. The terms may be defined as indicated in Table 12.5.

contamination (prion, e.g., TSE), contamination (particle), FBS out of specification, and FBS degraded. These hazards are caused by a variety of factors as shown in the "Cause" column of the hazard analysis table. The predominate cause is a potential lack of compliance by the vendor with established health authority regulations as noted in the following:

- Serum collection does not comply with applicable protocols.
- Serum collection by untrained personnel.
- Serum collected in contaminated vessels.
- FBS processing does not comply with cGMPs.
- Testing does not comply with European Pharmacopeia requirements.

Risk analysis: A hazard analysis is a tool that facilitates a high level assessment of risk. It typically looks at the classic risk parameters of severity and likelihood of occurrence of the harm. Detectability, often included in more extensive risk assessments, is not a factor in a hazard analysis. Both severity and likelihood were examined for the hazards identified and are shown in the hazard table. Both severity and likelihood were ranked on a three-level—high, medium, low—basis. If desired, a more detailed scale, for example, a 5-point or 10-point scale, may be utilized. An example of a three-point scale is shown in Table A.2.

The hazard analysis shows that all risks were assessed as either broadly acceptable or as low as reasonably practical given the current level of mitigation.

A.4 Risk Control

Risk reduction: Risk reduction involves identifying and implementing measures to minimize and controls the risks that have been identified. An underlying assumption is that mitigations will be reasonable and practical. A point of diminishing returns may be reached wherein mitigations are too costly or impractical to implement. At this point, one must determine if benefits outweigh the potential risks and decide if utilization of the material is warranted.

The mitigations in this example included supplier audits to assess compliance with Health Authority regulations and the inclusion of process steps, that is, filtration and irradiation to reduce particle load and microbial contamination. In addition, testing in compliance with regulatory requirements was identified as an additional mitigation to verify serum quality.

Risk acceptance: Risk acceptance is the act of determining if the risks remaining after all mitigations have been implemented is acceptable. In this example, after implementing the risk reduction activities, all risks were assessed as being either broadly acceptable or as low as reasonably practical. No further mitigations were identified as being possible.

Risk communication: Risk communication involves notifying key stakeholders of the conclusion of a risk assessment. The information communicated should include an assessment of residual risks and a recommendation of further action, that is, the acceptance or rejection of the material in question. In this example, the overall risk of using FBS was deemed acceptable and key stakeholders should be so notified.

Risk review: A review of the risk analysis should be performed if there is a change to the process or if new hazards are identified. Such changes may impact the mitigations previously identified and/or alter the residual risk inherent in the process.

REFERENCES

1. Federal Register 2008 Jul 2;73(128):37973.
2. ICH Q9, Quality Risk Management. 2005 Nov.
3. ICH Q8 (R2). Pharmaceutical Development. 2009 Aug.
4. ICH Q6B. Specifications: Test Procedures and Acceptance Criteria for Biotechnological/Biological products. 1999 Mar.
5. Nosal R, Tom Schultz, PQLI. Definition of criticality. J Pharm Innov 2008;3:69–78.
6. Beck G, Schenerman M, Dougherty J, Cordoba-Rodriguez R, Joneckis C, Mire-Sluis A, McLeod LD. Raw material control strategies for bioprocesses. BioProcess Int 2009;7(8):18–19.
7. EMEA/CVMP/743/00. Guideline on Requirements and Controls Applied to Bovine Serum used in the Production of Immunological Veterinary Medicinal Products. Available at http://www.emea.europa.eu/pdfs/vet/iwp/074300en.pdf. Accessed May 2011.
8. CPMP/BWP/1793/02. Note for Guidance on the Use of Bovine Serum in the Manufacture of Human Biological Medicinal Products. Available at http://www.emea.europa.eu/pdfs/human/bwp/179302en.pdf. Accessed May 2011.
9. EMEA/410/01, rev. 2: Note for guidance on minimizing the risk of transmitting animal spongiform encephalopathy agents via human and veterinary medicinal products adopted by the Committee for Proprietary Medicinal Products (CPMP) and by

the Committee for Veterinary Medicinal Products (CVMP). (2004/C24/03). Available at http://www.emea.europa.eu/pdfs/human/bwp/TSE%20NFG%20410-rev2.pdf. Accessed May 2011.
10. Geigert J. The Challenge of CMC Regulatory Compliance for Biopharmaceuticals. New York, NY: Kluwer Academic/Plenum Publishers; 2004.
11. ICH Q5A (R1), Viral Safety Evaluation of Biotechnology Products Derived From Cell Lines of Human or Animal Origin.
12. ICH Q5B, Quality of Biotechnological Products – Analysis of the Expression Construct in Cell Lines Used for Production of r-DNA Derived Protein Products.
13. ICH Q5D Quality of Biotechnological Products – Derivation and Characterization of Cell Substrates Used for Production of Biotechnological/Biological Production.
14. 5.1.7 Viral safety. European Pharmacopoeia (Ph. Eur.).
15. Food and Drug Administration 1993 Points to Consider in the Characterization of Cell Lines Used to Produce Biologicals, Rockville, MD.
16. U.S. National Archives and Records Administration. 2012, Code of federal regulations. Title 21. CFR 601.2(a). Applications for biologics licenses; procedures for filing.
17. Bates R. Downstream processing. In: Ozturk SS, Hu W-S, editors. Cell Culture Technology for Pharmaceutical and Cell-Based Therapies. Oxford, UK: Taylor and Francis Group, LLC; 2006. p 439–482.
18. Stanbury PF, Whitaker A, Hall SJ. Principles of Fermentation Technology. 2nd: Burlington, MA, Butterworth and Heinemann Publishers; 2000.
19. Seewoester T. cell separation and product capture. In: Ozturk SS, Hu W-S, editors. Cell Culture Technology for Pharmaceutical and Cell-Based Therapies. Oxford, UK, Taylor and Francis Group, LLC; 2006. p 417–438.
20. Cacciuttolo MA, Shane E, Oliver C, Tsao E, Kimura T. Scale-up considerations for biotechnology-derived products. In: Levin M, editor. Pharmaceutical Process Scale-Up. New York, NY, Marcel Decker, Inc; 2001. p. 95–114.
21. PDA Technical Report No. 15, Validation of Tangential Flow Filtration in Biopharmaceutical Application.
22. PDA Technical Report No.14, Validation of Column-Based Chromatography Processes for the Purification of Proteins.
23. Moreton C. Functionality and performance of excipients in quality-by-design world part 4: obtaining information on excipient variability for formulation design space. Am Pharm Rev 2009;12(5):28–33.
24. United States Pharmacopeia/National Formulary. Rockeville: United States Pharmacopeial Convention; 2009.
25. Merrell PH, I Silverstein. Inside IPEC–Americas: evaluating excipient stability. Pharm Technol 2009;33(9):74–75.
26. Cromwell MEM. Formulation, filling and packaging. In: Ozturk SS, Hu W-S, editors. Cell Culture Technology for Pharmaceutical and Cell-Based Therapies. Taylor and Francis Group, LLC; 2006. p 483–522.
27. DeGrazio FL. Parenteral Packaging Concerns for Drugs. Genetic Engineering & Biotechnology News. Vol. 25 21. New Rochelle, NY Mary Ann Liebert, Inc. publishers; 2005.
28. SEC. 501. [351] (a)(1)(3). Federal Food, Drug, and Cosmetic Act.

29. Food and Drug Administration. Guidance for Industry Container Closure Systems for Packaging Human Drugs and Biologics. 1999.
30. DeGrazio FL. The Importance of Extractables and Leachables Testing for Injectable Drug Delivery Systems. 2009 Available at http://www.ondrugdelivery.com. Accessed May 2011.
31. Laschi A., Natacha S., Antoine A., Beatrice B., Franscois C-M, Myriam D, Marc F, Stephanie G, Catherine L, Luc P, Christophe S. Container-Content Compatibility Studies: A Pharmaceutical Team's Integrated Approach. 2009. PDA; J Pharm Sci Technol 63:285–293.
32. Parenteral Drug Association (PDA). Technical Report 39, Revised 2007. Guidance for Temperature Controlled medicinal Products: Maintaining the Quality of Temperature-Sensitive Medicinal Products through the Transportation Environment, PDA; J Pharm Sci Technol 61(S-2).
33. Parenteral Drug Association (PDA). Technical Report 44, Revised 2008. Quality Risk Management for Aseptic Processes, PDA; J Pharm Sci Technol 62(S-1).
34. Ryan JA. Introduction to Animal Cell Culture. Technical Bulletin, Corning Incorporated; 2000.

13

RISK-BASED CHANGE CONTROL

William Harclerode, Bob Moser, Jorge A. Ferreira, and Christophe Noualhac

13.1 INTRODUCTION AND KEY POINTS

Integrating quality risk management (QRM) into pharmaceutical quality systems can be a daunting task. Even though the regulatory agencies have issued guidelines on QRM [1], each individual company must decide which specific risk assessment tools to use. The process is made even more difficult because there are many pharmaceutical quality systems and many different types of products. This chapter discusses an approach to risk-based change control for commercial products.

Integrating QRM into the change control system is essential to maintain risk management as a *living* process, but it can be especially challenging because change control covers so many areas in the pharmaceutical industry (equipment, facilities, utilities, processes, materials, computer systems, and documents, for example) and most of the product life cycle (technical transfer, commercialization, and product discontinuation). This chapter introduces some practical methods and tools that can be used to integrate QRM into an existing change control system.

13.1.1 Key Points

- QRM can increase both the quality and the speed of decision making, improve regulatory compliance, and increase efficiency of resource

utilization for implementation of changes. This ensures that patient safety is not adversely impacted because of the change.
- Risk assessment is typically used at two steps of the change control process (Fig. 13.1). First, a defined, preapproved risk-based approach can be used to perform initial risk assessment to determine whether they are in scope or out of scope. Second, a more detailed risk assessment can be performed for more complex changes, using formal tools such as failure modes and effects analysis (FMEA) or process hazards analysis (PHA).
- Patient risk is difficult to measure directly, without a medical professional's opinion. In the pharmaceutical industry, patient safety is also assured by meeting both product quality and regulatory compliance requirements, which are more easily measured. Therefore, product quality and regulatory compliance may be considered as acceptable surrogates for patient safety.
- Critical success factors for integrating QRM into change control include management commitment and support, a defined procedure for maintaining risk management in a *living* process (or a living document), a cross-functional change review team, and a process to ensure that any risk control actions identified for the change are completed before production and/or release of the product.

13.2 CHANGE CONTROL PROCESS

A formal change control process is a key component of a modern pharmaceutical quality management system. Change control is required to assure that any changes to established products, processes, or systems (such as equipment, facilities, utilities, materials, processes, or computers) are properly evaluated and implemented to protect product quality and ultimately ensure patient safety. Figure 13.1 provides a flow chart of a typical change control process and shows where QRM may be used.

QRM and knowledge management are two of the enablers used in implementing modern quality systems [2]. Knowledge management can assure that sound science and historical experience are used to evaluate changes. Regulatory agencies expect companies to know their products and to understand how changes might affect product quality and patient safety. QRM provides a proactive approach to identifying, evaluating, and controlling quality risks. Ideally, risk management for each product or process includes a living, controlled risk assessment document that serves as the standard against which all changes are compared. Otherwise, the risk assessment document must be prepared on a case-by-case basis.

The first step in a formal change control process is the initiation of the change. This starts with a written change request. The change request should include a description of the change, reason for change, justification for change, supporting documentation (including existing risk assessments), and a proposed risk-based implementation plan (including product quarantine requirements). The change

CHANGE CONTROL PROCESS

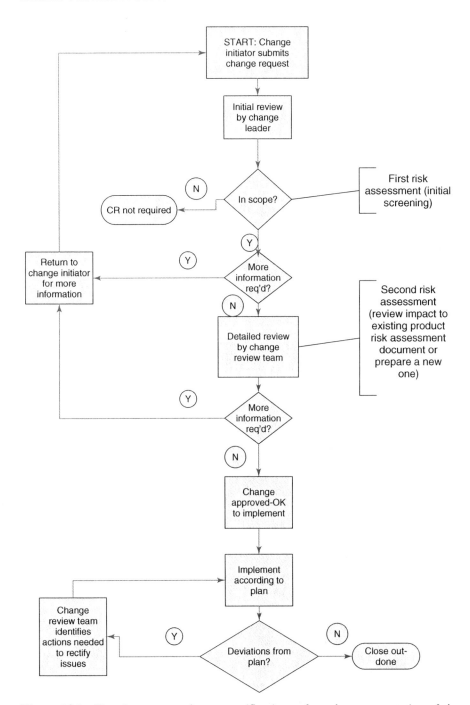

Figure 13.1 The change control process. (*See insert for color representation of the figure.*)

TABLE 13.1 Change Control Screening Criteria

Type of Change	Alternate Documentation System
Changes that do not impact a CGMP system	Maintenance work order
Changes that occur during commissioning, before completion of the initial system installation qualification (IQ)	Final IQ; ECR
Change that are allowed per established operating procedures (PM, calibration, etc.)	Equipment use log, maintenance work order, calibration record, PM record, ECR, etc.

request is then reviewed and the reviewer may request more information from the initiator.

The reviewer also performs an initial risk evaluation to screen out any change requests that do not need to be handled through this formal system. This step adds value, as handling changes through the formal change control system may consume more resources than handling them by other means.

Changes can be screened out from the formal change control system for two basic reasons (Table 13.1). First, changes that do not impact product quality or regulatory compliance, such as changes to non-GMP areas (such as nonprocessing areas, maintenance shop, and certain plant utilities such as electricity) are *out of scope*. Second, changes that are specifically allowed per approved standard operating procedures (SOPs) can also be eliminated, provided that these SOPs were developed using QRM concepts. One example is the calibration and preventive maintenance (PM) system. Established calibration procedures allow adjustment of the sensor, provided it was not outside of the preestablished tolerance. Maintenance procedures typically allow replacement of normal wear items using approved spares, and documentation may be performed through the maintenance work order system and equipment log books. A further example is change that occurs before the installation qualification of a system. As long as the user requirements specifications (URSs) have not changed, it may be acceptable to document these changes through the engineering change request (ECR) system, if the company SOPs permit. Table 13.1 shows the kinds of changes that might be processed outside of a formal change control system via procedural (SOP) controls.

Change requests that pass the screening criteria are then reviewed by a cross-functional change review team. The team performs a detailed review of the change request, supporting documentation (including any existing risk assessments), and especially the proposed change implementation plan, to assure that all considerations have been made to protect patient safety by maintaining product quality and regulatory compliance. The change review team members (which typically include representatives from the owning department, quality assurance (QA), quality control, operations, engineering, technical support, and regulatory affairs) look at the change request from their individual areas of expertise to assure that planned actions and documentation are sufficient. As part of this review, the team should verify the impact of the change on the existing risk assessment document for the impacted product or process. If this is not available, then a new risk assessment may have to be performed.

Once all items are deemed acceptable including the change implementation plan, the change request is approved as "OK to implement."

The change is then implemented according to the approved plan. Unanticipated events may prevent the implementation from going exactly according to plan. When this happens, the team (including QA) needs to be made aware of any significant deviations, and any further actions that may be required (such as deviations or Corrective and Preventive Action (CAPA)).

Lastly, when the change has been implemented and all significant deviations and actions addressed to the satisfaction of the change review team, the change request documentation can be reviewed and closed out.

13.3 BENEFITS OF USING QRM FOR CHANGE CONTROL

The main advantage of integrating QRM into the change control process is the quality and speed of the change review. The risk ranking format promotes objectivity in the decisions, and each change submitted to the Change Review Board can be addressed with a dependable and rational approach. After completing the risk assessment, the change can be implemented on the basis of a sound, scientifically documented rationale. This can provide a considerable improvement from the "tribal knowledge" decision making that is often used to determine change implementation plans.

When a change is addressed with QRM, the immediate benefit is the ability to make timely, science-based decisions. First, on the initial review of a submitted change, changes defined as *out of scope* can be immediately removed from the formal change control process. In addition, routine low risk changes that are already allowed through SOPs can be removed. The impact analysis of these low risk changes is predetermined and accepted, and no further discussion is required during a formal change control meeting. As a result, the process for implementing these types of changes is clear, and can be immediately launched.

Use of score-based risk tools (e.g., FMEA, PHA, etc.) can lead to faster assessment and scientific decision making because of a clearer understanding of relative risks associated with the change. The formality of the documented review process (using a risk assessment template) allows an immediate understanding and appreciation of the change and its impact on the product. The risk assessment discussion focuses on the risk resulting from the change and its impact on the identified critical parameters. The structured format and scientific approach for addressing the changes contributes to thorough and timely decision making.

Once the decision to accept the change has been made, the benefit of QRM is to reduce the time to implement the change. Activities adding no value to the product quality are avoided, and the change can be effected in a timely and efficient matter.

The second benefit of using QRM is more efficient management of resources. Qualification and validation activities for change controls require a significant amount of resources (manufacture, testing, and documentation). Focusing only on the parameters that are important to product quality (critical aspects) leads to a reduction in non-value-added activities, which provides a benefit to resources at all levels.

A third benefit is improved compliance. QRM provides a method by which product knowledge can be leveraged to deliver a rational and consistent approach for supporting change. It is a compliance benefit to have change control documents that can stand up to any future inspection by the regulatory authorities. A properly documented change request package does not rely on the memory of those involved to provide the rationale and justification for changes during agency inspections.

These resource and compliance benefits ultimately lead to improved efficiency. Short-term benefits may not be so obvious, because at first glance it may seem that the extra effort taken to document the risk management process is counterproductive. In the long run, however, this documentation process constitutes a small investment upfront, which yields a time savings later, in improved compliance and accessibility to information. Long-term benefits include avoiding the costs of later quality concerns by thoroughly addressing the change upfront. Over the long term, a good change control system will result in both improved quality and compliance, as well as avoiding unnecessary future costs.

13.4 ONE QRM TOOL: FMEA

FMEA is a method that can be used for analysis of potential failure modes within a system as a result of a change, in order to determine the effects on the system. The potential failure modes and effects are then quantified and analyzed in order to answer two main questions:

- Is the potential failure mitigated to an acceptable level?
- If not, what additional measures can be taken to further mitigate the potential failure?

In the context of change control for patient safety, the failures can easily be construed as any event that could affect quality and/or regulatory compliance. Mitigation is any control added to the system that aids in the prevention of the failure.

FMEA was formally introduced in the late 1940s for military use by the U.S. Armed Forces [3]. It was later used in the 1960s, during the space race, in order to safely put a man on the moon. Industry in the United States adopted FMEA in the 1970s, in part because of industrial disasters such as the chemical plant explosion in Flixborough, the United Kingdom, in 1974. One reason for the widespread popularity of the FMEA tool in industry may be the systematic approach used to aid in the mitigation of risk.

There are many other types of systematic analysis tools that are used to mitigate risks during change control [1]. These techniques are typically associated with hazard analysis [4]. These tools include, but are not limited to, the following:

- checklists;
- what-if analysis; and
- FTA (fault tree analysis).

A checklist method incorporates a list of qualitative, predefined system risks. A checklist is particularly useful when the variables are known, such as regulatory requirements. Checklists are routinely used as prescreening methods that identify known variables to reduce the amount of review time for more rigorous methods, such as FMEA.

In a "What-if" qualitative analysis, an individual and/or a team compiles a list of what-if questions designed to test the system; for example, "what if the power to a storage freezer is interrupted?" A potential consequence in this example may be that the product stored within the freezer is compromised. Potential safeguards (i.e., controls) may include high temperature alarms. The results of a what-if analysis are presented in the form of a table that includes the questions, potential consequences, safeguards in place, and recommendations.

The FMEA begins with the identification of a failure or a fault, whereas an FTA begins with a top risk; then all possible causes are evaluated in the form of a tree. The faults are quantified in accordance with each of their risks. The FTA is a visual tool (refer to Fig. 13.2 for an FTA example).

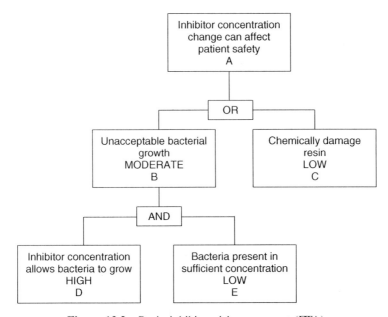

Figure 13.2 Resin inhibitor risk assessment (FTA).

The various risk assessment tools are used widely; each has its particular strengths and weaknesses. An FMEA is a versatile tool that can be used to leverage process knowledge and experience in an organized manner, in a team environment. Possible failure events and their causes are determined qualitatively. A second step then quantifies the risks associated with the failure event by assigning values for the FMEA risk elements of severity, occurrence, and detection. FMEA is one of the common methods that incorporate detection into the risk measurement equation. This analysis may lead to recommendations that decrease the probability that the failure will occur.

13.5 KEYS TO SUCCESSFUL IMPLEMENTATION

The importance of integrating QRM into change control may not always be recognized, because of the initial perception that it requires extra effort with no apparent benefit.

Not integrating QRM may ultimately lead to disparate failures, including product rejections, recalls, and regulatory agency observations associated with documentation or questionable rationales for addressing a change.

Building a solid cross-functional team and obtaining management commitment and support for the QRM program are the first, most obvious requirements for successfully integrating QRM into a change control system.

Successful change control teams must follow basic rules for conflict resolution, i.e., mutual respect, participating, and listening to each other in order to reach, ideally, consensus. A leader must facilitate the team's dynamics and promote active participation. Individual commitment is obviously crucial, as the final decision cannot be made if one or more of the members do not fulfill their role. Quality must be an active participant in the process, as patient safety is the main focus.

Although changes must be addressed in a timely matter, no artificial time constraint should be imposed to reach the final decision. Addressing the change using QRM should naturally lead to a quick decision and identification of appropriate actions.

Firms should have a documented approach to using risk management in change control. FDA's Q9 Guidance [1] does not mandate a specific QRM methodology. A formal scientific approach supplemented by comprehensive risk templates (FMEA, PHA, etc.) is recommended.

Finally, it is important to track the items in the agreed-upon implementation plan to completion.

13.6 USE OF A RISK ASSESSMENT TOOL IN CHANGE CONTROL

A step-by-step procedure for completing a risk analysis using the FMEA approach is presented here.

13.6.1 Step 1: Establish Definitions and Levels for Severity, Occurrence, and Detection and Risk Tolerance

One of the most important steps in FMEA is choosing the definitions for the different levels of severity (S), occurrence (O), and detection (D). These definitions should be standardized by each company to reflect the risk tolerance. Table 13.2 presents some example definitions used by one company.

The number of levels for severity, occurrence, and detection should be carefully chosen. If too many levels are used, then the team may spend excessive time determining which level to assign. Fewer levels generally will make decisions easier. A simple three-level approach (high, medium, low or 3, 2, and 1) may be a good starting point.

Severity is a measurement of the potential consequence of a hazard that could result from a change request, with respect to patient safety. As previously mentioned, product quality and compliance requirements are often used as measurements that assure patient safety. A medical professional's judgment should be sought when necessary.

Occurrence (or likelihood) of the failure event can be separated into categories ranging from frequent to almost never and may be defined in numeric (e.g., 1%, 10%, etc.) as well as qualitative (e.g., frequent, occasional, almost never, or 3, 2, 1) terms. When possible, numeric definitions should be used. Process capability index (CpK) is an especially useful measure. The presence of existing controls should be considered when measuring occurrence.

Detection measures the capability of existing controls to detect the failure event and to prevent a defective product from reaching patients. (A detection level of high means there is low detectability and thus high risk.) As with occurrence, detection may be defined in numerical as well as subjective terms. The presence of existing controls should be considered when measuring detection.

The change review team should also agree on its risk tolerance; i.e., the level of risk the team will accept without requiring further mitigation. For example, the team may decide that a risk priority ranking above a medium severity, medium occurrence, and high detection level would not be acceptable without mitigation.

13.6.2 Step 2: Define Critical Quality Attributes (CQAs) and Critical Process Parameters (CPPs)

It is important to identify the critical quality attributes (CQAs) for the change request, to understand the risk. CQAs are typically defined as those that have potential impact to patient health, product quality (for example, release specifications), and regulatory compliance (stability during shelf life, etc.).

The critical process parameters (CPPs) that ensure meeting the product quality requirements should also be known and understood. CQAs and CPPs are typically identified in a development report, technical transfer report, or validation report, and serve to help define the design space [5].

TABLE 13.2 Risk Ranking Definitions—Example

	Severity	Occurrence	Detection[a]
High	Direct (primary) impact to CQA or CPP of product, process, or utility; could result in critical (AQL) defect; potential for harm to patient health; microbial/foreign contamination; significant change to regulatory filing (preapproval) or master production record; direct product contact item; critical instrument	Potential exists for the event to frequently recur, process incapable (process capability CpK < 1.00)	Event is difficult to detect, or not frequently tested, or not tested; not likely to be detected before product release
Medium	Indirect (secondary) impact to CQA or CPP; could result in a major AQL defect; likely to result in customer dissatisfaction or product complaint; minor change to regulatory filing (changes being effected) or master production records	The event may occur but only occasionally; process is capable with tight control (CpK = 1.00 – 1.33)	Event may be detected by 100% human visual inspection or other administrative type of control; might be detected before product release
Low	No impact to CQA or CPP; may result in only a cosmetic complaint; no or minimal change to regulatory filing (annual report) or master production record; no impact to patient health; could result in minor or cosmetic AQL defect	The event is rare and unusual; process is capable (CpK > 1.33)	Event is 100% detected by engineering type of controls (electronic vision system, metal detection); defect is obvious; would be detected before product release

[a] Note that a detection level of high means there is low detectability, and thus high risk.

13.6.3 Step 3: List Failure Modes, Effects, and Causes

The potential impact of the proposed change is assessed with respect to failure events that affect product quality and regulatory compliance. When possible, the team should refer to failure modes already identified in existing *living* risk assessment documents. Where not available, a simple FMEA spreadsheet can be used such as the following (Table 13.3).

For each identified type of failure that could happen as a result of implementing the change request, list all failure modes, effects, and causes. For example, one failure mode for the tablet manufacturing process might be *incorrect direction of rotation for mixer*. The effect would be *failed content uniformity*. A specific cause could be *mixer wired incorrectly after motor change*.

Certain reasonable assumptions should be made to keep the hazard analysis manageable. Controls that already exist and that serve to mitigate the failure should be listed and included in the measurement for occurrence and detection. Other assumptions may be globally recognized (and do not need to be restated each time), for example, that existing quality systems are in place and functional (deviations, change control, CAPA, calibration, preventive maintenance, etc.).

13.6.4 Step 4: Assign Severity, Occurrence, and Detection Levels to each Failure Mode

The team should discuss each failure event and assign the appropriate levels to severity, occurrence, and detection. The level assignment should be based on objective data whenever possible.

13.6.5 Step 5: Calculate Overall Risk (RPR) for each Failure Mode

Using the assigned levels for severity, occurrence, and detection, calculate the risk priority ranking (RPR) for each failure event, using an agreed-upon method. The RPR can either be a number (if numbers were assigned) or a category (high–medium–low). Table 13.4 provides an example of one method of determining RPR based on high–medium–low categories.

TABLE 13.3 Sample Risk Assessment Spreadsheet

Process Step	Failure Mode	Effect	Severity	Cause	Occurrence	Current Controls	Detection	Overall Risk	Accept Risk?	Recommended Actions

TABLE 13.4 Calculation of Risk Priority Ranking (RPR)

RISK CLASSIFICATION

		Occurrence		
		Low	Medium	High
Severity	High	Level 2	Level 1	Level 1
	Medium	Level 3	Level 2	Level 1
	Low	Level 3	Level 3	Level 2

RISK PRIORITY RANKING

		Detection		
		Low	Medium	High
Risk classification	Level 1	MEDIUM	HIGH	HIGH
	Level 2	LOW	MEDIUM	HIGH
	Level 3	LOW	LOW	MEDIUM

13.6.6 Step 6: Mitigate Risk to an Acceptable Level

If the RPR exceeds the agreed-upon value in Step 1 (risk tolerance), the team decides what additional mitigation steps will be taken to reduce residual risk (risk remaining after remediation efforts have been completed) to an acceptable level.

In certain circumstances, the team may decide to accept risks that have been reduced to *as low as reasonable practicable (ALARP)*, based on a documented risk–benefit analysis [6].

The "law of unintended consequences" also applies to change control. Implementation of mitigation actions can create additional risks. It is therefore important to repeat the risk assessment process after mitigation has been implemented to verify that no new risks have been introduced.

13.6.7 Step 7: Communicate Acceptance of the Residual Risk to the Appropriate Level

The last step in the risk management process for change control is to make sure that the acceptance of any residual risk has been communicated to the appropriate level of management. Risk communication should include senior management for high risk rankings.

13.7 CASE STUDIES

To better illustrate how to integrate QRM in change control using the FMEA approach, four case studies are presented here. (The final step of evaluating risk after implementation of mitigation actions is not included in these examples.)

13.7.1 Case Study #1: Changing Magnesium Stearate from Bovine to Vegetable-Based

13.7.1.1 Change Description The animal grade magnesium stearate used in the formulation for a tablet is to be replaced by an equivalent vegetable grade to reduce risk of BSE/TSE contamination. (Magnesium stearate is added to the tablet's final blend in the manufacturing process to facilitate lubrication during tablet compression.)

The supplier proposes using an equivalent grade material with comparable particle size distribution and shape, to minimize the impact of the change. There is no change to the material release specification, supplier, or manufacturing site.

13.7.1.2 Potentially Impacted Critical Quality Attributes The final blend properties are sensitive to the hydrophobicity of magnesium stearate. Even though the material may meet the release specifications, a replacement material with slightly different hydrophobic properties could potentially impact the tablet's critical quality attributes, such as appearance, hardness, and dissolution (because of under- or over-lubrication).

Under-lubrication may cause adhesion of the powder to the punches during compression, leading to sticking and potential alteration of the embossing used to identify the tablet and/or the strength. An illegible imprint is considered a critical defect.

Over-lubrication, a result of excess coating of the granules with hydrophobic material, can delay dissolution or alter the cohesion between particles and adversely affect tablet hardness, producing soft tablets.

13.7.1.3 Calculated Risk The risk analysis (Table 13.5) led to unacceptable medium level risks and required a mitigation plan. It is assumed here that implementation of these recommended actions would not create any new risks to be evaluated.

13.7.1.4 Recommended Actions One verification batch will be produced with extensive stratified testing for appearance, hardness, and dissolution to ensure equivalency of the alternative Mg stearate in the formulation. Implementation of this action would reduce the risk to as low as reasonably practical, by providing

TABLE 13.5 Risk Assessment for Changing Source of Magnesium Stearate

Process Step	Failure	SEV	Cause/Process Failure	OCC	Current Controls	DET	RPR	Accept Risk?	Actions
Blend lubrication	Tablet fails appearance-embossing worn off	H	Under-lubrication (Filming, sticking, picking)	L	In-process testing (visual inspection) and release testing	M	M	No	Process verification (1 lot) with increased stratified in-process testing for appearance, hardness, and dissolution
Blend lubrication	Tablets fail hardness in-process spec not met (too soft)	H	Over-lubrication (tablet damaged during normal processing)	L	In-process testing (friability and hardness)	M	M	No	
Blend lubrication	Tablets fail dissolution spec.	H	Over-lubrication (excessive coating of the particle leads to retardation of product dissolution)	L	In-process testing (disintegration) and release testing	M	M	No	

CASE STUDIES 381

assurance (data) that future batches will meet the product quality and compliance requirements.

13.7.2 Case Study #2: Primary Component Change for a Sterile Product

13.7.2.1 Change Description Containers (glass vials) made by a particular supplier are being replaced. Alternate containers meeting current material specifications (dimensionally and chemically equivalent) are available from the other supplier at a different facility.

13.7.2.2 Potentially Impacted CQAs Incoming bioburden/contamination (existing cleaning and sterilization cycles are based on the now-discontinued vials) is one CQA (specification) that could be adversely impacted by using a different supplier.

Incoming bottle integrity (breaks or cracks can cause sterility failure) could also be adversely impacted.

13.7.2.3 Calculated Risk The overall risk for the identified risk events (Table 13.6) is generally high, primarily because the severity of a sterility failure and the detection of an occasional sterility failure are always high. (In normal (acceptable quality level) AQL testing, there is a low likelihood of detecting an occasional sterility failure.)

13.7.2.4 Recommended Actions Actions recommended to mitigate the risk include increasing incoming quality assurance (IQA) inspections, performing a vendor audit/qualification, reexecuting cleaning and sterilization cycle validation, and conducting at least one media fill run.

13.7.3 Case Study #3: Equipment Change

13.7.3.1 Change Description During the synthesis of a corrosive, temperature-sensitive pharmaceutical suspended solid, a 350-liter jacketed mixing vessel is used to heat and mix the suspension. It is being changed to a 700-liter vessel to increase batch size.

13.7.3.2 Potentially Impacted Critical Quality Attributes The attributes considered critical to meeting product specifications (and thus ensuring patient safety) were identified as follows:

- materials of construction (corrosion);
- mixing fluid dynamics (flow, shear, etc.); and
- temperature uniformity.

These characteristics and the integrity of the process need to be maintained in the design of the new equipment.

TABLE 13.6 Risk Assessment for Changing Bottle Manufacturer

Process Step, Change Control, Deviation, Failure Event, etc.	F	SEV	Causes	OCC	Current Controls	DET	RPR	Accept Risk?	Actions
Change supplier of product vials	Bottles are not adequately sterilized	H	Bottles have different dimensions and/or different materials of construction (MOC)	M	Vendor has confirmed will meet same dimensional specs and MOC; QA acceptance sampling	L	M	N	Increase IQA inspections and perform a vendor audit
Change supplier of product vials	Bottles not adequately sterilized	H	Bottles have higher incoming bioburden; inadequate GMP, manufacturing controls, etc.	M	Vendor states bottles will meet current bioburden specs; incoming IQA testing	H	H	N	Perform vendor GMP audit before purchase; perform increased IQA bioburden testing, perform media fill for first batch; perform cleaning validation studies
Change supplier of product vials	Bottles not adequately sterilized; residual film or presence of particulates	H	Bottles are contaminated; different mold release agent used	H	Vendor states release agent causes no problem with other pharma customers; IQA testing	H	H	N	Perform cleaning validation studies; perform media fill
Change supplier of product vials	Sterility failure	H	Bottles broken/cracked because of different packaging configuration or package size	M	Vendor states will provide vials adequately packaged to prevent breakage; IQA testing; final product testing	M	H	H	Place packaging requirements into purchasing spec; perform media fill, increase IQA testing for first three lots

CASE STUDIES

13.7.3.3 Calculated Risk The calculated risk was considered to be medium for product mixing and heating scale-up, primarily because the existing controls for this complicated process did not require a detailed mixing design analysis before purchasing the equipment. The risk of using the wrong materials of construction was considered low because existing controls are sufficient (Table 13.7). (Note that detection is low, because detectability was high.)

13.7.3.4 Recommended Actions The following recommendation was developed to assure that quality is maintained during this change:

- Perform a detailed 3D engineering study before designing and purchasing the new vessel. (The standard *best practices* analysis was not considered sufficient to mitigate the risk for this product, because the process was too complex.)

13.7.4 Case Study #4: Resin Inhibitor (FTA)

13.7.4.1 Change Description Consider a change in the concentration of an inhibitor used in a resin storage solution to prevent biological growth. The resin is used in the manufacture of a pharmaceutical product. It is desirable for process safety purposes to reduce the inhibitor concentration as low as possible.

FTA is a technique that can be used to evaluate the effect on patient safety because of a change in the pharmaceutical manufacturing process. An FTA is developed from the top–down. These fault trees are built using gates and events (blocks). The two most commonly used gates in a fault tree are the AND and OR gates. As a visualization example, consider the simple case described by the referenced FTA (Fig. 13.2). This example has been oversimplified for demonstration purposes.

In this example, consider two events (or blocks) comprising a top event (or a system).

A change in inhibitor concentration possibly affecting patient safety (A) can lead to

- unacceptable biological growth (B) or
- chemically altering the resin (C).

If occurrence of either event (B or C) causes the top event (A) to occur, then these events (blocks) are connected using an OR gate. If one of those events such as unacceptable biological growth (B) requires two other events to occur, such as

- inhibitor concentration too low allowing bacteria to grow (D); and
- bacteria present in sufficient concentration to grow (E).

Should they occur together causing the event above (B) to occur, they are connected by an AND gate.

TABLE 13.7 Risk Assessment for Changing Reaction Vessel

Process Step, Change Control, Deviation, Failure Event, etc.	Failure	SEV	Causes	OCC	Current Controls	DET	RPR	Accept Risk?	Actions
Increase vessel size	Product inadequately mixed; does not meet content uniformity	H	Agitator type and speed control; inadequate mix time	M	Current procedures require engineering study, design qualification, development batch, and process validation	L	M	N	Perform detailed 3D engineering/process simulation as part of design qualification before specifying new mixing equipment
Increase vessel size	Product does not meet release specs because of degradation (localized overheating)	H	Heat transfer system on jacket may not be as effective	M	Current procedures require engineering study, design qualification, development batch, and process validation; product release testing	L	M	N	Perform detailed 3D engineering/process simulation studies as part of design qualification before specifying new mixing equipment
Increase vessel size	Product is contaminated by corrosion	H	Inadequate MOC (materials of construction)	L	Current procedures require engineering study, design qualification, development batch, and process validation	L	L	Y	None required

The probability of the event is described by high, moderate, and low. The probability that a lower inhibitor concentration can allow bacterial growth (D) is considered high without further study. The probability that bacteria will be present in sufficient concentration to grow (E) is low because of cleaning procedures associated with the resin manufacture. Since both a low enough inhibitor concentration (D) and bacteria present in sufficient concentration (E) are needed to allow unacceptable growth, the overall risk is moderate. The probability that the resin will be chemically changed (C) is low, as it typically does not need the inhibitor to chemically preserve it.

The analysis of the FTA chart leads to the conclusion that key element is (D) to inhibit bacterial growth. A study may prove that a certain concentration of the inhibitor may be high enough to inhibit bacterial growth while being low enough to address the process safety concerns.

13.8 CONCLUSION

The benefits of integrating QRM into the change control system include improved risk communication, faster and more thorough science-based evaluations, and increased productivity.

Integrating QRM should not be difficult. Routine changes may be initially assessed from a risk perspective and then managed through administrative controls such as simple procedures. More complex changes can be assessed and managed using tools that are commonly accepted in industry, such as FMEA or FTA. Keys to successful implementation include a defined methodology, teamwork, and management commitment.

QRM ultimately focuses on risk to patients. Product quality and regulatory compliance may be used as acceptable surrogates for *risk to patients*. The use of QRM demonstrates to regulatory agencies that a company understands its products and has all of its processes under control.

REFERENCES

1. FDA Guidance for Industry: Q9 Quality Risk Management, June 2006.
2. FDA Guidance for Industry: Q10 Pharmaceutical Quality Systems, April 2009.
3. Procedure For Performing A Failure Mode Effect and Criticality Analysis, November 9, 1949, United States Military Procedure, MIL-P-16.
4. Process Safety Management, OSHA 29 CFR 1910.119.
5. FDA Guidance for Industry: Q8 Pharmaceutical Development, May 2006.
6. ISO 14971: Medical Devices-Application of Risk Management to Medical Devices.

INDEX

active pharmaceutical ingredient (API)
    development, 105, 114
    DOE for, 112
    influencing parameter, 110–111
adulterated drug, 353
American Society for Testing and Methodology (ASTM), 4
    E2500–07 Standard Guide, 4–5, 132–133, 168
aseptic processing, 228
    and absence of microorganisms, 243
    advances in, 229–230
    Agalloco–Akers (A–A) method, 234–235
    aseptic operator, 239
    culture-based microbial test methods, 265–268
    environmental controls, 237
    equipment and utensils, sterilization of, 237–238
    facilities for, 237
    factors influencing, 236
    FDA guidance, 231–232
    FMEA, 231
    future trends, 270–271
    hazards in, 254–259
    hierarchical relationship of risk and hazards, 260–263
    history of, 229–230
    ISO 14644-7 continuum, 230
    microbial contamination in, 230, 263–265
    microbiological monitoring methods, 230–231
    model for microbial ingress, 259
    Monte Carlo estimation in, 233
    number of microbes deposited on a product, calculation, 231
    and patient risk, 244–246
    PDA monitoring methods, 228–230
    quantitative risk assessment, 268–270
    risk assessment and management in, 246–254
    risk mitigation in, 236–239
    risk models for, 235
    risks and human assessment, 231–235
    risks associated with, 229
    sterility (safety) by design, 239–240
    sterilized products, delivering, 238
Aseptic Processing Guidance, 228
asymmetry of risk knowledge, 67
automated real-time risk calculation (ARC), 270

The Basics of FMEA 2nd Edition, 28
biases management, 70
bioburden ingress, risk of, 257, 264
Biologic License Applications (BLAs), 335

biopharmaceutical manufacturing process
  application of risk management tools to, 359–364
  approach to QRM, 325–326
  cell bank construction strategy, 332–335
  critical quality attributes and critical process parameters, 326–329
  distribution and cold chain supply, 355–357
  downstream processing, 340–342
  excipient control, 349–351
  extractables and leachables, impact of, 353–355
  extraction and purification process, 342–345
  fermentation/cell culture, 336–340
  packaging, 351–353
  raw material management, 329–332
  risk acceptability definitions, 358
  scale-up of production process, 345–349
  summary table of, 358

cause-and-effect relationship, 84–85
  diagrams (fishbone or Ishikawa diagrams), *see* Ishikawa diagram
cell banking, 332–335
  bacterial endotoxins, presence of, 335
  characterization and testing of cell banks, 333–335
  construction strategy, 332–333
  documentation of proper control procedures, 335
  QRM for, 333–335
  risk of contamination, 333–335
  storage of cell banks, 335
  tests for master cells banks (MCBs), 334
  tests for working cells banks (WCBs), 334
21 CFR 211.100 of U.S. CGMPs, 1
change control process, 368–371
  benefits of using QRM, 371–372
  use of risk assessment tool, 374–379
Charlton, Warren, 235
checklist analysis, 38
check sheets, 41
cleaning validation (CV), 180, 217–221
clinical trials, 114–124
  FMEA for pre-phase 3 manufacture and control, 123
  preliminary risk assessment for, 115–122
Clostridium botulinum, 227
Code of Federal Regulations (CFRs), 265
coincidences, 84–85
commissioning, 134–136
Communication from the Commission on the Precautionary Principle, 66
conditional probability, 81–82

confidence interval, 86
contamination control process, 173–174
  bracketing approach, 174
  in cell culturing and fermentation, 337
  change management and change control, 174–175
  failure investigation, 175–176
  family approach, 174
  gamma irradiation, 330
  project close out, 176
  supplier quality program, 330–331
contamination in aseptic processing
  Agalloco–Akers (A–A) method, 234–235
  aseptic operator as primary source of, 239
  containers/closures, 238
  Monte Carlo simulation of, 233
  sequence of activities, 238–239
continued process verification, 214–217
control charts, 43
control strategy, 91, 96–97
critical control points (CCPs), 2, 300–308
critical process parameters (CPPs), 95, 103, 180, 191–196, 327, 329
  identification of, 328
critical quality attributes (CQAs), 90–91, 93–97, 109, 154, 191, 195, 329
  identification of, 327
cross contamination risks, 217–221
culture-based microbial test methods, 265–268
Current Good Manufacturing Practices (CGMPs), 1, 256
  aseptic processing, 232
  for Finished Pharmaceuticals, 179

decision making, 3
  dual processes of, 63
design of experiments (DOE), 92, 110–113
  of API process, 112
  for matrix development, 112–113
design space, 91, 95–97
distribution processes for biopharmaceutical products, 355–357
downstream processing
  endotoxin contamination, risk of, 345
  inefficient harvest/recovery conditions, 342–344
  overview, 340–342
  purification conditions, 344
  purification of biotechnology-derived products, 340
  QRM, 342–345
  ultrafiltration/diafiltration (UF/DF) for, 344
  viral contamination, risk of, 344–345
dual-process theory, 63

INDEX

early-stage trials, risk management for, 114
endotoxins, risk of, 258–259, 345
Enterobacter sp., 266
Escherichia coli, 266
European Directive 2003/94/EC for quality assurance, 102–103
event tree analysis (ETA), 37–38
excipient control
  overview, 349–350
  QRM of, 350–351
extractables and leachables
  considerations, 354
  QRM of, 355
  risks associated with, 353
  testing for, 353–354
extrinsic hazards in aseptic processing, 256–258

failure mode effects analysis (FMEA), 92–93, 221, 328, 372–374
  in aseptic processing, 231
  benefits, 28
  description, 25–27
  elements of, 25–27
  examples, 28–29
  fitting into risk management program, 27
  limitations, 28
  10-point quantitative, 164–165
  for pre-phase 3 manufacture and control, 123
  risk ranking, 39
failure modes, effects, and criticality analysis (FMECA), 92–93
  benefits, 28
  description, 25–27
  elements of, 25–27
  examples, 28–29
  fitting into risk management program, 27
  limitations, 28
fault tree analysis (FTA), 92
  benefits, 33
  description, 32–33
  examples, 33
  limitations, 33
fermentation process, 336, 341, 346
  facilities and equipment for, 340
  provision of oxygen enrichment, 339
  scale-up considerations, 340
  sterilizing medium, 339–340
  suitable environment for, 337–339
first-in-human trials and risk management, 113
  preclinical studies, 113
fishbone diagram, *see* Ishikawa diagram for risk assessment
flowcharts/block diagrams, 41

flow-cytometric-based measurement platforms, 265

gel permeation chromatography (GPC), 341
global healthcare industry, 227
go/no go parameter, 108
Good Manufacturing Practice (GMP)—QRM, 54–55

harm, defined, 45
harvested cell culture fluid (HCCF), 341
hazard, defined, 45
hazard analysis and critical control point (HACCP), 2
  aim of, 295
  benefits, 33, 295–310
  critical control points (CCPs), 300–308
  decision tree analysis process, 303–306
  description for product and raw materials, 297
  documentation and record keeping, 309–310
  establishing critical levels, 306–307
  examples, 34–35, 37
  fitting into risk management program, 36
  hazard analysis chart, 300–301
  history of, 295
  identification of corrective actions, 308–309
  limitations, 35
  monitoring activities, 308
  phases for analysis, 298–299
  preventive measures, 299–300
  principles, 296
  process flow diagram, 297–298
  steps in, 33
  team members for, 296–297
  verification activities, 308–309
hazard operability analysis (HAZOP), 92
  benefits, 28
  definition, 28
  examples, 30–31
  fitting into risk management program, 30
  limitations, 30
  process flowchart for, 32
hazards in aseptic processing, 254–259
  extrinsic, 256–258
  intrinsic, 254–256
  ISO14971:2007(E) specification, 254
  risk of endotoxins, 258–259
health hazard evaluation (HHE), 41
heuristics, 64
histograms (bar charts), 43
hydrophobic interaction columns (HIC), 341

ICH (International Conference for Harmonization)
  Q8, 276

ICH (International Conference for Harmonization) (*Continued*)
  Q9, 114, 182, 325
  Q10, 96–97, 103
  Q6A, 103
  Q9 (guidance on quality risk management), 3, 10, 65–66
  Q8R1, 104
  Q8R2, 90, 94–96
industry guides
  Facilities Baseline Guides: Commissioning and Qualification, 4
  Guidance for Industry on the General Principles of Process Validation, 4
  ISPE Guide: Science and Risk-Based Approach for the Delivery of Facilities, Systems, and Equipment, 4
  Standard Guide for Standard Guide for Specification, Design, and Verification of Pharmaceutical and Biopharmaceutical Manufacturing Systems and Equipment, 4–5
  Technical Report No. 44, 4
integration of risk management, 67–70
  training, 69–70
interlukin-6 (IL-6), 258
International Society of Pharmaceutical Engineering (ISPE), 4
intrinsic hazards in aseptic processing, 254–256
intuition, 63
ion exchange column (IEC), 341
Ishikawa diagram, 42, 161
  for analytical method development, 112
  for risk assessment, 108, 110
  in tablet manufacturing, 107
ISO 14971 *Medical Devices—Application of risk Management to Medical Devices,* 66

joint probability, 82

Keynes, John Maynard, 76
Klebsiella pneumonia, 266

layer of protection analysis (LOPA), 37
lay–expert bias, 63
legal speed limit, 62
life cycle, 91, 97–98
lipopolysaccharide, 258–259
lipoteichoic acid, 258–259
luck, 76

mammalian products, 341–342
manmade disasters, 83
Maximin Principle, 66
MHRA, 54

QRM FAQs, 57–59
microbial contamination, 263–265
  at environmental levels, 264
  quantification of, 264–265
microbial ingress, model for, 259
microbial products, 341
microbiological control in nonsterile manufacturing, case study, 293–310
Monte Carlo simulation of contamination, 233, 269
6M's (Man, Machines, Measurement, Materials, Methods, and Management), 107–108
multidisciplinary team, 107

National Aeronautics and Space Administration (NASA), 45

occurrence rate, 86–87
out-of-tolerance parameter, 328
over-the-counter (OTC) oral dosage products, 3

packaging for biopharmaceuticals
  overview, 351–352
  QRM of, 352–353
Paradigm Change in Manufacturing Operations (PCMO*^{SM}), 98
Parenteral Drug Association (PDA), 3
  aseptic processing monitoring methods, 228–229
  Technical Report No. 15, 347
  Technical Report No. 44, 4
Pareto analysis, 43
patient safety, 3
peptidoglycan, 259
perception in risk, 65
perspectivism, theory of, 65
pharmaceutical and biopharmaceutical industries
  financial pressures, 2
Pharmaceutical cGMPs for the 21st Century—A Risk-Based Approach, 2
  initiatives and recommendations, 2
pharmaceutical product manufacturing processes, *see also* biopharmaceutical manufacturing process
  active pharmaceutical ingredients and excipients, management of, 281
  branding, 285
  cleaning program, 287–289, 291
  compression process, 284–285
  dispensing operation, 282–283
  drying step, 283–284
  effective risk management, 276–277
  equipments and instruments, 290–292
  facilities, 287–289
  film coating, 285

INDEX   391

finishing process, 284–287
fishbone diagram, 278
formulating a product, 282
granulation process, 283
level of environmental control, 289–290
list of potential risks, 313–320
maintenance and calibration practices, 291–292
material management, 287
milling process, 284
mixing/blending process, 284
packaging, 285–287
people interactions, assessment of risks, 293
personal hygiene, 292
process flow chart, 279
protective clothing and gear, 292–293
raw material management, 280
risk assessment, 277–293
storage conditions, 291
supplier management, 281–282
training programs for personnel, 293
use of risk assessment, 275–276
validation activities, 292
pharmaceutical products under development, basic rules, 102
Poisson distribution, 83–84
precautionary principle of risk, 65–67
preliminary hazards analysis (PHA), 92, 105–106
   approach to risk mitigation, 106–107
   benefits, 22
   description, 22
   example, 24
   fitting into risk management program, 25
   limitations, 22–24
   with stability data input, 106
probability
   addition rule, 79
   cautions, 78–79
   conditional, 81–82
   definitions, 77
   estimating, 85–87
   joint, 82
   luck and, 76–77
   multiplication rule, 81–82
   numerator and denominator in calculation, 78
   risk and, 79–82
   rules of, 77–78
process analytical technology (PAT), 90, 244
process capability analysis, 92
process design, validation of, 187–207
   cause/effect process parameter, 199–201
   critical material attributes (CMAs) and, 189
   critical process parameters (CPP), 191–196

   existing product, 192
   in-process hold stability studies, 203–205
   mixing process for solutions, example, 198–203
   new product, 192
   normal operating range (NOR), 193–195
   processing steps, 192–193
   proven acceptable range (PAR), 193–195
   quality by design (QbD) principles and, 189
   risk-based approaches, 189–191
   risk prioritization for commercial-scale experimental design, 205–207
   scaled-down models, 187
   tablet compression/coating process, example, 197
process mapping, 41
process parameter, 90–91, 95, *see also* critical process parameters (CPPs)
process performance qualification
   in multiproduct manufacturing, 212–214
   technology transfer activities, 209–211
process validation (PV)
   challenges in, 186–187
   general considerations, 183–186
   life cycle approach, 187–188
   of process design, 187–207
   risk-based life cycle approach, 180
   tools, 182–183
process validation (PV), guidance for, 130–131
   ASTM E2500 Standard Guide, 132–133
   FDA, 130–132
   for quality risk management, 181–182
product control strategy, 109–110
product isolation, 340
product polishing, 340
product purification, 340
Product Quality Lifecycle Implementation (PQLI®), 98
product realization, defined, 101
product specification file, 124–126
proven acceptable range (PAR), 112

qualification program, effective
   commissioning, role of, 134–136
   quality assurance, 133
   regulatory requirements, 130–132
   risk-based qualification, 133–136
   and risk management, 137–139
quality assurance, European Directive 2003/94/EC for, 102–103
quality by design (QbD), 244
   in aseptic processing, 239
   background, 89–90
   development of products using, 91–93

quality by design (QbD) (*Continued*)
　examples of approaches, 98
　steps in product development using, 94–98
quality risk management (QRM), 6, 49, *see also* biopharmaceutical manufacturing process
　agency-observed deficiencies, 51
　analysis of resin inhibitor (FTA), case study, 383–385
　auditing process, 55–61
　benefits, 371–372
　changing magnesium stearate from bovine to vegetable-based, case study, 379–381
　equipment change, case study, 381–383
　general system expectations, 51–52
　integrating into change control system, 367–368
　keys to successful implementation of, 374
　primary component change for a sterile product, case study, 381
　problems of subjectivity and uncertainty, 61–67
　regulatory expectations, 50–61
　regulatory expectations and practical considerations, 53–54
　risk communication and reporting, 52–54
　risk registers (risk master plan), 54–55
quality system requirement (QSR)s Design History File, 125
quality target product profile (QTPP), 94, 104–109, 112
　example, 104
　in R&D, 105
The Quality Toolbox Second Edition, 28
quality vision of pharmaceutical development, 101–104

rare event, 82–84
raw materials, risks associated with
　contamination of microorganism, 330
　control activity, 329–331
　QRM for, 331–332
　related to origin, 330
reason, 63
residual risk, 79
　evaluation, 44
responsibility of management, 50
restricted access barrier system (RABS), 229–230
risk, defined, 45
risk acceptance, 20–21
risk analysis, 19–20
　defined, 45
risk assessment, 3, 18–20
　of product quality, 130

risk assessment, defined, 45
risk-based approaches, 2
　life cycle approach, 180
　process design, validation of, 189–191
　qualification, 133–136
risk-based decision-making, 4–5
risk-based qualification plan, implementation of
　component-level impact assessments, 158–160
　computer system qualification, 171–172
　contamination control process, 173–174
　correlation between process steps and product quality attributes, 153–154
　criticality analysis, 161–163
　critical process parameters and operating conditions, analysis of, 169
　design review, 168–169
　fault tree analysis, 160
　FMEA assessment, 163–165
　impact assessment, 154–157
　installation qualification, 169–170
　Ishikawa analysis, 161
　level of qualification, 160
　notes, 168
　objective of plan, 153
　operational qualification, 170–171
　qualification tests, 154
　requalification, 172–173
　requirements, 152
　risk ranking, 163–164
　sequence, 155
　system-level impact assessments, 157–159
　system risk determination, 162
　test functions and acceptance criteria, 164–168
　three point qualitative risk priority rank determination, 167
risk-based qualification planning
　information management for, 151–152
　program-level alignment, 142–143
　project control system, 140–141
　project description, 139–140
　project performance metrics, 141
　qualification-related information, 146–151
　schedules, 141–142
　system boundaries, 143–145
risk communication, 109
risk control, 20–21
risk control, defined, 45
risk evaluation, 20
risk evaluation, defined, 46
risk identification, 18–19
risk information, 18
　sources, 44–45
risk management, 3
　components of risk, 8

defined, 46
ICH Q9 recommendations, 10
maturity, 69
of pharmaceutical manufacturing processes, 5–8
practical guide to, 8–11
process, 18–21
use of, 14
risk perception
heuristics, 70
influence factors, 64
risk prioritization and rationale, 111
risk prioritization number (RPN), 124
risk ranking and filtering, 92
risk ranking and filtering (RRF), 38, 40
benefits, 35
description, 35
examples, 35
limitations, 35
risk reduction, 20
risk registers (risk master plan), 54–55
risk review/communication, 21
risk training, 69–70

safety by design (SbD), 239–240
scale-up of production process
bioreactor operations, 346
cell culture and fermentation processes, 346
centrifugation, 347
chromatography, 347–348
harvest operations, 346
inappropriate facilities or equipment, risk of, 349
inoculum expansion, 346
medium preparation, 345–346
membrane/filtration operations, 347
overview, 345
purification techniques/process design, risk of, 348
QRM of, 348–349
viral contamination, risk of, 348
screening risk assessment, *see* preliminary hazards analysis (PHA)

Six Sigma programs, 83
social responsibility of a pharmaceutical manufacturer, 50
societal response, 66
societal trigger level, 66
statistical risk, 82
sterility assurance level (SAL), 227–228
sterility tests, 265–267
subject matter experts (SMEs), 168

t distribution, 87
tissue/cell culture, 336–337, 346
facilities and equipment for, 340
provision of oxygen enrichment, 339
quality and composition of culture media, 339
scale-up considerations, 340
sterilizing medium, 339–340
strain stability, 339
suitable environment for, 337–339
tools for risk management, 42
advanced, 23–24
applicability, 17–18
comparison, 43–44
preliminary hazards analysis (PHA), 22–25
total organic carbon (TOC) testing, 114
t-tests, 78–79
type II error (false negatives), 265
type II (two) error, 82
type I (one) error, 82

uncertainty, 76
unintended consequence of risk regulation, 67
U.S. Food and Drug Administration (FDA) regulations, 1, 3
Guidance for Industry on the General Principles of Process Validation, 4

Venn diagram, 79–80

what-if analysis, 41
5 why analysis, 43
Whyte, Dr., 232–233